THE HISTORY AND POETICS OF SCIENTIFIC BIOGRAPHY

Biographies of scientists carry an increasingly prominent role in today's publishing climate. Traditional historical and sociological accounts of science are complemented by narratives that emphasize the importance of the scientific subject in the production of science. Not least is the realization that the role of science in culture is much more accessible when presented through the lives of its practitioners. Taken as a genre, such biographies play an important role in the public understanding of science.

In recent years there has been an increasing number of monographs and collections about biography in general and literary biography in particular. However, biographies of scientists, engineers and medical doctors have rarely been the topic of scholarly inquiry. As such this volume of essays will be welcomed by those interested in the genre of science biography, and who wish to re-examine its history, foundational problems and theoretical implications.

Borrowing approaches and methods from cultural studies and the history, philosophy and sociology of science, the contributions cover a broad range of subjects, periods and locations. By presenting such a rich diversity of essays, the volume is able to chart the reoccurring conceptual problems and devices that have influenced scientific biographies from classical antiquity to the present day. In so doing it provides a compelling overview of the history of the genre, suggesting that the different valuations given scientific biography over time have been largely fuelled by vested professional interests.

About the Editor

Thomas Söderqvist is Professor in History of Medicine and Director of Medical Museion at the University of Copenhagen, Denmark.

Science, Technology and Culture, 1700–1945

Series Editors

David M. Knight
University of Durham

and

Trevor Levere
University of Toronto

Science, Technology and Culture, 1700–1945 focuses on the social, cultural, industrial and economic contexts of science and technology from the 'scientific revolution' up to the Second world War. It explores the agricultural and industrial revolutions of the eighteenth century, the coffee-house culture of the Enlightenment, the spread of museums, botanic gardens and expositions in the nineteenth century, to the Franco-Prussian war of 1870, seen as a victory for German science. It also addresses the dependence of society on science and technology in the twentieth century.

Science, Technology and Culture, 1700–1945 addresses issues of the interaction of science, technology and culture in the period from 1700 to 1945, at the same time as including new research within the field of the history of science.

Also in the series

Phrenology and the Origins of Victorian Scientific Naturalism
John van Wyhe

John Herschel's Cape Voyage
Private Science, Public Imagination and the Ambitions of Empire
Steven Ruskin

Jeremiah Joyce
Radical, Dissenter and Writer
John Issitt

The History and Poetics of Scientific Biography

Edited by

THOMAS SÖDERQVIST
University of Copenhagen, Denmark

ASHGATE

Published by
Ashgate Publishing Limited
Gower House
Croft Road
Aldershot
Hampshire GU11 3HR
England

Ashgate Publishing Company
Suite 420
101 Cherry Street
Burlington, VT 05401-4405
USA

Ashgate website: http://www.ashgate.com

British Library Cataloguing in Publication Data
The History and Poetics of Scientific Biography. – (Science, Technology and Culture, 1700–1945)
1. Scientists – Biography – Congresses. 2. Science – History – Congresses. 3. Scientists – Biography – Authorship – Congresses. I. Söderqvist, Thomas.
809.9'355'0922

Library of Congress Cataloging-in-Publication Data
The History and Poetics of Scientific Biography / edited by Thomas Söderqvist.
p. cm. – (Science, Technology and Culture, 1700–1945)
1. Science – Biography. 2. Science--History. 3. Scientists – Biography – Authorship – History. I. Söderqvist, Thomas.
Q141.H575 2007
509.2–dc22 2006029937

ISBN 978-0-7546-5181-9

This book is printed on acid-free paper.

Printed and bound in Great Britain by MPG Books Ltd, Bodmin, Cornwall.

Contents

List of Figures and Table

Figures

Table

Notes on Contributors

David Aubin is Associate Professor in the History of the Mathematical Sciences at the Institut de mathématiques de Jussieu, Université Pierre et Marie Curie-Paris 6. His most recent work includes the co-edition with Charlotte Bigg and Otto Sibum of a collected volume entitled *The Heavens on Earth: Observatory Techniques in the Nineteenth Century*. His contact address is <david.aubin@upmc.fr>.

Bernadette Bensaude-Vincent is Professor of History and Philosophy of Science at the University of Paris X. Her main research field is the history and philosophy of physics, chemistry and materials science. She can be contacted at <bensaude@u-paris10.fr>.

Charlotte Bigg is a research scholar at the Max Planck Institute for the History of Science in Berlin. Her recent publications include 'L'Optique de Précision et la Première Guerre Mondiale' in *Revue Suisse d'Histoire,* and 'The Panorama, or La Nature à Coup d'Oeil' in *The Osmotics of Romanticism* (ed. E. Fiorentini). She is currently working on Brownian motion research in early twentieth-century physics and chemistry, and can be contacted at <bigg@mpiwg-berlin.mpg.de>.

Chris Chilvers is Assistant Professor at the History of Technology Department, Technical Knowledge Centre, Technical University of Denmark. He is the British Society for the History of Science Programme Chair, and currently organises two international research programmes on the history of telecommunications and the political and social engagement of scientists. His contact address is <chc@dtv.dk>.

Jacalyn Duffin is a haematologist and historian who holds the Hannah Chair at Queen's University in Kingston, Ontario. In addition to biographies of René Laennec and James Miles Langstaff, she has written *History of Medicine: A Scandalously Short Introduction* (University of Toronto Press, 1999) and *Lovers and Livers: Disease Concepts in History* (University of Toronto Press, 2005), and she has edited an anthology of autobiographical essays by medical practitioners who are also historians (*Clio in the Clinic*, Oxford University Press, 2005). Her current research is on medical saints and miracles. Her contact address is <duffinj@post.queensu.ca>.

Patricia Fara is a Fellow of Clare College, Cambridge. Her books include *Sympathetic Attractions: Magnetic Practices, Symbols and Beliefs in Englightenment England* (Princeton University Press, 1996), *Newton: The Making of Genius* (Macmillan, 2002) and *Pandora's Breeches: Women, Science and Power in the Enlightenment* (Pimlico, 2004). She is currently working on a survey account from Babylon up to the present, provisionally called 'A Brief History of Science'. Her contact address is <pf10006@cam.ac.uk>.

Stephen Gaukroger is Professor of History of Philosophy and History of Science at the University of Sydney. He is the author of *Descartes: An Intellectual Biography* (Oxford University Press, 1995). He is presently involved in a research project on the persona of the philosopher in the early modern era. His contact address is <stephen.gaukroger@arts.usyd.edu.au>.

Thomas L. Hankins is Professor Emeritus at the University of Washington, Seattle. His previous writings on biography include *Jean d'Alembert: Science and the Enlightenment* (1970), *Sir William Rowan Hamilton* (1980), and the seminal article 'In defence of biography: The use of biography in the history of science' published in the journal *History of Science* in 1979. He can be contacted at <hankins@u.washington.edu>.

Signe Lindskov Hansen has an MA in Danish literature and a PhD from the University of Copenhagen. A revised version of her dissertation on biographies of Danish scholars and scientists from the mid-eighteenth century to the present is about to be published by Museum Tusculanum Press. She can be contacted at <s.hansen@lindskov.dk>.

Rebekah Higgitt recently completed her PhD at Imperial College London on biographical representations of Isaac Newton and the significance of his image to nineteenth-century British men of science. She is currently a Research Associate at the University of Edinburgh working on an ESRC-funded project on 'Geography and the British Association for the Advancement of Science, 1831–1939'. She is co-editor of *Early Biographies of Isaac Newton 1660–1885* (Pickering & Chatto, 2005). Her contact address is <rebekah.higgitt@ed.ac.uk>.

Helge Kragh is Professor of History of Science at the Steno Institute, University of Aarhus. He has written a biography of the physicist Paul Dirac, and in 2004 he published *Matter and Spirit in the Universe: Scientific and Religious Preludes to Modern Cosmology* (Imperial College Press, 2004). His present research interests focus on the history of modern physical science, including astronomy and chemistry. He can be contacted at <helge.kragh@si.au.dk>.

Beth Linker is Assistant Professor in the Department of the History and Sociology of Science at the University of Pennsylvania. She is currently finishing a book-length manuscript about medical welfarism and the physical rehabilitation of maimed soldiers in World War I America. Her next project will be a study of posture and fitness in the US, tentatively titled *Slouch: The Rise and Fall of American Posture, 1840–1980*. She can be contacted at <linker@sas.upenn.edy>.

Rena Selya is an historian focusing on the history of the life sciences in the nineteenth and twentieth centuries, with an emphasis on evolution, genetics and molecular biology. Her PhD dissertation at Harvard University, 'Salvador Luria's Unfinished Experiment: The Public Life of a Biologist in a Cold War Democracy', is being revised for publication. She is currently a visiting scholar at UCLA. Her contact address is <selya@post.harvard.edu>.

Vassiliki Betty Smocovitis is Associate Professor in the Departments of Zoology and History at the University of Florida, Gainesville. She is the author of *Unifying Biology: The Evolutionary Synthesis and Evolutionary Biology* (Princeton University Press, 1996). She is currently completing a biography of the plant evolutionary biologist G. Ledyard Stebbins. Her contact address is <bsmocovi@history.ufl.edu>.

Thomas Söderqvist is Professor of History of Medicine and Director of Medical Museion at the University of Copenhagen. His latest book, *Science as Autobiography* (Yale University Press, 2003), was an existential biography of the immunologist Niels K. Jerne. He is currently investigating representations of individuality in recent biomedicine and biotechnology. His contact address is <ths@mm.ku.dk>.

Liba Taub is Reader in History and Philosophy of Science, and Director and Curator of the Whipple Museum of the History of Science at the University of Cambridge. She is the author of *Ptolemy's Universe: The Natural Philosophical and Ethical Foundations of Ptolemy's Astronomy* (Open Court, 1993) and *Ancient Meteorology* (Routledge, 2003). She is currently working on a book on genres of scientific and mathematical discourse in the Greco-Roman world. Her contact address is <lct1001@cam.ac.uk>.

Series Editor's Preface

'Let us now praise famous men, and our fathers that begat us': so runs the biblical injunction, and that was how obituaries and éloges in scientific journals began. But the relationship of serious, critical history to biography (not only where science is concerned) remains awkward, as these essays show. Biography is a popular genre, and those who want to curl up with a book or beguile a long journey delight in it. Scientific biographies, written by more or less well-informed journalists or professional writers, today sell well; but heavyweight lives by academics anxious about context do not, and university presses nowadays look askance at proposals for biographies.

I was myself drawn by an enthusiastic editor at Blackwell's into editing a series of scientific biographies in the 1980s and early 1990s; but he moved on, the company lost interest, drew the line at a scientist as obscure as Justus Liebig, and were happy to unload the series on to Cambridge University Press – at that point, keen to take it. There again, editors changed, the goalposts moved further so that only scientists of extreme eminence could be considered, and the series ground frustratingly to a halt. The test of acceptability had become, I was told, to ask around the office if anybody had heard of the subject proposed. Simply to generate further lives of the exceedingly well-known seemed unworthy of a university press, and of only moderate interest to historians of science.

Biography thus seems in mainstream academe to be marginal, popular (a dreaded word), and amateurish: it is accessible, it involves empathy and writing skills, appreciation of tragedy, and even poetic imagination. Like the alchemist, the biographer is creating life; hoping with Tennyson that the dead man will touch him from the past and be led back to the land of the living. This makes biography rather (but not necessarily altogether) different from the sort of professional writings, using learned words and copious references to theory, that are supposed to score points in Research Assessment Exercises, and before Promotion Committees. There, biography is too often despised and rejected, in favour of much more tedious tomes where learning is worn heavily; and authors sometimes have to disguise biography as something else, like micro-history.

The biographer is not only, like the body-snatcher, a resurrection man; modern biography starts with James Boswell on Samuel Johnson, and living scientists have had their Boswells – we meet one here. Boswell waited until his subject was dead before writing, thereby facing competition from John Hawkins, but it looks as if he was wise. Certainly obituaries, which can be works of art, depend upon personal knowledge, but are not quite the same as biographies. Whether scientific and medical biography is really a distinct genre, quite different from writing lives of other kinds of intellectuals, or indeed of politicians, lawyers, admirals or footballers, remains uncertain: all biographers face problems of balance, of maintaining interest, of conveying something of the day-to-day life (and perhaps drudgery) of the subject, and his or her significance in the bigger scheme of things.

Despite the supercilious glare of those committees, these essayists are not glum about the significance and prospects of scientific biography. They exude enthusiasm. Certainly, 'lives' attract students into history of science. I was myself moved to enter the field as a graduate student under Alistair Crombie by the question 'What made Humphry Davy tick?' – which, thirty years on, with the blandishments of that Blackwell's editor, led to my writing his biography. In the 1830s, he had had two double-decker lives. J.A. Paris's was mildly debunking, and had social mobility as its guiding principle: its frontispiece was an engraving from Sir Thomas Lawrence's swagger portrait of Davy. The President of the Royal Academy depicted the President of the Royal Society in a commanding pose, with the safety lamp in view: both men had risen from very humble origins to high positions in the brittle Regency world, where portrait painting and applied chemistry were in demand. John Davy's was adulatory, depicting the philosopher from the point of view of the younger brother, who in the spirit of family piety went on to edit Humphry's collected works so that he could rank alongside Boyle and Newton. His frontispiece showed handsome young Humphry, with his sparkling eyes, at the outset of his glittering career at the Royal Institution.

Paris's, on the whole, seemed, a century and a half on, the more promising line; but there were naturally all sorts of concerns that I had that he had not: what is new about a new biography is, after all, the biographer. To be the subject of two big biographies within ten years of one's death was unusual, but if the Regency was the age of personality, the Victorian era saw the flourishing of biography: of men of science indeed, but of Dean Stanley as well as Roderick Murchison, Thomas Carlyle and Frederick Temple as well as Michael Faraday. All were in a 'Life and Letters' mould, the letters usually bowdlerised: later writers do not need to consider the sensibilities of the survivors in the same way. Science was only just becoming professional, and its practitioners did not seem to require any very special approach. We live in a different world.

What the essays in this book show is that biography is an attractive and important component of the history of science, that biographers need and possess gifts that would be useful to more analytic historians of science and medicine, and that to be healthy, relevant and interesting, the history of science must include close study of individuals as well as of institutions and social trends. We may hope that university committees and presses will take note; and like this Ashgate series, delight in good biographies. In the mean time, enjoy these stimulating and readable essays.

David M. Knight
December 2006

Preface

The idea for this volume came in 2001, when I was doing research in the wonderful holdings of biographies in the Science Museum Library in London. An ad hoc programme committee, consisting of Janet Browne, University College London; Geoffrey Cantor, University of Leeds; Richard Yeo, Griffith University, Brisbane, and myself invited a select group of scholars to a two-day meeting in May 2002 at the Magleås conference centre in the luscious decidious forests north of Copenhagen.

I am grateful to my fellow organisers, the authors and the other workshop participants for investing so much enthusiasm and critical energy in the discussions. The Danish Research Councils provided generous economic support to the meeting and the subsequent publication of this volume.

The Introduction and the 15 chapters speak for themselves. There is nothing more to add. Time to go to print.

Thomas Söderqvist
University of Copenhagen
September 2005

Introduction

A New Look at the Genre of Scientific Biography

Thomas Söderqvist

The genre of scientific biography enjoys a paradoxical status in today's discourse about science's past.[1] On the one hand, life writing constitutes a literary corpus with a strong presence and a strong impact on the public understanding of science; on the other hand, the genre is hardly discernable in the discussion about the historiography of science, technology or medicine (and invisible in the science studies literature). The purpose of this volume is to address this paradoxical presence versus absence of biography in today's discourse about science, technology, and medicine of the past.[2]

First the presence. A bibliographical overview reveals that about four to five thousand biographies of scientists, engineers and medical men and women have been published, in book format, in the Latin, French, German, English, Italian, Dutch and Scandinavian languages over the last four hundred years; in addition several hundred titles have come out in the Spanish, Hungarian and Russian languages and so on.[3] The first handful of *vitae* of the men of the 'new science' (Copernicus, Tycho, Peiresc and Galileo) appeared in the early and mid-seventeenth centuries, and in the following two centuries a couple of hundred different scientific lives were published, often in Latin, but increasingly in the vernacular. By the end of the nineteenth century the number of titles had risen to about ten a year, often in the 'life and letters' format,

1 The terms 'scientific biography' and 'genre of biographies of scientists' are used synonymously here. With full respect for the vagueness of the term 'scientist', I restrict myself to biographies of those kinds of actors who are listed in the *Dictionary of Scientific Biography*: that is, natural philosophers and natural historians, mathematicians, physicists, chemists, biologists and so on, plus scientifically trained engineers and medical doctors – but not inventors, general practitioners, social scientists, economists or scholars in the humanities, law or theology.

2 The genre of scientific autobiography has not been included in this volume, in spite of the fact that it is perhaps even more unacknowledged.

3 The estimate is based on Oettinger (1854), Howsam (1997), Morton and Moore (1994), and *Isis Cumulative Bibliography*, 1913–65, 1966–75, 1976–85 and 1986–95; in addition, the collections and catalogues of the Science Museum Library, The Wellcome Library and the Royal Society Library, all in London, and the history of science collection in the Niedersächsische Staats- und Universitätsbibliothek, Göttingen, have been surveyed. Biographies published in several smaller European languages (for example, in Finnish, Czech, Greek and Portuguese), are not included in this estimate. Nor have I had an opportunity to survey biographies written in non-European languages.

and in the early post-Second World War period to some twenty per year. In the 1990s about seventy-five biographies of scientists, including engineering and medical scientists, were published each year. In addition to all these monographs, seemingly innumerable short lives have appeared, from the late Renaissance to the present, for example in the form of entries in biographical dictionaries,[4] as *éloges* in academy publications (starting with the Leopoldian Academy in the 1690s),[5] as obituaries in newspapers, journals and magazines, as memorial articles in scientific journals, and so forth. Thus from a quantitative point of view, scientific biography constitutes a most impressive metascientific genre all the way through modernity.

Compared to other genres concerned with the history and present state of science – like philosophy of science, historiography of science, technology and medicine, sociology and anthropology of science and so on – biographies have had a strong popular appeal. Quite a few have been issued in several (often revised) editions, and some have been publishers' delights, selling tens and even hundreds of thousands of copies. For example, Adrian Desmond and James Moore's *Darwin* (1991), an 800-plus-page masterpiece of historical scholarship and writing, has been translated into four languages (German, Italian, Japanese and Portuguese) and found its way into six book clubs; a conservative estimate is a hundred thousand copies sold worldwide in total.[6] Biographies from other language areas have sold well too. To take a random example from the German and Russian literatures: Johannes Hemleben, *Galileo Galilei in Selbstzeugnissen und Bilddokumenten*, a popular yet serious biography, was reprinted several times between 1969 and 1977, selling altogether 33,000 copies;[7] the original Russian edition of Pelageya Kochina's *Love and Mathematics: Sofya Kovalevskaya* ([1981] 1985) is said to have sold a staggering 117,000 copies.[8]

No studies of the cultural impact of scientific biography exist, and one can only speculate about the genre's historical importance. How has it contributed to the recruitment and socialization of young scientists? To the dissemination of scientific virtues and philosophies? To the self-understanding and identity formation of scientists, engineers and clinicians? Or to the maintenance of the moral fabric of science, technology, and medicine? We don't know. Yet the existence of thousands of published volumes of scientific lives indicates that the public understanding of science and its practices may have been significantly shaped by the genre. It is not too far-fetched to suggest that scientific biography as a whole may in fact have had a

---

4 Bayle's *Dictionnaire historique et critique* (1697) carried biographical articles, for example on Copernicus. Jöcher's *Compediöses Gelehrten-Lexikon* (1733), was the first biographical dictionary that specialised in scholars, including mathematicians, astronomers, natural philosophers and natural historians.

5 The best-known are of course the *éloges* of l'Academie des sciences; see Paul (1980).

6 Jim Moore, personal communication (13 July 2005). Fourteen years after its first publication, *Darwin* still held a sales rank of around 57,000 on <www.amazon.co.uk> – few history of science books ever reach this rank, even when they are fresh from the press.

7 Hemleben (1969); print run according to title page.

8 Kochina (1985); print run according to dust cover.

stronger cultural and political impact than any other genre of metascientific writing in the last four hundred years.[9]

Yet, in spite of this formidable presence, the genre is not very visible in today's discourse about science as a historical, social and cultural phenomenon. Historians of science are remarkably silent about biography as a product of and contributor to the social and political formation of modern science and its identities. Biography as a genre of writing and a cultural phenomenon is largely absent from treatises of the historiography of science (and totally absent from books dealing with science studies methodology).[10] Almost every imaginable aspect of science, technology and medicine in society, culture and history – for example, the role and importance of scientific societies and academies, scientific disciplines and specialties, scientific meetings, expeditions and travels, the role of archives and manuscripts, diaries and letters, museum exhibitions and popular science magazines, scientific practice in the laboratory, the clinic and the field, science in novels and the movies, and so forth – has been treated extensively and in depth. But not scientific lives and their literary vehicle, scientific biography. General cultural historians, literary historians or book historians do not have much to say about the genre either. And the biographers themselves have usually continued to practise their craft without reflecting much upon its history or place in culture. Individual lives – either embodied life trajectories or literary representations of such trajectories – do not have a real habitation in the professional reflection about science in history and culture; biography seems to be the ugly duckling in today's discussions about historiographical approaches and science studies methodologies.

A recent handbook – the *Reader's Guide to the History of Science* (2000), edited by historian of science Arne Hessenbruch – illustrates this paradoxical presence/absence status of biography in today's discourse about science, technology and medicine. It contains some 125 evaluative and for the larger part well-researched and well-written bibliographical essays on the biographical literature on important individuals in the history of science, technology and medicine. This metabiographical approach – that is, focusing on the evaluation of already published biographies about a person rather than writing yet another short biographical essay – reflects the publisher's (Fitzroy Dearborn, formerly St James Press, now swallowed by Taylor & Francis) long tradition for this kind of biographical dictionaries.[11] Despite this commendable tradition, however, the entries in the *Reader's Guide* are not matched by a corresponding awareness of biography as a genre. The handbook covers a number of interesting topics relating to science in modern culture, such as 'popularisation', 'literature and science', 'reading culture and science' and 'history of science: general

9 With the caveat that blockbusters such as Thomas Kuhn's *The Structure of Scientific Revolutions*, which still, forty years after its first publication in 1962, ranks around 2,600 on <www.amazon.com> (accessed 13 July 2005), have contributed significantly to the cultural impact of history and philosophy of science.

10 Kragh (1987) is an exception by devoting a whole chapter to scientific biography.

11 This innovative format that has so far not been adopted by other scientific, technological or medical dictionaries, including the recently published *Dictionary of Medical Biography* (Bynum and Bynum, 2006).

works' – but the reader will look in vain for an entry on 'biography' in general, or on 'scientific biography' as a phenomenon in the history of modern culture, or anything along these lines.

The *Reader's Guide to the History of Science* is not an exception, but rather a symptom of the status of biography in the discourse about metascientific methodologies today. Particular biographies – especially biographies of 'dead white males', like Galileo, Newton, Darwin, Huxley and Richard Feynman (and occasional 'dead white females' like Marie Curie and Dorothy Hodgkin) – have acquired an established, if not necessarily favoured, place in the awareness of most historians of science and students of science studies. But biography as a *generic* phenomenon, as a *kind* of writing and a constitutive part of scientific culture throughout modernity, is much less visible. Other biographical genres, particularly literary biography (and to some extent art biography) have been the topic of quite a few studies, both historical and formal-literary, some of which have been rather sophisticated from a theoretical and methodological point of view, to the extent that biography, including autobiography, is now a respected topic in literary studies.[12] But not scientific biography – yet.

Metabiographical Comments in Prefaces and Reviews

Although the genre of scientific biography has not yet acquired its proper place on the agenda of today's discourse about science in history and culture, there have none the less been a few attempts to reflect on it. Authors of biographies have occasionally discussed their craft in prefaces and introductions. Furthermore, reviewers of biographies in newspapers and magazines, in the general scholarly literature and in specialised journals of history of science, technology and medicine, have occasionally dropped a few general remarks. And finally, a small but steadily increasing number of scholarly commentaries have appeared, especially over the last decades. Such reflections make it possible to reconstruct the contours of a history of metabiographical discourse which, in turn, might provide the backdrop for a discussion of the status and role of the genre today.

Comments by authors of biographies are actually fairly rare. In fact, one of the very first 'scientific' biographers, Pierre Gassendi, prefaced his *vita* of Tycho Brahe with a long explanation for his undertaking, saying, among other things, that he wished to present Tycho's astronomical work in details in the same way as his classical predecessors had focused on the public achievements of generals and statesmen (Gassendi, 1654). But such acts – whether of decorum or sometimes perhaps authorial whim – have always been rare among biographers. Most life writers simply embark on the narrative after a short preface and introduction, but usually without any reflections on their own writing process, even less on the genre as a whole. Ronald W. Clark, one of the most prolific authors of scientific biographies

12 See Backscheider (1999) and France and St Clair (2002), and, of course, the quarterly journal *Biography*. There is even a masters' programme in 'life writing' in the University of East Anglia, Britain's premier literary school (see <http://www.uea.ac.uk/eas/admissions/courseprofiles/q325t>, accessed 11 July 2005).

in the 1960s and 1970s, did not write a single word about his craft, neither in the books themselves nor in separate articles. He, like most biographers of scientists, seem to have been more devoted to telling a straight story rather than indulging in self-reflection.

Nevertheless, a few authors do indeed (like Gassendi) comment on their craft, at least in passing, and such occasional remarks in prefaces and introductions give some insight into the shifting authorial sensibilities with respect to the genre and its uses. For example, a number of late nineteenth- and early twentieth-century scientific life writers expressed the dominant biographical empiricism of the time. The ideal of these biographers was to downplay their role as authors by fashioning themselves as merely collectors of biographical facts. The reader should do the interpretation, the 'author' was just a transparent medium between subject and reader. 'Prepared by', 'collected and arranged by' and 'compiled and edited by' were some of the euphemisms of late Victorian authorship. Bernard J. Harrington introduced his *Life of Sir William E. Logan* (1883) with a description of himself as a mere 'compiler', and added that his aim had not been 'to write a eulogium or yet a lengthy criticism of Sir William, but rather to bring together such of his own words as will recall him to the minds and hearts of old friends, or enable … [others] to form for themselves an estimate of his character and work' (p. v).

Such apparently humble authorial attitudes were related to a strong belief in the strength and independence of the biographical sources: 'It has been the author's desire to let documents and letters speak as far as possible for themselves', commented Silvanus P. Thompson in *The Life of William Thomson, Baron Kelvin of Largs* (1910, p. x), the standard biography of Kelvin until the 1980s. The German biographer Felix Auerbach added, in *Ernst Abbe*, that 'a biography stands and falls with the full completeness [*sic*!] and trustworthiness of the source material' (1918, p. vi).[13] Such biographical empiricism lived well into the first half of the twentieth century. Carl van Doren, the author of *Benjamin Franklin* (1938), a classic among Franklin biographies, considered himself a writer who simply collected the numerous facts from 'Franklin science' (that is, the 'Darwin industry' of the time). Van Doren emphasised he had no explicit agenda, not even a moral lesson to tell: 'No effort has been made to cut his nature to fit any simple scheme of what a good man ought to be. Here, as truly as it has been possible to find out, is what Franklin did, said, thought, and felt' (p. viii).

This empiricist ideal stands in stark contrast to the recent tendency to see the biographer as an active and reflective constructor of a life in science. Compare, for example, Adrian Desmond's cinematic view of himself as the 1990s *auteur* of Thomas Henry Huxley with Leonard Huxley's self-effacing attitude when portraying his father: 'So far as possible, I have made his letters, or extracts from them, tell the story of his life' (Huxley 1900, p. vi). Desmond seductively draws the reader into the skilfully crafted persona of Huxley, almost as if he were the director of the movie, whereas his predecessor expresses his wish to let the readers make their own conclusions. Such temporal changes in authorial intentions not only illustrate the

13 'Eine Biographie steht und fällt mit der restlosen Vollständigkeit und Zuverlässigkeit des Quellenmaterials.'

changing sentiments towards biographical writing from (hypocritical?) self-effacing traditionalism to narcissistic postmodernism; it also provides food for thought on how, today, to write a biography which does not just add to the ephemerae of the book market.

Book reviews provide another window into temporally and culturally shifting biographical and historiographical sensibilities. In many respects, reviews have not changed much. Most reviewers focus on a descriptive summary of the narrative, and end with a couple of sentences about how good (or bad) a read the book is. But rarely do reviewers expand on the composition, style or other textual, literary or aesthetic qualities of the book; in other words, biographies rarely become objects of literary criticism. Few reviewers relate the work at hand to other scientific biographies, not to mention the genre of biography as whole. It is as if biographies by default are classified under the rubric of 'history of science', and are supposed to be read for their informative and historical contextualising usefulness only – or their entertainment value.

Occasionally, however, biography reviewers rise above the nitty-grittiness of the particulars to reflect on the genre, its structure, functions, uses and historicity. 'This is the age of biography', proclaimed the anonymous reviewer of Samuel Smiles' *Industrial Biography: Ironworkers and Toolmakers* in *The Times* of 28 December 1863, and continued with a reflection on the status of biography of the time: 'there never was a time when personal memoirs or biographies proper have been so much studied as in the present century, it appears that the new intellectual movement which dates from the period of the French Revolution has this for one of its chief characteristics that it is essentially biographical'. The great biographies had so far dealt only with 'soldiers and sailors, and statesmen, poets and artists, and philosophers'. But one should not forget, the reviewer continued, those who were responsible for the 'material grandness' of the time. Hence 'the lives of our engineers are peculiarly worthy of being written'.

Few reviewers are as programmatic as Smiles'. Usually generic remarks have to be distilled from the texts: 'If the first object of a biography is to enlist the sympathy of the reader for the man then the book is a signal success', wrote the reviewer of Hugh Robert Mills' *The Life of Sir Ernest Shackleton* (Debenham, 1923, p. 124), thereby expressing a changing attitude to scientific heroes in the early twentieth century – not uncritical worship, but humanistic sympathy. A related topic among reviewers in the first half of the twentieth century was the empathetic relation of biography authors towards their subjects. The reviewer of J.M.D. Olmsted's *Charles-Édouard Brown-Séquard: A Nineteenth Century Neurologist and Endocrinologist* (1946), himself a scientist and an accomplished biographer, opined that a good biography demands that 'the biographer must have been in close spiritual rapport with his subject' (Fulton, 1948). This was how Boswell had related to Johnson, and Harvey Cushing to William Osler; but when it came to Olmsted and Brown-Séquard, the reviewer felt that 'author and subject were not spiritually attuned'. So even if Olmsted gave a 'just, critical appraisal' of his subject's scientific contributions, the portrait 'lacks color in its descriptive passages' (Fulton 1948, p. 187). Others, in contrast, warned against the spectre of empathy: the author of *Lawson Tait 1845–1899* was accused of having been 'captivated by his spectacular subject' (Faulconer, 1949, p. 525).

Today, few reviewers are interested in the spiritual rapport between biographer and biographee. The most striking feature of biography reviews in history of science journals from the last decades is the frequency of the adjectives 'hagiographical' versus 'contextual' as standard derogatory and praising criteria respectively. The rejection of a life narrative as 'hagiographical' is not a particularly recent phenomenon – for example, in his review of a Rutherford biography, the Cambridge physicist C. G. Darwin warned of the danger of 'the degeneration into a dreary hagiography' to which biographers of great scientists were exposed (Darwin, 1940, p. 324)[14] – but in the last decades 'hagiographical' has become the most common reason for debunking a biography, and vice versa, 'contextual' (or biography's usefulness for the social and historical contextualisation of science) the most common reason for approval. The two criteria sometimes appear together, as in historian of science Mary Jo Nye's praise of Bernadette Bensaude-Vincent's *Langevin: science et vigilance* (1987) both for substituting 'a valuable critical biography for hagiography' and for contributing to the understanding of history of science 'in a broad intellectual and social context' (Nye, 1989, pp. 338–9). Such is the power of these two signal words that they often have replaced a closer analysis of the composition or style or other textual qualities of the biographies under review.

Earlier Metabiographical Scholarship

A third kind of reflection on scientific biography is provided by specific metabiographical commentaries. Whereas occasional authorial remarks on the genre can be traced back to its very beginnings in the seventeenth century (and while reviews of biographies began to appear in the mid-nineteenth century), systematic reflection on the genre is of much younger date.

The rapidly growing number of biographies in the first half of the twentieth century gave rise to review essays of subject-specific biographical traditions – Faraday, Darwin, Einstein biographies and so on – which also contained observations on the changing patterns of biographical scholarship about the specific scientist in question, but with few exceptions (Fullmer, 1967) they restricted themselves to the problems involved in writing about 'their man' and were not overtly concerned with the general problems involved in writing scientific lives.[15] The first attempts

14 He also pointed to the danger of 'the psychological account of the (usually Freudian) thoughts which the biographer thinks that his subject ought to have been thinking' (Darwin 1940, p. 324).

15 For example, Henry Guerlac (1954) wrote on Lavoisier and his biographers, Dorinda Outram (1976) on the tradition of Cuvier *éloges*, David Cassidy (1979) on biographies of Einstein, Marilyn Marshall (1980) on Fechner studies, Steven Jacyna (1983) on the succession of nineteenth-century images of John Hunter, Frederick Churchill (1982) and Marjorie Greene (1993) on the Darwin biographical tradition, L. Pearce Williams (1991) on Faraday biographies, and so on; in addition A. Rupert Hall wrote a whole book on eighteenth-century biographies of Newton (1999). It is interesting to note that the entries on individual scientists in the *Reader's Guide to the History of Science* (Hessenbruch, 2000) mentioned above follows the same pattern.

to address the generic problems of scientific life writing appeared in the 1960s and 1970s. In a short article in the *Bulletin of the Atomic Scientists* of 1963, Harold I. Sharlin complained about the fact that scientific biographies were mostly written by other scientists, 'the result too often being not biography, but a laboratory report on a specimen that seems never to have been alive' (Sharlin, 1963, p. 27). June Fullmer followed up with a discussion of biographies of medical scientists, and pointed to the fact that the genre is highly diverse, and that the cause of this diversity might be that 'biographers, like all writers, bend both consciously and unconsciously to the trends of taste pursued by their contemporaries' – an observation that is probably still valid (Fullmer, 1971, pp. 173–4). The most influential commentator on the genre was Thomas L. Hankins, whose article 'In defense of biography' (1979) was published a year before his magisterial *Sir William Rowan Hamilton*.[16] Hankins's aim was to rescue the genre of scientific biography from its many detractors by demonstrating its usefulness for contextual historical studies. Scientific life writing, said Hankins, 'gives us a way to tie together the parallel currents of history at the level where the events and ideas occur. ... We have, in the case of an individual, his scientific, philosophical, social and political ideas wrapped up in a single package' (p. 5).

The number of commentaries grew rapidly in the late 1980s and 1990s, probably reflecting the changing (and now again more positive) attitudes to biography among historians of science. For example, Robert M. Young (1988) claimed a central place for biography as 'the basic discipline for human science' because of its powers to introduce contextualism and historicity in understanding human matters, including science. Helge Kragh included a chapter about biography in his textbook *An Introduction to the Historiography of Science*, but took a more traditional position than Young when suggesting that the genre is useful for writing about 'the aristocrats' of science while 'the thousands of less important or less exciting scientists will remain beyond the reach of biography' (Kragh, 1987, p. 173). Kragh's elitist view surely resonates with L. Pearce Williams, who made a passionate *plaidoyer* for biography as a genre that specialises in the study of 'great men', in contrast to social studies of science that are primarily interested in the 'swamp' (Williams, 1991). Others plainly noted that the possibilities for writing about 'great men' were not yet exhausted. In a review essay aptly titled 'Has the social history of medicine come to age?', Ludmilla Jordanova was struck by 'the restricted range of genres and topics tackled' in the history of medicine and public health, adding that 'one of the most dramatic examples of this is the almost total absence of scholarly biographies', and continuing: 'Even for the really big names celebrated by whig history, few have been the recipients of sustained biographical treatment'; not even Edward Jenner has received any substantial modern treatment (Jordanova, 1993, p. 438; cf. Söderqvist, 2004). Other commentators again went beyond the concern with 'great men' by raising more genre-specific problems; for example, Susan Sheets-Pyenson (1990) reminded her fellow historians of science that biography is not only a historical craft, but also a literary art.

16 The fact that Hankins's paper was published in the journal *History of Science* probably contributed to its status as a minor 'citation classic'.

The 1990s saw a number of conferences and collections reflecting on the genre. In 1991, Australian historians of science Michael Shortland and Richard Yeo solicited a collection of essays published under the title *Telling Lives: Essays on Scientific Biography* (1996). Four years later, a symposium at the Oregon State University in Corvallis was devoted to the life and work of Linus Pauling within the framework of the art of biography in general (Krishnamurti, 1996). Antonella La Vergata at the University of Bologna edited a special 180-page issue on 'Le biografie scientifiche' in the Italian history of ideas journal *Intersezioni* (issue no. 1, April 1995), and a year later the editors of the *Canadian Bulletin of Medical History* took the initiative to publish an interview with two leading Canadian medical biographers, Michael Bliss and Jackie Duffin, followed up by a discussion symposium in a later issue of the journal. In addition to occasional sessions and panels at major professional history of science meetings, there have been conferences on biographies of scholars in Stockholm in 1997 (Badou, 1998) and in Copenhagen in 2001, and at least another collection of articles has been published in the Dutch history of science journal *Gewina* (vol. 23, 2000). A conference at Newnham College, Cambridge in 1999 dealt with biography as a tool to bridge the history of science and the history of women in science (Govoni 2000), and occasional postgraduate courses have dealt with scientific biography.[17] Significantly, such reflections on biography were made almost exclusively by historians of science and historians of medicine; with few exceptions, historians of technology have shown little interest in the genre (Buchanan, 1996); scholars in science studies and philosophy of science have so far not expressed any interest in biography as a genre of writing about science past.

In sum, there has been a growing scholarly interest in discussing scientific biography in the last two decades – a tradition which apparently reflects the renaissance of the genre in the 1980s and 1990s. This fledgling metabiographical tradition is fragmented, however, and some contributions have been published in languages (Swedish and Dutch) which are not immediately accessible for international readers. Furthermore, much of the literature is either essayistic – even anecdotal – or normative and programmatic, or restricted to the authors' more or less idiosyncratic experiences of writing one or two biographies. Much is repetitive; the same tricks of the trade or methodological points are repeated from article to article, and the discussions start from scratch at each meeting and in each collection of papers. In other words, the genre has not been the subject of any systematic, scholarly treatment, either methodological or historical.

Chapter Summaries

This collection of essays may not have entirely transcended such idiosyncratic, essayistic, fragmented and anecdotal approaches. But it expresses an ambition to bring the growing interest in scientific biography a step further and thereby contribute

17 For example, a graduate course at Oregon State University, taught by Mary Jo Nye in 2004.

to the resolution of the presence/absence paradox that permeates this rich and varied genre.

It is hardly meaningful to speak about 'scientific biography' before the rise of modern science in the seventeenth century. As Liba Taub points out in Chapter 1, there was no genre of scientific biography in the Greco-Roman world, yet there were lives (*bioi*) of mathematicians and philosophers who sought to explain nature (*physis*). By analysing the three extant *bioi* of Pythagoras (by Diogenes Laertius, Porphyry and Iamblichus), Taub concludes that these accounts were not biographies as we understand the term today. Their purpose was rather to contribute to the history of a philosophical tradition, especially by relating the interactions between teachers and their students, and to serve as guides for the readers on how to live and how to be more divine. Ancient *bioi* of Pythagoras were intended to be edifying.

So were many seventeenth-century *vitae*. In Chapter 2, Stephen Gaukroger points out that seventeenth-century natural philosophers like Francis Bacon and René Descartes thought that the project of a new science needed a wholly new kind of philosopher. The personal qualities involved in this project did not define an individual in the twentieth-century scientific-biographical sense, but a wholly new kind of persona. The study of this kind of person is not biography in the present sense, but is nevertheless closer to our present understanding of the genre of biography than the history of scientific thought.

In Chapter 3, David Aubin and Charlotte Bigg raise the question of the self-fashioning of the (outstanding) scientist – that is, how the self is perpetually elaborated, perceived and reflected. Implicitly drawing on the classical Plutarchian model of comparing parallel lives, they analyse how the various identities – scientist, explorer, discoverer (British or *français*) and so on – were combined in different ways to form the selves of the two late nineteenth-century astrophysicists Jules Janssen and Norman Lockyer. This focus on dynamic self-fashioning has also led the authors to revise received views on historical material, including commemorative medals, which they suggest can be used to supply insights into such processes of self-construction.

Patricia Fara, in Chapter 4, is also interested in self-fashioning, but her focus is on visual strategies. Analysing portraits of Isaac Newton, Joseph Banks and other scientific practitioners, she discusses the sitters' strategies of pictorial self-fashioning as well as the artists' involvement in creating role models to be emulated. Through considering the interactive triangular relationships between sitters, artists and viewers, Fara examines how the production, reproduction and reception of portraits lead to shifting interpretations and affect the biographical information they can provide.

In Chapter 5, Thomas L. Hankins explores a key paradox in our understanding of the individual in scientific culture, namely that biographers concentrate on how science is related to personal, cultural and social forces, while tenure committees and peer reviewers are expected to keep all such aspects of science securely out of sight. This irony raises the question of the relation between biography and the reward system of science, which Hankins approaches in an innovative way, that is, by investigating patents and copyrights as a reward system. Both biographies and patents are awarded to individuals, and they are both based on narratives of discovery. The history of patents shows how this particular reward system has determined scientific and technological development; the historiography of science can, suggests Hankins, benefit by raising the same issues in writing biographies of scientists.

Stephen Gaukroger uses seventeenth-century tragedy as a guide to understanding contemporary conceptions of the persona of the early modern philosopher. Similarly, in Chapter 6, Christopher Chilvers suggests that the poetic categories of tragedy can be employed to understand the role of the individual in the history of science. His essay is part of a growing body of work that develops explicitly 'politically engaged' dimensions in the history of science. Chilvers analyses the life of the Soviet physicist Boris Hessen, better known today as a leading Marxist historian of science in the 1920s and 1930s, who was arrested and shot during the Stalin purges in 1936. His defence of quantum mechanics and relativity can be interpreted as his *hamartia*, whereas the *perepeteia* (reversal) began with the party philosophers' attack on independent interpreters of Marxism.

Biographical accounts of a particular scientist are, of course, culturally and historically variant. But there are also factors that diminish the amount of narrative variation, for example biographers' dependence on a common foundation of documentary evidence. In Chapter 7, Helge Kragh suggests yet another factor: the biographical tradition. Biographies of particular scientists often build upon earlier exemplars that serve as models for later biographers. Kragh traces the development of biographies of the Danish astronomer Tycho Brahe, from Pierre Gassendi's *vita* in 1654 to John Christianson's *On Tycho's Island* (2000). Gassendi's shadow was looming large in the Tycho-biographical tradition until the late twentieth century, when his biographers freed themselves from the exemplar, largely because new historiographical sensibilities opened up for a wider array of contextual approaches.

In Chapter 8, Signe Lindskov Hansen takes another biographical tradition – that of the Danish theologian and anatomist Niels Stensen (a.k.a. Steno) – as her point of departure for a discussion of the function of biography as a genre. Hansen argues that biographies have a programmatic function, an underlying agenda, by which biographers promote religious, political or aesthetical views. This agenda manifests itself in certain story-telling strategies (and not least those of tragedy) that are visible in a variety of rhetorical manoeuvres aimed at seducing the reader.

In Chapter 9, Rebekah Higgitt also addresses questions of biographical rhetoric; she is concerned with how the various nineteenth-century depictions of Newton were presented and made persuasive. It was not the case, as one might believe, that biographers always attempted to establish their subject as a hero. Some tried indeed to retain the idealised eighteenth-century image of the great man, but others sought to undermine the myth. Despite the different attitudes, however, all Newton biographers stressed the need for an inductive and factual approach to the sources as a means of giving their biographical accounts authority and the appearance of impartiality.

Four authors draw on their own experiences in writing biographies to raise crucial questions concerning the use of the genre. Two of them deal with securely dead men. Bernadette Bensaude-Vincent, who has published books on Antoine-

Laurent Lavoisier and Paul Langevin, discusses in Chapter 10 the problem of taking the impact of the public image of the individual into account. The memories of Lavoisier and Langevin are both visibly present in the French urban landscape, and both names are surrounded by popular clichés. How to deal with the popular myths – shall they simply be debunked or included as part of the biographer's project? And more generally, how to handle the fact that biography is a genre that mediates between commemoration and history?

In Chapter 11, Jacalyn Duffin draws upon her experience of publishing the lives of René Laennec, one of the most famous medical scientists in Europe in his time, and James Miles Langstaff, a then completely unknown country doctor. Publishers' attitudes towards biographies of medical scientists were still cautionary in the early 1990s: why didn't she write social history instead? 'Can't you make it look like it is not a biography', one publisher responded to Duffin's Laennec manuscript. Eventually, *To See With a Better Eye* was published in 1999, and just a year later it was honoured with the Hannah Medal of the Royal Society of Canada.

The experience of writing about living scientists is dealt with in two other essays. Through her biographical account of molecular biologist Salvador Luria, Rena Selya examines in Chapter 12 the relationship between science and democracy to gain a deeper understanding of the history of molecular biology and late twentieth-century American history in general. But in doing so she had to confront her protagonist's strong autobiographical voice, and the whole array of personal, and more or less mythical, narratives that surround the rise of molecular biology. This kind of experience reinforces a certain cautionary attitude on the part of the biographer: How to avoid hagiography? How to maintain a separate voice from that of her subject? How to be more rigorously historical? The interpersonal relationship does not have to be a threat to historiographical rigour, however.

In an autobiographical account of her thirteen-year-long *pas de deux* with evolutionary botanist and science celebrity Ledyard Stebbins, in Chapter 13 Vassiliki Betty Smocovitis discusses the pros and cons of having a close rapport with one's subject. Working with Stebbins sometimes tested her emotional resilience and social skills, but also provided abundant gratifications; for example, Stebbins opened the door to his friends and colleagues, who became a network of living witnesses. Does the death of the subject then lead to a more 'objective' view of the biographical subject? Smocovitis claims that it brings forth detachment and a more vivid portrait, but not necessarily a more 'objective' or 'neutral' account.

Finally, two essays investigate the relation between writing history and life-writing as two distinct genres of writing about the past. In Chapter 14, Beth Linker takes her point of departure in the irony that in their effort to purge their historical accounts of 'elitism', social historians of medicine have come close to producing a heroic portrait of their 'founding father', the Swiss-American physician and medical historian Henry Sigerist. Furthermore, Sigerist would have been a reluctant participant in their ideological crusade against biography, since his methodological pluralism included medical biography alongside social history. Linker suggests that social historians of medicine will only be able to accept biography if they reassess their own history, including their own Sigerist myth.

In the concluding Chapter 15, Thomas Söderqvist discusses the history of the relationship between biography and historiography of science, especially in the twentieth century. He distinguishes three phases of historians' attitude to scientific biography. First, historians of science like George Sarton and Donald McKie thought of biography as a genre that could stem the tide of anti-scientism by cultivating and accentuating the human aspects of science; then followed a 'Cold War generation' of historians of science who wished to professionalise their craft by weeding out the personal aspects of science, and finally, the last two decades have witnessed a return of biography in the guise of an *ancilla historiae*, that is, as a tool for contextualising the history of science. It is time to write the history of scientific biography from a truly historicist point of view, to see the genre in all its forms and varieties without being unduly influenced by the present standards for what constitutes good history of science.

Acknowledgements

I am grateful to Betty Smocovitis and Liba Taub for comments on this introductory essay.

Bibliography

Anon. (1863), [Review of Samuel Smiles, *Industrial Biography*, 1863], *The Times*, 28 December (Palmer's Full Text Online version, <http://www.proquest.co.uk/products/palmers.html>; p. 5, col. a).

Auerbach, Felix (1918), *Ernst Abbe: Sein Leben, sein Wirken, seine Persönlichkeit, nach den Quellen und aus eigener Erfahring geschildert*, Leipzig: Akademische Verlagsgesellschaft.

Backscheider, Paula R. (1999), *Reflections on Biography*, Oxford: Oxford University Press.

Badou, E. (ed.) (1998), *Forskarbiografin. Föredrag vid ett symposium i Stockholm 12–13 maj 1997*, Stockholm: Kgl. Vitterhetsakademien.

Bayle, Pierre (1697), *Dictionnaire historique et critique*, 2 vols, Rotterdam.

Bliss, Michael and Jacalyn Duffin (1996), 'Life writing in medical history [interview]', *Canadian Bulletin of Medical History*, **13**, 123–37.

Buchanan, R. Angus (1996), 'Theoretical aspects of engineering biography', *ICON: Journal of the International Committee for the History of Technology*, **2**, 53–8.

Bynum, W.F. and Helen Bynum (eds) (2006), *Dictionary of Medical Biography*, 5 vols, Westport, CT: Greenwood Press.

Cassidy, David C. (1979), 'Biographies of Einstein', in H. Nelkowski et al. (eds), *Einstein Symposium, Berlin, aus Anlaß der 100. Wiederkehr seines Geburtstages, 25. bis 30. März 1979*, Berlin: Springer-Verlag, pp. 490–500.

Churchill, Frederick B. (1982), 'Darwin and the historian', in R.J. Berry (ed.), *Charles Darwin: A Commemoration 1882–1982: Happy is the Man that Findeth Wisdom*, London: Academic Press, pp. 45–68

Darwin, C.G. (1940), [Review of A.S. Eve, *Rutherford*, 1939], *Nature*, **145**, 324–5.

Debenham, F. (1923), [Review of H.R. Mills, *The Life of Sir Ernest Shackleton*, 1923], *Nature*, **112**, 124–5.

Desmond, Adrian (1997), *Huxley: From Devil's Disciple to Evolution's High Priest*, Reading, MA: Addison-Wesley.

Desmond, Adrian and James Moore (1991), *Darwin*, London: Michael Joseph.

Faulconer, Robert J. (1949) [Review of H. Flack, *Lawson Tait 1845–1899*, 1949], *Bulletin of the History of Medicine*, **23**, 524–5.

France, Peter and William St Clair (2002), *Mapping Lives: The Uses of Biography*, Oxford: Oxford University Press.

Fullmer, J.Z. (1967), 'Davy's biographers: Notes on scientific biography', *Science*, **155**, 285–91.

Fullmer, J.Z. (1971), 'Medical lives and medical letters: A chapter in the history of scientific biography', in Edwin Clarke (ed.), *Modern Methods in the History of Medicine*, London: Athlone Press, pp. 173–93.

Fulton, John F. (1948), [Review of J.M.D. Olmsted, *Charles-Édouard Brown-Séquard*, 1946], *Isis*, **39**, 187–8.

Gassendi, Pierre (1654), *Tychonis Brahei, equitis Dani, astronomorum coryphæi vita*, Paris.

Govoni, Paula (2000), 'Biography: A critical tool to bridge the history of science and the hhistory of women in science. Report on a conference at Newnham College, Cambridge, 10–12 September 1999', *Nuncius*, **15**, 399–409.

Greene, Marjorie (1993), 'Recent biographies of Darwin: The complexity of context', *Perspectives on Science*, **1**, 659–75.

Guerlac, Henry (1954), 'Lavoisier and his biographers', *Isis*, **45**, 51–62.

Hall, A. Rupert (1999), *Isaac Newton: Eighteenth-Century Perspectives*, New York: Oxford University Press.

Hankins, Thomas L. (1979), 'In defence of biography: The use of biography in the history of science', *History of Science*, **17**, 1–16.

Harrington, Bernard J. (1883), *Life of Sir William E. Logan, Kt., LL.D, F.R.S., F.G.S. &c., First Director of the Geological Survey of Canada. Chiefly Compiled From His Letters, Journals and Reports*, Montreal: Dawson Brothers.

Hemleben, Johannes (1969), *Galileo Galilei in Selbstzeugnissen und Bilddokumenten*, Hamburg: Rowohlt.

Hessenbruch, Arne (ed.) (2000), *Reader's Guide to the History of Science*, London: Fitzroy Dearborn.

Howsam, Leslie (1997), *Scientists Since 1660: A Bibliography of Biographies*, Aldershot: Ashgate.

Huxley, Leonard (1900), *Life and Letters of Thomas Henry Huxley*, 2 vols, London: Macmillan.

Jacyna, L.S. (1983), 'Images of John Hunter in the nineteenth century', *History of Science*, **21**, 85–108.

Jöcher, Christian Gottlieb (1733), *Compendiöses Gelehrten-Lexikon*, 3rd edn, Leipzig: Gleditsch.

Jordanova, Ludmilla (1993), 'Has the social history of medicine come of age?', *The Historical Journal*, **36**, 437–49.

Kochina, Pelageya ([1981] 1985), *Love and Mathematics: Sofya Kovalevskaya*, Moscow: Mir.

Kragh, Helge (1987), *An Introduction to the Historiography of Science*, New York: Cambridge University Press.

Krishnamurti, Ramesh S. (ed.) (1996), *The Pauling Symposium: A Discourse on the Art of Biography*, Corvallis, OR: Oregon State University Libraries.

Le Vergata, Antonello (ed.) (1995), 'Le biografie scientifiche', *Intersezioni: Rivista di storia delle idee*, **15**, 1–183.

Marshall, Marilyn E. (1980), 'Biographical genre and biographical archetype: Five studies of Gustav Theodor Fechner', *Storia e critica della psicologia*, **1**, 197–210.

Morton, Leslie T. and Robert J. Moore (1994), *A Bibliography of Medical and Biomedical Biography*, 2nd edn, Aldershot: Scolar Press.

Nye, Mary Jo (1989), [Review of B. Bensaude-Vincent, *Langevin*, 1987], *Isis*, **80**, 338–9.

Oettinger, Edouard-Marie (1854), *Bibliographie biographique universelle: Dictionnaire des ouvrages relatifs a l'historie de la vie publique et privée des personnages célèbres de tous les temps et de toutes les nations, depuis le commencement du monde jusqu'a nos jours*, Brussels: J.J. Stienon.

Outram, Dorinda (1976), 'Scientific biography and the case of Georges Cuvier: With a critical bibliography', *History of Science*, **14**, 101–37.

Paul, Charles B. (1980), *Science and Immortality: The Éloges of the Paris Academy of Sciences (1699–1791)*, Berkeley, CA: University of California Press.

Sharlin, Harold I. (1963), 'The scientist in biography', *Bulletin of the Atomic Scientists*, **19** (November), 27–8.

Sheets-Pyenson, Susan (1990), 'New directions for scientific biography: The case of Sir William Dawson', *History of Science*, **28**, 399–40.

Shortland, Michael and Richard Yeo (eds) (1996), *Telling Lives: Essays on Scientific Biography*, Cambridge: Cambridge University Press.

Söderqvist, Thomas (2004), 'Why are there so few scholarly biographies in the history of medicine and public health?', *Michael: Publication Series of the Norwegian Medical Society*, **1**, 119–29.

Thompson, Silvanus P. (1910), *The Life of William Thomson, Baron Kelvin of Largs*, 2 vols, London: Macmillan.

Van Doren, Carl (1938), *Benjamin Franklin*, New York: Viking Press.

Williams, L. Pearce (1991), 'Faraday and his biographers', *Bulletin for the History of Chemistry*, **11**, 9–17.

Young, Robert M. (1988), 'Biography: The basic discipline for human science', *Free Associations*, **11**, 108–30.

Chapter 1

Presenting a 'Life' as a Guide to Living: Ancient Accounts of the Life of Pythagoras

Liba Taub

There was no genre of 'biographies of scientists' in the Greco-Roman world, yet information about natural philosophers and mathematicians was communicated in many ancient works. In some cases, the biographical details provided seem to be offered only incidentally. So, for example, we get snippets about incidents in the life of Thales, often celebrated as the first natural philosopher, from Plato (*Theaetetus* 174a). He relates a story, credited to a female servant, that Thales fell into a well while star-gazing. This claim to intellectual unworldliness is somewhat balanced by the account given by Aristotle (*Politics* 1259a5ff.) relating how Thales cornered the market in olive presses in Miletus.

Biography was not a clearly demarcated genre for the ancient Greeks and Romans.[1] The boundaries with other genres, including the eulogies or *encomia* used to praise heroes and important citizens,[2] are not distinct and may be somewhat artificial (Harvey, [1937] 1974, p. 158). Christopher Pelling has suggested that ancient biography should be understood as a set of 'overlapping traditions, embracing works of varying form, style, length, and truthfulness' (1996, p. 241).[3] The *bios* ('life'; plural = *bioi*) as an account or celebration of a particular life can be found in a range of ancient writings, and may also include a discussion of the opinions (or *doxai*) of the individual being described; such accounts often carried an ethical or religious message. Those writings known as 'doxographical' concentrate on presenting opinions, without necessarily providing biographical details of the life of the individual (Mansfeld, 1999, pp. 17–19; Runia, 1999). While biography itself was not a rigidly defined genre for the Greeks and Romans, the *bioi* are linked by the desire to celebrate individuals.

---

1 This essay is part of a larger project, tentatively titled *Genres of Ancient Science*. For the most part (but not always), I have adopted a 'latinised' spelling of Greek names and terms (for example, Iamblichus, rather than Iamblichos), to conform to general usage.

2 Harvey ([1937] 1974), p. 158, defines the encomium (*egkōmion*) as 'a Greek choral hymn … in celebration, not of a god, but of some man'. He explains that 'the word means a song "at the kōmos" (here the revel at the end of a banquet), and thus suggests a eulogy of the host'.

3 Söderqvist (Chapter 15 in this volume) points to the difficulties in speaking about a homogenous genre of biographies of scientists in the modern period.

If we look for evidence of biographies of scientists in Greek and Roman writings, there are difficulties in using the term 'scientist' in this context. The term 'scientist' is modern; William Whewell coined it in 1840 (vol. 1: cxiii);[4] arguably, there is no exact analogue in other periods to that professional label. Nevertheless, there are a number of the ancients whom few historians would question should be included in the study of the history of science. These would include those who sought to explain nature (*physis*) and the natural world. Many such thinkers have been called 'philosophers', both by ancient writers as well as modern. Mathematicians should also be included, since those Greek and Roman writers who thought about such matters often considered mathematics, like philosophy, to be a branch of theoretical knowledge.

But ancient understandings of what constitutes mathematics and philosophy do not map on to modern conceptions in a straightforward manner. Just as today there is no unanimity regarding what constitutes science and scientific practice, so in antiquity people did not always agree about the categories in which they placed those pursuits which might today fall under the historian's rubric of 'science'. As G.E.R. Lloyd has noted, 'a distinction between "philosophers" and "scientists" is in general hard to draw in Greco-Roman antiquity. Natural science is a domain that straddles both those disciplines as *we* perceive them' (Lloyd, 1991, p. 301).[5]

Aristotle is often credited with generating interest in the writing of *bioi* that reflected intellectual and ethical concerns (Pelling, 1996, p. 241; Momigliano, 1993, p. 119). A member of Aristotle's Lyceum, Aristoxenus of Tarentum (b. *c.* 370 BCE), is credited as the author of *bioi* of at least four philosophers, Pythagoras (b. mid-sixth century BCE), Archytas (fl. *c.* 400–350 BCE), Socrates (469–399 BCE) and Plato (*c.* 429–347 BCE); these may have formed part of a series of *bioi* of philosophers (Barker, 1996, 169–70; cf. Pelling, 1996, pp. 241–2). A number of authors, including Sotion of Alexandria (active between 200 and 170 BCE), are also said to have produced such a series. These sorts of collections of *bioi* became a standard form for presenting intellectual history; 'the "succession" of teachers and pupils was a helpful idiom for explaining influences' (Pelling, 1996, p. 242).[6]

Many of the accounts of the lives of ancient 'scientists' – natural philosophers and mathematicians – reflect these two aims: to celebrate the individual and also to present intellectual history and lineage. These two strands of ancient biographical writing are not entirely separate. More generally, many *bioi* – not only those

---

4 Cited in *The Compact Edition of the Oxford English Dictionary* (1971), under 'scientist'.

5 The ancient Greeks were themselves very interested in classifying human activities; on the classification of knowledge and expertise, see Tatarkiewicz (1963), pp. 231–40, and Kühnert (1961). There was no one term, either in Greek or in Latin, which carried the meaning of the modern word 'science'. While some have suggested that the Greek work *episteme* is equivalent to the Latin *scientia*, this is open to discussion. See also Söderqvist (Introduction to this volume) on the choices he has made regarding who to include as a 'scientist'.

6 Autobiography is not treated in this volume, but it is worth noting that Galen's *On My Own Opinions* (1999) can be considered part of intellectual history. Momigliano (1993) considers the role of the Peripatos in shaping Hellenistic biography; in particular, see Ch. 4, 'From Aristotle to the Romans'.

concerned with natural philosophers and mathematicians – emphasise the heroic character of the individual whose life is being described. Plutarch (b. before 50 CE, d. after 120), in his *Parallel Lives* of Greek and Roman heroes, was particularly interested in ethical and moral matters (cf. Lamberton, 2001, pp. 69–74). Some of the later *bioi*, for example the *Life of Apollonius* by Philostratus (d. between 244 and 249 CE), appear almost hagiographical. Furthermore, as a number of scholars have noted, 'the Christian Gospels have points of contact with the Greek tradition, with their charismatic hero and their anecdotal narrative texture' (Pelling, 1996, p. 242; cf. Talbert, 1977).

A 'hero' was a member of a class of beings worshipped by the ancient Greeks. Heroes were generally conceived of as the powerful dead, and understood as forming a class intermediate between the gods and human beings. As Emily Kearns has noted, from the fourth century BCE onwards, there was in practice great variation in the types of honours offered to heroes. She explains that 'at one end of the spectrum it could have a strong resemblance to the offerings given to a dead relative; at the other, it might be barely distinguishable from worship paid to a god'; 'there was a tendency in many parts of the Greek world for mourners to depict the ordinary dead in heroic forms, to call them "hero", and even on occasion to establish regular heroic cults and a priesthood' (Kearns, 1996, pp. 693–4; Kearns, 1989).[7]

In 1965, Moses Hadas and Morton Smith published *Heroes and Gods: Spiritual Biographies in Antiquity*, an examination of what they called ancient 'aretalogy'. In their preface, they noted that 'aretalogy is not recognised as a word in our dictionaries, nor is the type of literature it designates treated as a separate genre' (Hadas and Smith, 1965, p. xiii). To some extent, 'aretalogy' could be understood as the celebration of *arete*, that is, virtue. But Hadas and Smith use the term to describe something more specific: 'a formal account of the remarkable career of an impressive teacher that was used as a basis for moral instruction'.[8] Hadas and Smith were concerned with literary images of a particular kind of person, namely a hero or god. They defended their focus on literary images by arguing that:

> For the effect exerted upon the course of history the authorized image of the hero is more important than his historical personality. It is upon the image rather than the person that reverence is bestowed, whether formally in an organised cult or informally in popular tradition, and it is the cult, formal or informal, that ensures the survival of the image. (Hadas and Smith, 1965, p. 4).

Hadas and Smith argue that biography, as a constructed literary image, carries historical significance; they focused their study of ancient biography on aretalogical

---

7 Diogenes Laertius (5.91) relates how Heraclides of Pontus attempted to organise heroic honours for himself.

8 Hadas and Smith (1965, p. 60) acknowledged that 'we have no complete text surviving from the past specifically labelled aretalogy'. Some scholars, including Burridge (1992/1995, pp. 17–19) have argued against 'aretalogy' as a separate genre, and prefer to use the term 'aretalogical' as an adjective to describe other literary forms; see also Tiede (1972), pp. 1–13, on the 'problem' of aretalogies, and Talbert (1977), pp. 12–13, who rejects the suggestion that the gospels are examples of aretalogical biography.

accounts of heroic teachers as moral exemplars, whose activities and teachings may also have religious import, as well as ethical influence. Turning to the *bioi* of ancient natural philosophers and mathematicians, there are often important overlaps between the intellectual histories told through a series of *bioi* of philosophers and the aretalogical accounts that focus on a particular heroic teacher.

Bioi of Pythagoras

Pythagoras holds a special place in the history of science, with many natural philosophers, mathematicians and scientists claiming descent from his intellectual line (Kahn, 2001, pp. 156–72). So, for example, Kepler wrote in a letter to Galileo dated 13 October 1597 that 'you are following the lead of Plato and Pythagoras, our true masters' (Kepler, 1951, pp. 40–41; cf. Kepler, 1945, p. 145).[9] In the twentieth century, Einstein argued that a scientist may be a '*Platonist* or *Pythagorean* insofar as he considers the viewpoint of logical simplicity as an indispensable and effective tool of his research' (cited by Kahn, 2001, p. 172).[10] Alfred North Whitehead, in his 1925 Lowell Lectures, published as *Science in the Modern World*, credited Pythagoras with 'founding European philosophy and European mathematics' (Whitehead, 1925, p. 54).

Three ancient accounts of the life of Pythagoras survive from late antiquity, and while a detailed analysis and comparison of the texts is not possible here, they provide a useful focus to explore various strands of ancient biography. The three 'lives' of Pythagoras can all be loosely dated to about the third century CE, and were written by Diogenes Laertius (dating uncertain, but probably first half of the third century CE), Porphyry (234–*c.* 305 CE) and Iamblichus (*c.* 245–*c.* 325 CE); all three were composed significantly after the death of their subject (fifth century BCE). The life of Pythagoras was celebrated not only in textual accounts; Samos, his birthplace, honoured this favourite son by representing him on coins (see Figure 1.1).

9 See also Hallyn (1990), p. 62, note 25, on the assessment of the importance of Pythagoreanism to Copernicus, and Jardine and Segonds (1999), pp. 222ff., with regard to Kepler's interest in Pythagoreanism.

10 For the complete passage, see 'Reply to Criticisms', in Einstein (1949), p. 684. I am grateful to Tom Ryckman for help in locating this.

Figure 1.1 Pythagoras, depicted on a coin from Samos dating from the reign of Trajan Decius (249–251 CE), in the collection of the British Museum (BMC 351). © The Trustees of The British Museum.

The focus here will be on the three extant biographies, for even a brief examination of these ancient biographies of Pythagoras should indicate directions for exploring elements of intellectual history and aretalogy included in modern histories of science, which are often also motivated by similar urges to establish an intellectual heritage and to celebrate heroes. The *bioi* of Pythagoras served to establish an intellectual history for thinkers who aligned themselves with Pythagoras and Plato. In looking at ancient biographies of natural philosophers, some of those works that survive should be understood as an attempt to formulate an intellectual history; so, for example, Porphyry's *Life of Pythagoras* was part of his larger project on history of philosophy (O'Meara, 1989, p. 25). Some biographies of natural philosophers and mathematicians may be regarded as part of a tradition of writing about divine beings – that is, heroes. In certain cases, the subject of the *bios* is presented as a moral exemplar, whose life and teachings provide a guide for living. Significantly, some of these ancient biographies combine elements that are ethical and/or religious with intellectual history.

The complicated nature of ancient biographies of 'scientists' – *bioi* concerned with natural philosophers and mathematicians – is especially apparent when we look at the biographies of Pythagoras. In 1962, in his pathbreaking book *Weisheit und Wissenschaft: Studien zu Pythagoras, Philolaos und Platon*, Walter Burkert argued that Plato and his school had largely created a conception of Pythagorean philosophy (known to us through late ancient sources) to equip themselves with an intellectual heritage.[11] According to Burkert, much of what we think we know about Pythagoras and those who Aristotle referred to as 'the so-called Pythagoreans' is, arguably, an example of the desire to construct an intellectual history.

Pythagoreanism is generally understood as having two rather distinct forms or schools after the fifth century BCE. The 'scientific' or philosophical form (whose advocates were the so-called *mathematikoi*) manifested itself in the fourth century BCE in the thinking of Philolaus and Archytas of Tarentum and the Pythagoreans whom Plato knew and succeeded.[12] The other was a religious, sectarian form, whose adherents were known as *akousmatikoi*, those following certain oral teachings (Graf, 1996, pp. 1284–5; cf. Burkert, 1972, pp. 192–208). These two strands of Pythagoreanism, the scientific/mathematical and religious/ethical, are both reflected in the extant biographies of Pythagoras, and so make the untangling of these two traditions problematic. To attempt to disassociate the two traditions is probably a mistake, for this may give a skewed view of the range of teachings that were offered under the umbrella of Pythagoreanism. These two traditions may also, to some extent, mirror the somewhat overlapping desires of the biographers to produce both intellectual-historical and aretalogical accounts.

Each of the three extant 'lives' of Pythagoras was part of a larger work: Diogenes Laertius' *Lives of Eminent Philosophers*, Porphyry's history of philosophy (*Philosophical History*) and Iamblichus' compendium *On Pythagoreanism*.[13] Diogenes Laertius' biography of Pythagoras is one of a large number of *bioi* he included in his *Lives of Eminent Philosophers*. Porphyry was a scholar who wrote on a wide variety of topics, including philosophy, grammar, rhetoric and religions. He produced over sixty works, including commentaries on works by Plato (including the *Timaeus*) and Aristotle (including the *Physics*), as well as *Homeric Enquiries*. Porphyry wrote a history of philosophy from Homer to Plato. His *Life of Pythagoras* is an excerpt from this work (Smith, 1996).[14] Iamblichus probably studied with

11 See also Kahn (2001), pp. 2–3, on Burkert's contribution.

12 On Plato's relationship to the Pythagoreans, see Kahn (2001), pp. 39–62.

13 The relationship between the three works has been the topic of some scholarly debate; see, for example, Burkert (1972), pp. 97ff.

14 Porphyry was responsible for arranging the work of his teacher, Plotinus, into the sets of nine treatises known as the *Enneads*. His treatise on vegetarianism, *On Abstinence*, is well known, but his own writings on metaphysics are almost entirely lost. Of particular relevance to historians of science are the fragments of his commentaries on Plato's *Timaeus*, Aristotle's *Physics*, Ptolemy's *Harmonics*, as well as his introduction to Ptolemy's *Tetrabiblos* and a treatise on the entry of the soul into the embryo (a work formerly attributed to Galen, but probably by Porphyry).

Porphyry in Rome or Sicily before founding his own school, possibly in Apamea (modern Syria). Many of his writings are now lost, including commentaries on works by Plato and Aristotle. He compiled a compendium of Pythagorean philosophy, incorporating extracts derived from earlier writers; the first four books survive, amongst which *On the Pythagorean Life* is the first (O'Meara, 1996).[15] These three surviving accounts of Pythagoras' life incorporate material from earlier sources. The first 'life' of Pythagoras was apparently that written by Aristoxenus (now lost); Charles Kahn has suggested that many of the marvellous and moralistic features found in the later accounts by Diogenes Laertius, Porphyry and Iamblichus find their source in his work (2001, p. 69).

Diogenes Laertius' *Bios* of Pythagoras

Any discussion of ancient biographies of 'scientists' must include Diogenes Laertius, because of his work *Lives of Eminent Philosophers*. Almost nothing is known about Diogenes Laertius, including his dates, where he lived, and with whom he studied; he is often disparaged as a 'late' source, but he is our only source of information for many of the philosophers about whom he wrote. His work, probably written in the first half of the third century CE, appears to pre-date the other extant biographies of Pythagoras.

In his *Lives*, Diogenes Laertius provides biographies and summaries of the opinions of the ancient Greek philosophers, beginning with Thales (sixth century BCE) and ending with Epicurus (341–270 BCE). Diogenes Laertius himself may have been an Epicurean since he celebrates Epicurus' divinity (Mejer, 1978).[16]

Diogenes Laertius' work incorporates the two approaches to biography distinguished above. It provides an intellectual history, but is also motivated by an ethical desire to provide exemplars of the good life (Gardiner, 2003). He makes it clear that Epicurus is to be regarded as a god; the religious and moral message of this 'life' is explicit. But as we read Diogenes Laertius' account, we also get a detailed look at Epicurus' natural philosophy; in fact, Diogenes Laertius is responsible for the preservation of much of what we know of Epicurus' ideas, through the quotation of three of his letters. In other words, Diogenes Laertius' account of Epicurus is concerned not only with his providing an account of his life, but also with the details of his work.

Diogenes Laertius organises the lives of the philosophers into two 'successions', a method of organisation developed by Theophrastus as well as Sotion of Alexandria (Long, [1925] 1972, p. xx). The first 'succession' is the Ionian, beginning with Thales, continuing through Socrates, and then branching into three 'schools'. The 'Italian' succession begins with Pherecydes and continues through Pythagoras.

---

15 The other surviving books of the compendium are the *Protrepticus* (believed to contain material from Aristotle's lost work by the same name), *On General Mathematical Science* and *On Nicomachus' Arithmetical Introduction*.

16 Some scholars have suggested that Diogenes Laertius was a Sceptic, or he may have been an adherent of no particular school; cf. Long (1970), Sharples (1996) and Barnes (1992), p. 4243–4.

Some of the scholars who have worked on ancient *bioi* have examined their structure, arguing that certain features seem to be 'standard'. Richard A. Burridge, as part of his study of the Christian gospels and Greco-Roman biography, identified the following as the usual topics covered: ancestry, birth, boyhood and education, great deeds, virtues, death and consequences ([1992] 1995, pp. 145–7, 178–80, 207–9). An earlier scholar, Armand Delatte, identified general areas according to which Diogenes Laertius organised the biographical material on each of his subjects. Delatte recognised that Diogenes Laertius was a biographer specifically interested in relating his subjects' intellectual role and contributions; he described the following as the usual topics covered (1922; cited in Long, [1925] 1972, pp. xxi–xxii): origin, education (including philosophical training and travels), place in a succession or founding of a school, character and temperament (illustrated by anecdotes and sayings), important life events, anecdotes relating to the subject's death and epigrams, chronological data, works, doctrines, documents (for example, last will, letters), other men of the same name, miscellaneous notes (including lists of followers, inventions and political activity). The order of topics identified by Delatte only loosely applies to Diogenes Laertius' *bios* of Pythagoras; nevertheless, there is almost a sense that Diogenes Laertius was following a checklist of themes to be covered. The founding of a school and one's ideas, teachings and writings are particularly relevant for intellectual history.

Diogenes Laertius' *bios* of Pythagoras appears at the beginning of his examination of the philosophy of Italy (Book 8), having ended his treatment of the 'Ionian' succession with his discussion of the philosopher Chrysippus (*c.* 280–207 BCE; treated in Book 7). The opening passage reads as follows:[17]

> Having now completed our account of the philosophy of Ionia starting with Thales, as well as of its chief representatives, let us proceed to examine the philosophy of Italy, which was started by Pythagoras, a son of the gem-engraver Mnesarchus, and according to Hermippus, a Samian, or, according to Aristoxenus, a Tyrrhenian from one of those islands which the Athenians held after clearing them of the Tyrrhenian inhabitants. (Diogenes Laertius, [1925] 1972, vol. 2, p. 321).

Here, as elsewhere, Diogenes Laertius names different sources and points to differences in their accounts.[18] Throughout the *Lives*, Diogenes Laertius proudly names his sources, which number well over two hundred (Hope, 1930, pp. 59–60; cf. Long, [1925] 1972, p. xix); the breaks and shifts that occur at various points in his account are very likely due to his moving from one source to another.

Continuing, he briefly recounts (8:2–3) Pythagoras' education and travels (which offered him opportunities for study abroad), then reports (8:4–5) what Heraclides of Pontus (fourth century BCE) claimed Pythagoras used to say about himself, regarding his former lives.

---

17 All translations of Diogenes Laertius in what follows are by Hicks, in Diogenes Laertius ([1925] 1972).

18 In his Prologue in Book 1 (1.15), Diogenes Laertius names Pherecydes as the first in the Italian school, followed by Pythagoras, and then his son, Telauges. We are told elsewhere (8.2) that Pythagoras was a pupil of Pherecydes of Syros.

Diogenes Laertius considers Pythagoras' activities as an author (8:6–9), noting that 'there are some who insist, absurdly enough, that Pythagoras left no writings whatever', and points to evidence, credited to Heraclitus (fl. *c.* 500 BCE), that Pythagoras did write a number of works ([1925] 1972, vol. 2, p. 325). Amongst the several treatises said to have been written by Pythagoras was one with a title shared by other writings attributed to pre-Socratic authors, *On Nature*.

Here, Diogenes Laertius emphasises Pythagoras' identification and standing as a physicist (*physikos*). Furthermore, the title of the work, *On Nature*, was that which would have been expected of physicists, and therefore attests to Pythagoras' reputation as someone who philosophised about nature. Other works attributed to Pythagoras (such as *On Education* and *On Statesmanship*) are discussed, as is the report, following an account given by Sosicrates (probably mid-second century BCE), that Pythagoras answered Leon the Tyrant of Phlius' question about who he was by answering 'a philosopher', because, as Pythagoras went on to explain, the philosopher seeks truth (8:8). The section ends with a brief summary (8:9) of the contents of Pythagoras' three treatises, noting that he forbade prayer, on the grounds that we do not know what will help us, called for temperance in drinking and eating, and offered cautions about the harmful effects of too much sexual activity.

A number of Pythagoras' accomplishments are then reported, in a seemingly random order. So, for example, we are told (8:11) that once, disrobed, his thigh appeared to be gold, and that crossing the river Nessus a number of people heard the river welcome him. Then (8:11–12) we learn that he first led geometry to perfection, spending most of his time on the arithmetical aspect of geometry and also discovering the musical intervals on the monochord, while 'not neglecting even medicine'. Diogenes Laertius reports that: 'We are told by Apollodorus the calculator that he offered a sacrifice of oxen on finding that in a right-angled triangle the square on the hypotenuse is equal to the squares on the sides containing the right angle' ([1925] 1972, vol. 2, p. 331). Following the report of this discovery, we learn (8:12) that Pythagoras was the first to suggest that athletes have a diet of meat, rather than dried figs and soft cheese, but Diogenes Laertius notes that this may have been the work of a different Pythagoras, one who was an athletic trainer. In any case, Diogenes Laertius takes this opportunity (8.13) to discuss *our* Pythagoras' views on diet, including his vegetarianism. In this account, we find the mingling of details relating to wonders and marvels associated with Pythagoras, as well as accounts of his mathematical accomplishments. Diogenes Laertius specifically mentions some of his sources, and even questions some of the information he transmits.

We then (8:14) learn something of Pythagoras' views on reincarnation, and that, according to Aristoxenus, he 'was the first to introduce weights and measures into Greece'; moreover, 'it was he who first declared that the Evening and Morning Stars are the same, as Parmenides maintains'. Immediately following this information, Diogenes Laertius reports that 'so greatly was he admired that his disciples used to be called "prophets to declare the voice of God", besides which he himself says in a written work that "after two hundred and seven years in Hades he has returned to the land of the living"' ([1925] 1972, vol. 2, p. 333). Here we have an intellectual history, that is, a history of ideas and intellectual discoveries, intertwined with an aretalogical account, which (although it is not much developed) includes references to disciples as well as a resurrection story.[19]

We are told (8:15–16) that 'down to the time of Philolaus [*c.* 470–390 BCE] it was not possible to acquire knowledge of any Pythagorean doctrine, and Philolaus alone brought out those three celebrated books which Plato sent a hundred minas to purchase' ([1925] 1972, vol. 2, p. 335). We learn a little bit more about Pythagoras' teaching and lecturing and his views on education, and also his gift for friendship. A number of his 'watchwords' or 'precepts' follow (at 8:17); Diogenes Laertius glosses the meaning of these sometimes cryptic sayings. He discusses Pythagoras' diet, appearance and habits (8:19–20), his dealings with oracles and priestesses (8:21), and his ethical advice to his disciples (8:22–4).

Diogenes Laertius then names Alexander (Polyhistor; first half of the first century BCE, who reported what he found in the Pythagorean memoirs or notebooks, *Pythagorikai hypomnēmata*)[20] as his source for a rather lengthy list of Pythagoras' ideas on a range of topics. These topics are those which, traditionally, interest historians of science and mathematics: 'the principle of all things is the monad or unit'; 'the sun, the moon, and the others stars are gods; for, in them, there is a preponderance of heat, and heat is the cause of life'; 'the moon is illumined by the sun' ([1925] 1972, vol. 2, pp. 341–3). This section is carefully demarcated by Diogenes Laertius, both at the beginning and at the end, as information coming from Alexander (8:24–36).

At 8:36, Diogenes Laertius notes that 'what follows is Aristotle's'. Diogenes repeatedly makes an effort to distinguish and name the source of his information; he then provides a series of anecdotes from a variety of sources (8:36–8), along with several different accounts of Pythagoras' death (8:39–40), an anecdote attributed to Hermippus regarding Pythagoras and his mother (regarding a journey into Hades), and information about Pythagoras' wife Theano and his son Telauges (his successor, and possibly a teacher of Empedocles) (8:42–4). Diogenes Laertius (8:45) also offers information about his date of flourishing (60th Olympiad) and the longevity of his school (nine or ten generations). We also learn about other men of the same name living around the same time at no great distance from one another (8:46–8). A few brief tributes are then presented (8:48): 'Favorinus says that our philosopher used definitions throughout the subject matter of mathematics'; 'further, we are told he was the first to call the heaven the universe and the earth spherical' ([1925] 1972, vol. 2, pp. 351 and 365). A letter from Pythagoras to Anaximenes is quoted (8:49), and Diogenes Laertius concludes his *bios* of Pythagoras.

Throughout the *Lives of Eminent Philosophers*, as in his *bios* of Pythagoras, Diogenes Laertius aimed to provide an intellectual history. The organisation of the work into 'schools' or 'successions' makes this clear. In addition, Diogenes Laertius' work is clearly a compilation, and he is proud to name his sources at various points.

---

19 Resurrection is not a standard feature of Diogenes Laertius' accounts of philosophers' lives. Accounts of discoveries, sometimes referred to as 'heurematology' or 'heurematography', may constitute another genre. I am grateful to Nick Jardine for sharing the literature on this topic; a useful starting point is Copenhaver (1978).

20 See Kahn (2001), pp. 74ff., on this work.

His account of the life of Pythagoras is not presented as a continuous narrative, but rather a collection of information about his subject, who is only one among many in the *Lives of Eminent Philosophers*. But Pythagoras holds a special place in Diogenes Laertius' history of philosophy: he was the founder of a philosophical succession.

Porphyry's *Life of Pythagoras*

The status of the philosopher as hero, moral exemplar and even divinity is crucial in the works under examination here. Scholars have pointed to the rising influence of Christianity as important in shaping the *bioi* of Pythagoras produced by both Porphyry and Iamblichus. Porphyry and his student Iamblichus were Neopythagoreans or Neoplatonists, depending on how one interprets these labels.[21] In their accounts, Pythagoras is, as Kahn (2001, p. 134) has noted, 'the paradigm of the sage as divine man'. The marvels and wonders associated with Pythagoras are emphasised by Porphyry and Iamblichus; there are indications that both authors may have written their accounts with the specific goal of providing an alternative to the Christian gospels (Kahn, 2001, p. 134; Dillon and Herschbell in Iamblichus, 1991, pp. 25–6; cf. O'Meara, 1989, p. 214).

Porphyry's *Life of Pythagoras* was part of his *Philosophical History* from Homer to Plato; it is not complete, and our version ends abruptly. (Except for short extracts, it is the only part of the *Philosophical History* to survive.) Dominic O'Meara has noted (1989, pp. 25–6) that Porphyry's '*Life* reads like a learned compilation of source materials concerning Pythagoras'. His account shares some similarities of organisation with that of Diogenes Laertius, and some content, but contains more information and is more of a continuous narrative. Like Diogenes Laertius, Porphyry begins with a consideration (1–5) of Pythagoras' origin and family (including his offspring), naming a number of sources that offer divergent accounts. But Porphyry, unlike Diogenes Laertius, does not begin his 'life' of Pythagoras by placing him within a particular intellectual tradition. In his opening passages, while commenting on Pythagoras' childhood precociousness, Porphyry focuses on his family background and parentage: 'It is agreed by most that Pythagoras was the son of Mnesarchus, but there has been disagreement concerning Mnesarchus' race. Some say he was a Samian, but Neanthes in the fifth book of his work on Pythagoreanism says he was a Syrian.' He indicates that there are differing versions of Pythagoras' origins, and relates that:

> Apollonius, in his book *On Pythagoras*, also gives the name of Pythagoras' mother, Pythaïs, a descendant of Ancaeus, the founder of Samos. Apollonius says, too, that some declare Pythagoras was by procreation the child of Apollo and Pythaïs, and only nominally the child of Mnesarchus. As a matter of fact, one of the Samian poets does say, 'Pythagoras,

21 See, for example, Gorman (1979), p. 2, and O'Meara (1989), pp. 4–5, on the appropriation of Pythagorean ideas by Neoplatonic philosophers, and Kahn (2001), p. 133, on the absorption of neo-Pythagoreanism into neo-Platonism. Porphyry and Iamblichus shared a teacher–student relationship, but Iamblichus is often described as a rival to Porphyry; see O'Meara (1989), p. 214.

> the friend of Zeus, whom Pythaïs, most beautiful of the Samians, bore to Apollo.' Finally, that he was a pupil not only of Pherecydes and Hermodamas but of Anaximander too, is also stated by Apollonius. (Porphyry, 1965, pp. 107–8).[22]

Following his recounting of the various accounts of Pythagoras' ancestry, like Diogenes Laertius, Porphyry then moves on to the topic of Pythagoras' education (6–13), again naming his sources. Several anecdotes are offered (14–15), including one regarding the training of athletes and recommendations regarding diet (again meat, rather than cheese and figs).

Porphyry's order of exposition corresponds more closely to that described by Delatte than that of Diogenes Laertius. Porphyry recounts Pythagoras' travels once he left Samos for Italy, and the founding of his school (16–22). The description of Pythagoras' character is accomplished through illustrative anecdotes; Porphyry assures us (28–29) that 'ten thousand other things yet more marvelous and more divine are told about the man, and told uniformly in stories that agree with each other'; indeed, he says, 'to put it bluntly, about no one else have greater and more extraordinary things been believed' (1965, p. 116).

Information about his listening to the harmony of all things, including the harmony of the spheres, is reported (30–31), as are matters related to his daily life, including his diet and sacrificial practices (32–6). His teaching methods and ideas are discussed at some length (36–53), including their development within his school. Porphyry reports (46–7) that 'he practiced a philosophy of which the object was to deliver and set free of such fetters and bonds [as incarnation] the mind that had been separated from the cosmic mind for incarnation in us'; 'accordingly, it uses mathematics and the sciences dealing with the borderland between bodies and the bodiless to train in advance the eyes of the soul' (1965, p. 122).

In contrast to Diogenes Laertius presentation, Porphyry's account provides information about Pythagoras's ideas and teaching in a more unified manner; the references to Pythagoras' ideas and teaching are not scattered throughout the *bios* of Pythagoras, as they are in Diogenes Laertius'. As noted above, Diogenes Laertius' ordering and changes in topic may reflect his own shifting between sources. Porphyry's account seems to be more digested, coherent and unified, and may allow readers to focus more readily on Pythagoras' intellectual achievements.

Similarities between Porphyry's account of Pythagoras' work and the Christian synoptic gospels have been examined by Morton Smith (Hadas and Smith, 1965, pp. 101–4). Indeed, a number of twentieth-century scholars have pointed to parallels between the extant *Lives* of Pythagoras and the gospels (Lévy, 1927; Dillon and Herschbell in Iamblichus, 1991).

Iamblichus' *On the Pythagorean Life*

Iamblichus presents by far the lengthiest account of Pythagoras' life, but his work *On the Pythagorean Life* is arguably not only a biography, but rather a preliminary guide

22 All translations of Porphyry in what follows are by Smith, in Porphyry (1965).

to Pythagorean philosophy. It was part of his larger work, *On Pythagoreanism*,[23] and is somewhat different from those presented by Diogenes Laertius and Porphyry. Kahn (2001, p. 5) credits Eduard Zeller (1880–92) with being the first to point out that 'the further a document is from Pythagoras' own time, the fuller the account of Pythagoras becomes'; however the length and detail of an account does not reflect its accuracy.

Iamblichus' account is more concerned with divinity than the others. It may be deliberately anti-Christian, and may also have been consciously modelled on and in competition with the Christian gospels (Clark, 1989, pp. ix–xiii; O'Meara, 1989, pp. 214–15). Furthermore, Iamblichus' account focuses not only on Pythagoras himself, but offers a good deal of information concerning his followers and the members of his 'school'. So, for example, *On the Pythagorean Life* ends with a list of the names of the 'most famous' Pythagorean men and women.

From the very beginning of the work, Iamblichus invokes a completely different style from that of both Diogenes Laertius and Porphyry. He begins (Ch. 1) with the explanation that:

> At the start of every philosophical investigation, it is after all the custom, at least for all who are sound-minded, to invoke God. But at the outset of that philosophy rightly believed to be named after the divine Pythagoras, it is surely all the more fitting to do this; for since this philosophy was at first handed down by the gods, it cannot be comprehended without the gods' aid. Moreover, its nobility and greatness exceed human ability to understand it immediately: only when the goodwill of the gods leads the way, can someone with gradual approach slowly appropriate something from it.[24]

Iamblichus makes it clear from the start that philosophy is not only an intellectual endeavour: it is dependent upon its link to the divine for its accomplishment. Indeed, Iamblichus proclaims that Pythagoras himself will be the divine guide through whom his philosophy can be comprehended: 'And after the gods, we shall choose as our leader the founder and father of this divine philosophy' (1991, p. 31). In other words, the divine philosopher Pythagoras himself will be our teacher.

Iamblichus' account of Pythagoras' life and his school which follows is a highly developed aretalogical account of an heroic, indeed divine, teacher; some might argue that it verges on the hagiographical. Iamblichus does mention, somewhat briefly, Pythagoras' ideas and 'discoveries' in *On the Pythagorean Life*, but the purpose of this first book of *On Pythagoreanism* was not to expound and explain Pythagorean philosophy in detail (O'Meara, 1989, pp. 33–4); this was the purpose of the later books, namely the *Protreptic to Philosophy* (Book II), *On General Mathematical Science* (III), *On Nicomachus' Arithmetical Introduction* (IV), *On Arithmetic in Physical Matters* (V), *On Arithmetic in Ethical Matters* (VI), *On Arithmetic in Theological Matters* (VII), *On Pythagorean Geometry* (VIII), *On Pythagorean*

23 See O'Meara (1989), pp. 30ff., on the place of the *Life* within the larger work, and a consideration of its possible titles.

24 Here and in what follows, translations of Iamblichus are by Dillon and Herschbell, in Iamblichus (1991).

Music (IX) and (probably) *On Pythagorean Astronomy* (Book X). (Only the first four books of *On Pythagoreanism* survive; there are only excerpts of Books V–VII.)

O'Meara (1989, p. 39) has argued that *On the Pythagorean Life* is itself an exhortation or 'protreptic to Pythagorean philosophy through the illustration of the spiritual credentials of the founder of that philosophy'. As Gillian Clark (1989, p. xv) has explained, once students of Pythagoreanism

> had read the life of Pythagoras, and become convinced that Pythagoras was a divine soul sent to reveal the truth and teach human beings how to live, they were to continue with the *Protrepticus*, 'Exhortation to Philosophy', which offers Pythagorean sayings and philosophers side by side with extracts from Plato and Aristotle. Thus encouraged, they advanced to a series of highly technical works on aspects of Pythagorean mathematics: that is, mathematics understood as the study of the structure of reality.

It is only after the prologue describing the divinity of Pythagoras that Iamblichus begins his discussion of his origins (in Ch. 2). Unlike Diogenes Laertius and Porphyry, Iamblichus starts his account of Pythagoras' life not by focusing on his home town or biological parents, but instead describing Pythagoras' divine lineage: 'The story goes, then, that Ancaeus who dwelt in Same in Cephallenia was sired by Zeus'; 'the tradition is that Mnesarchus and Pythais, Pythagoras' parents, were from the household and family started by Ancaeus who founded the colony' (1991, p. 33). Thus, Pythagoras was celebrated as the descendant of a god, namely Zeus.

Iamblichus provides us with details of the way in which Mnesarchus learned that his son would possess divine gifts, and so re-named his wife and chose an appropriate name for his child, referring to the Pythian Apollo. He rejects suggestions that Apollo himself impregnated Pythagoras' mother. Nevertheless, Iamblichus (Ch. 2) assures us that 'no one would dispute, judging from his very birth and the all around wisdom of his life, that Pythagoras' soul was sent down to humans under Apollo's leadership, either as a follower in his train, or united with this god in a still more intimate way' (1991, p. 35).

After the account of Pythagoras' divine lineage, Iamblichus begins his description of his life, touching on his childhood, education and travels, prior to his return to Samos at about age 56 (Ch. 2–4). Following details of his life in Samos (5), we learn of his move to Italy and receive a brief description of his character and philosophy (6). Iamblichus is occasionally repetitive and sometimes contradictory; he tends not to name his sources, but in many instances offers a wealth of detail. So, for example, we learn (2) that Pythagoras travelled to see several of the pre-Socratic philosophers; 'and as he visited each in turn, the result of association with him was such that all cherished him and admired his character, and made him a partner in their discourses. And what is more, Thales gladly accepted him as a student, and admired his difference from other youths' (1991, p. 39).

Iamblichus discusses Pythagoras' speeches in Italy (Ch. 7), his visit to and teachings in Croton (8–11), devoting one chapter (10) to outlining Pythagoras' advice to children ('never start a fight') and another to his address to the women of Croton (11), in which he commended the feminine practice of lending clothes and jewellery when needed without requiring a witness. Following his explanation of why Pythagoras was the first person to call himself a philosopher (12), he offers examples of his ability to teach animals, and his views regarding transubstantiation (13–14). The importance of music in education and restoring the soul (15), his views on friendship (16), his relationships with his followers are described, and some of his maxims reported (17–18). Iamblichus (19) explained that Pythagoras 'discovered many ways of teaching and training, and transmitted the appropriate portion of wisdom according to each one's own nature and ability' (1991, p. 115); the practices of Pythagorean philosophy are then described (20–24). Iamblichus returns to the topic of education through music (Ch. 25), and Pythagoras' discovery of the principles of harmony (26). The political activities of and civic benefits bestowed by Pythagoras and his followers on humanity are described in some detail (27). Various marvels and miracles associated with Pythagoras, as well as his piety, are then celebrated (28); this is a particularly lengthy chapter. The wisdom of Pythagoras, including his work in physics, ethics and logic, as well as his study of geometry, are described briefly (29). His contributions to justice (30), his practice of self-control (31), his precepts on courage (32) and his teachings on friendship (33) are recounted, followed by customs of the Pythagoreans (34), the various political problems encountered by the Pythagoreans, and the way the sect continued to operate (35). *On the Pythagorean Life* concludes with a list of his successors, including the 17 'most famous' Pythagorean women (Ch. 36).[25]

Iamblichus' concern here was not to provide a full account of Pythagorean philosophy; he did that in the other books of *On Pythagoreanism*. Rather, *On the Pythagorean Life* is an introduction to living as a Pythagorean; the account of the life and teachings of Pythagoras is a crucial part of that introduction. Furthermore, in Iamblichus' account, Pythagoras' followers play a particularly important role; many details of his relationships to his disciples, including his teaching practices, are offered, as are the names of many Pythagoreans.

Iamblichus' *bios* of Pythagoras is a highly developed 'aretalogical' account of the heroic teacher as moral exemplar. Pythagoras' students, followers and disciples are especially present, serving to emphasise his role as a teacher, as well as the founder and leader of a community committed to a particular way of life. Iamblichus celebrated Pythagoras as an heroic teacher and divine guide to living.

But, of course, Pythagoras was not the only such guide to living available during the third century. Diogenes Laertius promoted Epicurus as a divine ethical teacher. Christianity developed with the aim of promoting Jesus. The Christian threat to pagan philosophers was certainly known to Porphyry and Iamblichus; Porphyry wrote a critique of Christianity, *Against the Christians*. In composing accounts of the life of Pythagoras, our authors combined intellectual history with moral, ethical and, in some cases, apparently religious concerns. As Moses Hadas explained, aretalogical accounts could be shaped by 'the level of the audience to which the aretalogy might be addressed, and the general religious climate that fostered its growth' (Hadas and Smith, 1965, p. 62). Our authors provided accounts of Pythagoras as a philosopher, but their accounts were shaped by various factors, including their own, differing, philosophical allegiances.[26]

25 See Clark (1989), pp. xvi–xviii, on the place of women within Pythagoreanism.

26 While these must be noted, they are too complicated to be treated fully here.

Iamblichus' account may well have been shaped by the Christian climate in which it was written. O'Meara (1989, p. 215) has suggested that 'Iamblichus' successors could of course come to consider his Pythagorean theology as a revelation of much greater antiquity and purity than that of the Christians.' Nevertheless, as he has indicated:

> it would be difficult to show, on the basis of the extant remains of *On Pythagoreanism*, that Iamblichus had Christianity specifically in mind as a target against which his Pythagoreanizing programme was to be directed. At most one could point to structural parallels between his figure of Pythagoras (his divine authority, attributes, mission, words and deeds among men) and Christ. (p. 214)[27]

John Dillon and Jackson Herschbell have argued that *On the Pythagorean Life* is not really a biography, and should not be understood as an example of that genre. As they point out, 'much of the central part of the work seems hardly excusable in a biography: it concerns not Pythagoras himself, but the Pythagoreans in general'. They make the case that 'the work is, in fact, a dramatised study of a way of life, with a strong protreptic purpose', exhorting and urging the reader towards Pythagorean philosophy and the Pythagorean way of life. Dillon and Herschbell (Iamblichus, 1991, p. 25) go so far as to suggest that 'if it is permissible (that is, if it be correct to recognise it as a genre transcending the strictly Christian milieu), it seems best to classify [the work] as a gospel'.[28]

The survival of Iamblichus' *On the Pythagorean Life*, and the loss of most of the books of *On Pythagoreanism*, may skew our perception of the emphasis intended on Pythagorean ideas. O'Meara has argued that 'the succession of books in Iamblichus' work is … related to a pedagogical progression from the general and common to what is more difficult, higher, and specifically Pythagorean. The progression is a "protreptic": it is designed to lead the soul up to "greater things".' The student of Pythagoreanism will study, first, Pythagoras and his school, then general and Pythagorean philosophy, then Pythagorean mathematics. The programme for Pythagorean mathematics included general mathematical science, arithmetic itself, as well as arithmetic in physics, ethics, and theology, geometry, music and (probably) astronomy (O'Meara, 1989, pp. 34–5). In O'Meara's view, it was Iamblichus' stress on the role of mathematics in the personal achievement of divinity that distinguished his account of Pythagoras, his teachings and his school; Iamblichus emphasises that the 'specifically Pythagorean revelation consists in the "most scientific", unerring forms of knowledge, those having to do with pure, immaterial unchanging realities, namely mathematics and the study of true being and of the divine'. Indeed, 'mathematics can train and prepare the soul … for higher wisdom' (O'Meara, 1989, p. 89). Yet in *On the Pythagorean Life*, the philosophical and mathematical ideas are

27 Clark (1989), pp. ix–xv, has argued that a secret convert to paganism, the Emperor Julian (331–363 CE, emperor 361–363), intended Iamblichus' work to be used to train pagan priests.

28 Whereas some other scholars have pointed to similarities to the synoptic gospels, Dillon and Herschbell (in Iamblichus, 1991) suggest that Iamblichus' account most closely resembles the gospel of John.

only briefly mentioned; in this work, it is the divine character of Pythagoras that is highlighted.

Within the 'aretalogical' literature, the desire to describe the philosopher's ideas and teachings interacted with the aim of promoting the divinity of the heroic teacher. Indeed, as Hadas noted, accounts of 'the teacher for whom supernatural claims are made asks for something like religious conversion'. He suggested that the image of such teachers survives 'because those who cherish the legend constitute something like an organised cult, with leaders concerned for its propagation and its adaptation to new climates and conditions'. The dual aims of institutionalising teachings while promoting divine status interact with each other; 'mystics are endowed with rational doctrine and rational thinkers with mystique' (Hadas and Smith, 1965, p. 32).

Conclusion

What are these three *bioi* of Pythagoras about? Pythagorean philosophy has often been understood as emphasising the value of mathematics.[29] But the ancient *bioi* considered here cannot be regarded simply as biographies of an important figure in the history of science. Rather, the three ancient biographies of Pythagoras belong to a genre born in a specific time and place. These accounts are not biographies as we understand the term, even while recognising that biography is a diverse genre; nor are they the gospels of a failed religion. Their purpose was to provide the history of an intellectual tradition, relating the interactions of teachers and their students, and also to celebrate the achievements of heroic philosophers, whose 'lives' are meant to serve as guides for others on how to live, how to benefit from philosophy, and how to be more divine.

Acknowledgements

I am grateful to Niall Caldwell, Clare Drury, John Drury, Katie Eagleton, Frances Gardiner, Nick Jardine, Jeremy Morris, Martin Rees, Tom Ryckman, Thomas Söderqvist, Laurence Totelin, Karin Tybjerg and Frances Willmoth for their generous help and suggestions. I dedicate this article to the memory of my dear friend, Clare Drury.

Bibliography

Barker, Andrew D. (1996), 'Aristoxenus', in Simon Hornblower and Antony Spawforth (eds), *The Oxford Classical Dictionary*, 3rd edn, Oxford: Oxford University Press, pp. 169–70.

29 Whitehead (1925), pp. 53–4, pointed to 'old Pythagoras, from whom mathematics, and mathematical physics, took their rise'. He suggested that Pythagoras had endowed these fields 'with the luckiest of lucky guesses', but then asked: 'or, was it a flash of divine genius, penetrating to the inmost nature of things?'

Barnes, Jonathan (1992), 'Diogenes Laertius IX 61–116: The Philosophy of Pyrrhonism', *Aufstieg und Niedergang der Römischen Welt*, 2.36.6, pp. 4241–4301.

Burkert, Walter (1972), *Lore and Science of Ancient Pythagoreanism*, trans. Edwin L. Minar Jr, Cambridge, MA: Harvard University Press; original publication *Weisheit und Wissenschaft: Studien zu Pythagoras, Philolaos und Platon*, Nuremberg: Verlag Hans Carl, 1962.

Burridge, Richard A. ([1992] 1995), *What are the Gospels? A Comparison with Graeco-Roman Biography*, Cambridge: Cambridge University Press,

Clark, Gillian (1989), *Iamblichus: On the Pythagorean Life, Translated Texts for Historians*, vol. 8, Liverpool: Liverpool University Press.

Copenhaver, Brian P. (1978), 'The Historiography of Discovery in the Renaissance: The Sources and Composition of Polydore Vergil's *De inventoribus rerum*, I–III', *Journal of the Warburg and Courtauld Institutes*, **41**, 192–214.

Delatte, A. (1922), *La Vie de Pythagore de Diogene Laërce*, *Mémoires*, Deuxième Série 17, Brussels: Maurice Lamertin, Académie Royale de Belgique, Classe des Lettres et des Sciences Morales et Politiques.

Diogenes Laertius ([1925] 1972), *Lives of Eminent Philosophers*, trans. R.D. Hicks, 2 vols, *Loeb Classical Library*, Cambridge, MA: Harvard University Press.

[Einstein, Albert] (1949), *Albert Einstein: Philosopher-Scientist*, ed. P.A. Schilpp, Evanston, IL: The Library of Living Philosophers.

Galen (1999) *On My Own Opinions*, ed. and trans., with commentary, by Vivian Nutton, Berlin: Akademie Verlag.

Gardiner, Frances (2003), 'Diogenes Laertius: The Man behind the Text', unpublished undergraduate dissertation, Department of History and Philosophy of Science, University of Cambridge.

Gorman, Peter (1979), *Pythagoras: A Life*, London: Routledge and Kegan Paul.

Graf, Fritz (1996), 'Pythagoras (1), Pythagoreanism, 1. Pythagoreanism (Religious Aspects)', in Simon Hornblower and Antony Spawforth (eds), *The Oxford Classical Dictionary*, 3rd edn, Oxford: Oxford University Press, pp. 1,284–5.

Hadas, Moses and Morton Smith (1965), *Heroes and Gods: Spiritual Biographies in Antiquity*, London: Routledge and Kegan Paul.

Hallyn, Fernand (1990), *The Poetic Structure of the World: Copernicus and Kepler*, trans. Donald M. Leslie, New York: Zone Books.

Harvey, Paul (ed.) ([1937] 1974), *The Oxford Companion to Classical Literature*, Oxford: Clarendon Press.

Hope, Richard (1930), *The Book of Diogenes Laertius, its Spirit and its Method*, New York: Columbia University Press.

Iamblichus (1991), *On the Pythagorean Way of Life*, trans. John Dillon and Jackson Herschbell, Texts and Translations 29, *Graeco-Roman Religion Series* 11, Atlanta, GA: Scholars Press.

Jardine, Nicholas and Alain Segonds (1999), 'Kepler as reader and translator of Aristotle', in Constance Blackwell and Sachiko Kusukawa (eds), *Philosophy in the Sixteenth and Seventeenth Centuries: Conversations with Aristotle*, Aldershot: Ashgate, pp. 206–33.

Kahn, Charles (2001), *Pythagoras and the Pythagoreans: A Brief History*, Indianapolis, IN: Hackett Publishing.

Kearns, Emily (1989), *The Heroes of Attica*, Bulletin Supplement 57, London: University of London, Institute of Classical Studies.

Kearns, Emily (1996) 'hero-cult', in Simon Hornblower and Antony Spawforth (eds), *The Oxford Classical Dictionary*, 3rd edn, Oxford: Oxford University Press, pp. 693–4.

[Kepler, Johannes] (1951), *Johannes Kepler: Life and Letters*, trans. Carola Baumgardt, New York: Philosophical Library.

Kepler, Johannes (1945), *Briefe 1590–1599*, in Max Caspar (ed.) (1938–88), *Gesammelte Werke*, Bd XIII, Munich: C.H. Beck.

Kühnert, Friedmar (1961), *Allgemeinbildung und Fachbildung in der Antike*, Berlin: Akademie Verlag.

Lamberton, Robert (2001), *Plutarch*, New Haven, CT: Yale University Press.

Lévy, Isidore (1927), *La Légende de Pythagore de Grèce en Palestine*, *Bibliothèque de l'École des Hautes Études* 250, Paris: Librairie Ancienne Honoré Champion.

Lloyd, G.E.R. (1991), 'Introduction' to 'Observational error in later Greek science', in G.E.R. Lloyd, , *Methods and Problems in Greek Science: Selected Papers*, Cambridge: Cambridge University Press, pp. 299–302.

Long, Herbert S. (1970), 'Diogenes (6) Laertius', in N.G.L. Hammond and H.H. Scullard, *The Oxford Classical Dictionary*, 2nd edn, Oxford: Clarendon Press, pp. 348–9.

Long, Herbert S. ([1925] 1972), 'Introduction' to Diogenes Laertius, *Lives of Eminent Philosophers*, trans. R.D. Hicks, 2 vols, *Loeb Classical Library*, vol. I, Cambridge, MA: Harvard University Press, pp. xv–xxvi.

Mansfeld, Jaap (1999), 'Sources', in K. Algra, J. Barnes, J. Mansfeld and M. Schofield (eds), *The Cambridge History of Hellenistic Philosophy*, Cambridge: Cambridge University Press, pp. 3–30.

Mejer, Jørgen (1978), *Diogenes Laertius and His Hellenistic Background*, Hermes Einzelschriften 40, Wiesbaden: Steiner.

Momigliano, Arnaldo (1993), *The Development of Greek Biography*, expanded edn, Cambridge, MA: Harvard University Press.

O'Meara, Dominic J. (1989), *Pythagoras Revived: Mathematics and Philosophy in Late Antiquity*, Oxford: Clarendon Press.

O'Meara, Dominic (1996), 'Iamblichus (2)', in Simon Hornblower and Antony Spawforth (eds), *The Oxford Classical Dictionary*, 3rd edn, Oxford: Oxford University Press, p. 743.

Pelling, Christopher (1996), 'biography, Greek', in Simon Hornblower and Antony Spawforth (eds), *The Oxford Classical Dictionary*, 3rd edn, Oxford: Oxford University Press, pp. 241–2.

Porphyry (1965), *The Life of Pythagoras*, trans. Morton Smith, in Moses Hadas and Morton Smith, *Heroes and Gods: Spiritual Biographies in Antiquity*, London: Routledge and Kegan Paul.

Runia, David T. (1999), 'What is Doxography?', in Philip J. van der Eijk (ed.), *Ancient Histories of Medicine: Essays in Medical Doxography and Historiography in Classical Antiquity*, *Studies in Ancient Medicine* 20, Leiden: E.J. Brill, pp. 33–55.

Sharples, Robert W. (1996), 'Diogenes (6) Laertius', in Simon Hornblower and Antony Spawforth (eds), *The Oxford Classical Dictionary*, 3rd edn, Oxford: Oxford University Press, pp. 474–5.

Smith, Andrew (1996), 'Porphyry', in Simon Hornblower and Antony Spawforth (eds), *The Oxford Classical Dictionary*, 3rd edn, Oxford: Oxford University Press, pp. 1226–7.

Talbert, Charles H. (1977), *What is a Gospel? The Genre of the Canonical Gospels*, Philadelphia, PA: Fortress Press.

Tatarkiewicz, W. (1963), 'Classification of the arts in antiquity', *Journal of the History of Ideas*, **24**, pp. 231–40.

The Compact Edition of the Oxford English Dictionary (1971), 2 vols, Oxford: Oxford University Press.

Tiede, David Lenz (1972), *The Charismatic Figure as Miracle Worker*, *Dissertation Series* 1, Missoula, MT: Society of Biblical Literature for The Seminar on the Gospels.

Whewell, William (1840), *The Philosophy of the Inductive Sciences Founded upon their History*, 2 vols, London: John W. Parker; Cambridge: J. and J.J. Deighton.

Whitehead, Alfred North (1925), *Science and the Modern World: Lowell Lectures, 1925*, New York: Macmillan.

Zeller, Eduard (1880–92), *Die Philosophie der Griechen in ihrer geschichtlichen Entwicklung*, I–III, Leipzig: O.R. Reisland.

Chapter 2

Biography as a Route to Understanding Early Modern Natural Philosophy

Stephen Gaukroger

The kind of contextualisation needed to understand the personae of natural philosophers in the sixteenth and seventeenth centuries is different from that needed for the understanding of modern figures. I think this difference is reflected in the mechanisms for self-understanding exhibited in early modern drama, writings on the passions, on morality, and more specifically on the 'station' of the philosopher. This difference requires different biographical techniques, or at least different emphases, from those needed in writing a biography of a twentieth-century figure. One of these differences provides the topic of this paper. I want to focus on something that would usually be ignored in accounting for the development of early modern natural philosophy, but which is of particular importance in the biography of early modern natural philosophers, and to show how this enables us to understand something of fundamental importance to how the natural-philosophical enterprise was conceived. I am referring to the construction of the persona of the philosopher/natural philosopher.

By way of background to this question, consider the case of seventeenth-century tragedy. Actors who have been brought up in a naturalistic tradition notoriously have difficulty in playing parts from the classical, and the sixteenth- and seventeenth-century, repertoire. They avoid simply playing the lines by attempting to play the character, and the way in which they do this is to try to capture the psychology of the character. Using this to build up an image of how the character thinks and behaves, they take on the persona of the character and play the character. While this may be appropriate for nineteenth- and twentieth-century drama, however, it does not always work with the drama of earlier periods. In the classical Greek tragedy, for example, the thing that primarily motivates the character's behaviour will be a violation of the natural order of things, and this is still an ingredient in seventeenth-century drama. Here we have something whose meaning is destroyed if it is translated into a purely psychological struggle. Now seventeenth-century tragedy is, I suggest, a good guide to seventeenth-century conceptions of responsibility, identity, moral integrity, self-awareness and the other factors that play a role in one's persona. Drama deals with fictional characters who can represent all kinds of things, whereas biography deals with actual historical figures, but the raw materials are the same, and there is a fundamental connection between how one conceives of what motivates one and how one represents motivations dramatically.

Put in broad terms, my claim is that the factors that make up one's persona – one's sense of oneself, one's conceptions of one's responsibility, one's identity, one's moral integrity and so on – are historically and culturally variant. One extreme historical case that has been investigated in detail from the early decades of the twentieth century onwards by a number of writers such as Onians (1954), Snell (1960), Adkins (1970), Vernant (1965), Vernant and Vidal-Naquet (1972) and Detienne (1967) is the contrast between archaic and classical Greek conceptions of the person. Here we find a very radical shift in what might be called forms of consciousness, so radical that it is difficult for us to apply any of our notions of moral responsibility and identity.[1] These extreme differences make serious biography impossible, where by 'serious biography' I mean something that goes beyond mere chronology of a life, and which does not just extrapolate from our own understanding of ourselves on the assumption that our own understanding of what our identity and motivation consist in can be universalised. One alternative to the latter is to try to go native, to attempt to recapture what motivates the seventeenth-century psyche. But this is impossible, and even if it were possible, it would be as futile as the attempts of the actor who thinks that to play Macbeth, what one must do is capture his psychology, ignoring the cosmically significant, fatalistic context which provides the possibility of the drama in the first place. Rather, what we need to understand is the context within which thought about what we would now refer to as questions of personality arise, and how they are articulated within this context.

The history of biography is revealing here. In working on Descartes' biography, it soon became clear to me that his own autobiographical accounts were not at all like the kinds of things that one expects from an autobiography in our post-Lytton Strachey era. The autobiographical account of the *Discours de la Méthode* does not contain an accurate warts-and-all account of Descartes' personal and intellectual development. It comprises a moral tale in which a central character, Descartes, exemplifies the qualities that he sees as necessary to the life of a philosopher (Gaukroger, 1995, pp. 106–11). The aim is not to capture the particular frame of mind he was in at particular times, but to identify the appropriate states of mind and behaviour in a philosopher. This is most striking in his account of the famous dreams in the *Discours*, where they herald not merely an intellectual discovery, but a major turning point in his intellectual life, the point at which he becomes a philosopher of a new kind. However, the reports by his first biographer, Baillet, which give a description of the dreams and Descartes' own state of mind at the time from a now lost account by Descartes himself, and other extant sources such as the contemporary *Olympica*, give us a completely different story. But Descartes is not glossing over or rationalising an episode in his life; on the contrary, he is giving it meaning in terms of a standard trope (standard from classical antiquity onwards) of dreams heralding discovery.

---

1 This aspect of the situation is pursued in Marcel Mauss's path-breaking 1938 lecture on the self – Carrithers, Collins and Lukes (1985) contains a translation of the original essay, and essays discussing and developing its themes. Mauss sees the concept of a person as essentially a Roman development.

Biography

Descartes' sense of what a biography does and ours are different. I don't think Descartes' 'autobiography' has a completely different aim from the modern biography. I think its aims, while not wholly coincident with ours, do nevertheless overlap with ours, and if this is the case, then what is interesting is that the path taken to these aims is so different. We find it hard, from the perspective of biography, to see the point of Descartes' autobiography, which has no interest in recording how he felt and worked, but rather with how someone in this position *should* have felt and worked. But Descartes may well have seen our interest in biography, and the kind of aims we try to achieve in biography, as mere self-indulgence, as an amoral form of narcissism, a genre useless for moral or personal guidance.

It is not just that what Descartes wants to capture from the autobiographical genre is different from what we would want to capture from this genre; rather, what Descartes sees as important about a person is, in certain respects, different from what we see as important. This, finally, takes me to the point from which I want to start my investigation. It is this: in writing biographically about Descartes, we write about him, at one important level, as a person. But what Descartes – and more generally early modern European intellectual culture – sees as being important to being a person is different from what we see as being important. To put the point more strongly, what early modern European intellectual culture sees as constituting a person is different from what we see as constituting a person.

This, then, is my starting point, and I want to look at one particular aspect of the question. I want to look at a particular type of person – the early modern philosopher – and I want to look at an event that highlights, in a very striking way, what is involved in thinking through what it means to be this kind of person, namely someone who is a philosopher. The event in question is a conscious and concerted attempt to change how philosophers think of themselves, how they think of their aims and motivations.

The question of changes in how philosophers think of themselves plays no part in standard histories of philosophy: it is simply assumed that what motivates and drives philosophers falls outside the domain of philosophical investigation. But it doesn't, for it is a key ingredient in early modern philosophical thought. The kinds of tools that one develops for studying early modern philosophers at a biographical level are, I want to suggest, exactly the kinds of tools needed to open up this question. For investigation of how philosophers think of themselves and what they are doing – at the level of detail and depth that interests the biographer but finds no place in the standard history of philosophy or history of science – provides the model for their constructions of the archetypal natural philosopher, for this is exactly what they are concerned with in their own self-reflections. Ultimately, one thinks of, or explores, oneself by measuring oneself against an archetype of the kind of person one thinks one should be. If one is a philosopher, one thinks of oneself by measuring oneself against an archetype of the philosopher. In the case we shall looking at, this archetype of the philosopher has collapsed, so that a new canonical image of the philosopher has to be constructed. It is here, at one of the formative moments of early modernity, that we get a rare insight into the early modern idea of what goes into the construction of a philosophical persona, and therefore into just what makes up this kind of philosophical persona.

The Persona of the Philosopher

I am going to be looking at Bacon and Descartes, as the two formative figures in the construction of a new philosophical persona, but since they are not thinking in a vacuum, let me just draw attention to three important streams of what we might term personal construction which form part of the background to what they are doing.

The first, and the oldest, derives from the fact that the philosophers since Greek and Roman antiquity had seen one of their chief aims as being to provide guidance for how life was to be lived.[2] The philosopher was seen as the archetypical Sage, and the Stoic ideal of indifference to calamity and misfortune, developed by Cicero and Seneca in Roman literature, is perhaps the most easily recognisable of the images of the character of the philosopher. Achieving this goal was not considered easy, however, and it required philosophers to undergo training to prepare themselves for a life quite different from that of their fellows. Such questions were paramount throughout antiquity, and at least from Socrates onwards, the philosopher took on or fostered a distinct persona and attitude, depending on the philosophical doctrine or school. For Plato, for example, the persona of the philosopher fitted him for kingship, whereas for Diogenes the Cynic, on the other hand, it fitted him to the life of a beggar or a slave.

This fostering of a philosophical persona is particularly marked in the Hellenistic era, where *ataraxia*, peace of mind, was explicitly the aim of all the major schools, and where regulation of the passions played a major role for Epicureans and Stoics alike in attaining the state of mind, and corresponding behaviour, worthy of or appropriate to a member of their philosophical school. This philosophical self-fashioning was pursued in a different way in the Christian era, and it is nowhere clearer than in the disputes of the 1260s and 1270s arising from the attempts to introduce an autonomous philosophical system, that of Aristotle, into Christian thought. The Averroist Boethius of Dacia, one of the strongest advocates of philosophical 'self-affirmation', argued that it was easier for the philosopher to be virtuous than for anyone else, and that whoever does not live the life of a philosopher does not live rightly or virtuously (Boethius, 1987, pp. 32–5; Wieland, 1982). Tempier responded in the 1277 Condemnation by condemning the propositions 'that no station in life is to be preferred to the study of philosophy' and 'philosophers alone are the wise men of the world'.

No less striking is the idea of philosophical self-fashioning that pervaded the Renaissance thought, with Pico della Mirandola's eulogy on 'the dignity of man' being in fact an attempt to redefine the office of the philosopher as the paradigm Sage, and to set out a programme for the attainment of this goal. Works like Montaigne's *Essais* and Burton's *Anatomy of Melancholy* show that the genre was alive and well

---

2 The literature on this is extensive, but the key work remains Pierre Hadot (1995). See also Hadot (1993) and Rabbow (1954).

in the early modern era, and although the way in which it was pursued differed, a theme that runs through all the literature is the mastery of the passions. In general terms, philosophers in antiquity, in the Middle Ages and in the early modern era were able, with varying degrees of success, to construct images of themselves as paradigmatic bearers of moral, aesthetic and intellectual responsibility. Whatever deep philosophical quarrels they may have had among themselves, it was important to establish that the philosophical view was not simply one kind of opinion among others. What was required to establish this was the construction of a philosophical persona capable of bearing and displaying this authority: an authority which was very different from that borne and displayed by theologians and statesmen, for example, whose claims on moral, natural-philosophical and other questions may have overlapped with, and perhaps competed with, those of philosophers. The question raised here is one about the relation between philosophy and the behaviour appropriate for the philosopher, or at least the philosophically educated: what kind of persona philosophy does or should shape or encourage.

The second form of personal construction is not peculiar to philosophers, though it certainly helps shape the philosophical project: it is the application of what were originally monastic values to the population in general. In the sixteenth and seventeenth centuries we find what can be called an 'internalisation' of religion, a shift from something public to something private. This is a feature common to both Catholicism and Protestantism, as part of a shared, widespread sense that Christianity needed to be renewed, given focus, strengthened, and indeed ultimately transformed in quite a radical way. As Delumeau had shown, the key ingredient in this process is a *contemptus mundi*, something originally developed and refined in the monasteries and later transmitted to the whole of society, in the first instance through the mendicant orders, as 'a self-evident truth' which has three components – hatred of the body and the world, the pervasiveness of sin, and a sharp sense of the fleetingness of time (Delumeau, 1988, p. 144).[3] There is a concern with self-reform, motivated by feelings of guilt and repentance, which we can find not merely in the devotional literature, but which is also reflected in the philosophical literature from Montaigne onwards, especially in writing on the passions and ethics.

The third phenomenon is a concern with the values of civilisation, which takes the form of concern with the self. Although it is initially political rather than religious in orientation, it shares much ground with the second development, above all in that its concern is with developing internal motivation rather than external coercion. The importance of the move from 'external into internal compulsion' in the 'civilising process' of the early modern era has been stressed in the pioneering work of Norbert Elias (1982, 1983) This civilising process begins, at least in France, in court society as a means of subordinating the interests of the nobles to those of an absolutist monarch, but from the late fifteenth century onwards, it has a far broader significance, exemplified in the numerous manuals which appeared in the sixteenth century, describing in detail how one should behave – that is, regulate one's behaviour – in a variety of circumstances. In an extremely popular and influential series of manuals that Erasmus published between 1500 and 1530, for example,

3 See also Delumeau (1978) and (1983), as well as Ariès (1975).

there are set out rules for how to behave in church, in bed, while at play, while eating and so on; the manuals are exhaustive, covering everything from dress, deportment and gestures to facial expressions and demeanours.[4]

There are other developments that one would have to look at were one pursuing these questions in detail, but these three will serve to indicate how a number of streams overlap at certain junctures, providing a rich if complex background to the concerns of Bacon and Descartes. These concerns hinge on the collapse of the dominant sixteenth-century notion of the philosopher, and the unsatisfactory nature of the alternatives provided by Renaissance humanism, practitioners of the mechanical and practical arts, and so on. Bacon and Descartes attempt to shift the image of the paradigm philosopher from that of the Scholastic cleric to the intelligent man of the world. The skill comes in identifying just where the problems lie in the former and what qualities are required of the latter. This skill is manifested in the adequacy of the persona that results.

Bacon

A crucial part of Bacon's project for the reform of natural philosophy was a reform of its practitioners. One ingredient in this was the elaboration of a new image of the natural philosopher, an image that conveyed the fact that the natural philosopher is no longer an individual seeker after the arcane mysteries of the natural world, employing an esoteric language and protecting his discoveries from others, but a public figure in the service of the public good, that is, the crown.

I have argued in detail elsewhere that we can detect here a transformation of traditional humanist concerns into a natural-philosophical context (Gaukroger, 2001). Renaissance humanists raised the question of the responsibilities appropriate to the humanist, in particular whether the life of activity in affairs of state (*negotium*) should be preferred to that of detachment and contemplation (*otium*). The answer almost invariably given – not least by Bacon himself, in the seventh Book of *De Dignitate* – was that *negotium* should be preferred to *otium*.[5] Once this question had been decided, the issue then became not just the appropriate learning but also, given the practical nature of the programme, the appropriate behaviour of such a practical humanist. The appropriate education, attitude of mind, behaviour, and even demeanour, were the questions that informed thinking on these matters. Indeed, this is nowhere more marked than in *De Dignitate*, where Bacon turns the Socratic dictum 'know thyself' into a question of behaviour, telling us that 'it is not enough for a man only to know himself; for he should also consider the best way to present himself to advantage; to disclose and reveal himself; and lastly, to turn and shape himself according to the occasion'.[6]

The choice, then, is in the first instance, between the active or practical life and the contemplative life, where philosophers had traditionally fallen in the latter

4 There is a representative selection of these writings in translation in Section 2 of Rummel (1990). On the role of civility in England, see Barker (1948).

5 Bacon (1857–74), vol. 1, pp. 713–44, and vol. 5, pp. 3–30.

6 Ibid., vol. 1, p. 779, and vol. 5, p. 66.

category. The explicit shift to the defence of the active or practical life places new requirements on philosophy, for philosophers now had to show that they were able to live up to the aims of the active or practical life. What Bacon effectively does is to transform philosophy into something that comes within the realm of *negotium*. This is completely at odds with the conceptions of philosophy of classical antiquity and the Christian Middle Ages. Promoted through the rhetorical unity of *honestas* and *utilitas*, philosophy is presented as something good and useful, and thus as intrinsic to the active life. Indeed, it starts to become a paradigmatic form of *negotium*, and in this way it can usurp the claims made for poetry by writers such as Philip Sidney, who argued that poetry can move one to act virtuously, whereas philosophy cannot do this.

In the humanist thought that makes up the source from which Bacon derives much of his inspiration, moral philosophy figures very predominantly. There are two respects in which the model of moral philosophy is important here. First, philosophical self-fashioning had always turned on the moral question of the understanding and regulation of the passions, and because of this they have a peculiar centrality, for they have not merely been one object of study among others for philosophers, but something which must be understood if one is to be 'philosophical' in the first place. Mastery of the passions was, in one form or another, not only a theme in philosophy, but a distinctive feature of the philosophical persona from Socrates onwards. Renaissance and early modern philosophers pursue the theme of self-control with no less vigour than had the philosophers of antiquity. This is the model around which Bacon wishes to shape his new practitioner of natural philosophy.[7] It is a model inappropriate to the artisan, and it gives the new practitioner a dignity and standing that the collective nature of his work would not otherwise suggest.

Second, in a humanist dimension, being virtuous and acting virtuously are the same thing: there is no separate practical dimension to morality. Indeed, this forms the basis for much humanist criticism of traditional moral philosophy; Sidney, for example, in stressing the superiority of the active, practical life over the contemplative one, draws what he takes to be the consequences for moral thought, namely that teaching the nature of virtue is not the same thing as, and indeed is no substitute for, moving people to practise virtue, and that all philosophy has managed is the former. Sidney and Bacon both want to obliterate the distinction between being moral and acting morally. Moreover, it is interesting to note here that Bacon stresses in the *Advancement of Learning* that moral philosophy is a cognitive enterprise, one in which the practical outcome is constitutive of the discipline.[8] If, as I am suggesting, we see natural philosophy as being in some respects modelled on moral philosophy, something which is natural enough in a humanist context, and which is reinforced in the shift from *otium* to *negotium*, then we may be able to delve a little more deeply into why Bacon famously claims that the aim of the natural philosopher is not merely to discover truths, even informative ones, but to produce new works.

7 This is particularly evident in Bacon's account of his scientific utopia, *New Atlantis*, where self-respect, self-control and internalised moral authority are central.

8 Bacon (1857–74), vol. 3, pp. 432–4.

For Bacon, the natural philosopher is not simply someone with a particular expertise, but someone with a particular kind of standing, a quasi-moral standing, which results from the replacement of the idea of the Sage as a moral philosopher with the idea of the Sage as a natural philosopher, whose paradigm is Solomon. And just as the Sage as moral philosopher cannot be such unless his grasp of morality is manifest in his behaviour, so the Sage as natural philosopher cannot be such unless his grasp of nature is manifest in his behaviour, and the only way in which it can be manifest is in the production of new works.

How does one become such a Sage? In the most general terms, at least one ingredient in the answer is a very traditional one: the purging of the emotions. But Bacon puts a distinctive gloss on this. The Sage for Bacon must purge not just affective states, but cognitive ones as well. In his doctrine of the 'Idols' of the mind, Bacon provides an account of the systematic forms of error to which the mind is subject, and here the question is raised of what psychological or cognitive state we must be in to be able to pursue natural philosophy in the first place. Bacon believes an understanding of nature of a kind that had never been achieved since the Fall is possible in his own time because the distinctive obstacles that have held up all previous attempts have been identified, in what is in many respects a novel theory of what might traditionally have been treated under a theory of the passions, one directed specifically at natural-philosophical practice.

Bacon argues that there are identifiable obstacles to cognition arising from innate tendencies of the mind (Idols of the Tribe), from inherited or idiosyncratic features of individual minds (Idols of the Cave), from the nature of the language that we must use to communicate results (Idols of the Market-Place), or from the education and upbringing we receive (Idols of the Theatre). Because of these, we pursue natural philosophy with seriously deficient natural faculties, we operate with a severely inadequate means of communication, and we rely on a hopelessly corrupt philosophical culture. In many respects, these are a result of the Fall and are beyond remedy. The practitioners of natural philosophy certainly need to reform their behaviour, overcome their natural inclinations and passions and so on, but not so that, in doing this, they might aspire to a natural, prelapsarian state in which they might know things as they are with an unmediated knowledge. This they will never achieve. Rather, the reform of behaviour is a discipline to which they must subject themselves if they are to be able to follow a procedure which is in many respects quite contrary to their natural inclinations.

The reform of one's persona is needed because of the Fall: after the Fall, it is lacking in crucial ways. Whereas earlier philosophers had assumed that a certain kind of philosophical training would shape the requisite kind of character, Bacon argues that we need to start further back, as it were, with a radical purging of our natural characters, in order to shape something wholly new.

Descartes

It might seem that Descartes has significant problems in the construction of the persona of the philosopher that Bacon does not have. Bacon's purging is targeted very precisely in his doctrine of Idols, and his understanding of what is needed to build on the newly cleared foundations is not abstract and metaphysical, but something psychological and practical: in keeping with his conception of the reformed philosophical enterprise. But in fact Descartes' approach is very similar to Bacon's in general outline.

To understand how, it is crucial that we distinguish between two kinds of enterprise. The first, which I see as being largely legitimatory, is set out in the *Principia*, and the route it follows is that of a radical purging of the mind of anything that can conceivably be doubted, establishing clarity and distinctness (manifested paradigmatically in the *cogito*) as the only criterion by which to establish the veridicality of our ideas, and then, having established that our understanding of the natural world must begin with quantitatively and mechanistically formulated ideas, building up a novel cosmology. This is the way to establish the truth of Cartesian natural philosophy, but Descartes does not claim that it is the way to pursue this natural philosophy. It is not the route followed by the Cartesian natural philosopher. This route, and the requisite state of mind and character of the natural philosopher who wishes to pursue it, are formulated in quite different terms, ones that involve psychological and moral considerations as much as epistemological ones.[9]

Descartes' discussion of this route occurs in *La Recherche de la vérité par la lumière naturelle*, which contrasts the fitness for natural philosophy of three characters: Epistemon, someone well versed in Scholasticism; Eudoxe, a man of moderate intelligence who has not been corrupted by false beliefs, and Poliandre, who has never studied, but is a man of action, a courtier and a soldier (as Descartes himself had been). Epistemon and Poliandre are taken over the territory of sceptical doubt and foundational questions by Eudoxe, but in a way that shows Poliandre's preparedness for, or capacity for, natural philosophy, and Epistemon's lack of preparedness. Preparedness here is in effect preparedness for receiving instruction in Cartesian natural philosophy. The *honnête homme*, Descartes tells us:

> came ignorant into the world, and since the knowledge of his early years rested solely on the weak foundation of the senses and the authority of his teachers, it was close to inevitable that his imagination should have been filled with innumerable false thoughts before his reason could guide his conduct. So later on, he needs to have either very great natural talent or the instruction of a very wise teacher, to lay the foundations for a solid science. (Descartes, 1974–86, vol. 10, p. 496)[10]

The thrust of Descartes' discussion is that Poliandre has not had his mind corrupted, because, in his role as an *honnête homme*, he has not spent too much time on book learning, which 'would be a kind of defect in his education'. The implication is that Epistemon has been corrupted in this way, and so is not trainable as the kind of natural philosopher Descartes seeks. It is only the *honnête homme* who can be trained, and it is Poliandre whom Eudoxe sets out to coax into the fold of Cartesian natural philosophy, not Epistemon. It is true that we might think of the procedure of radical doubt and the purging that results as a way of transforming everyone into

9 For a detailed discussion, see Gaukroger (2002).

10 On *La Recherche*, see Ranea (2000).

an *honnête homme*, and to some extent it is, although in his account of the passions Descartes makes it clear that, once we leave the programmatic level, ridding ourselves of prejudices and preconceived ideas is not so simple, and it requires the cultivation of a particular mentality, which is really what we witness in *La Recherche*.

In the *Recherche*, the *honnête homme* alone is identified as the kind of person who uses his natural faculty of clarity and distinctness to the highest degree: or, at least, it is he who, when called upon, uses it to the highest degree. This does not mean that the *honnête homme* alone is able to put himself through the rigours of hyperbolic doubt and discover the true foundations of knowledge: in theory, everyone is able to do that, Scholastics included. After all, hyperbolic doubt erases our beliefs (everyone's beliefs) to such an extent that everyone becomes a natural-philosophical *tabula rasa*:

> An examination of the nature of many different minds has led me to observe that there are almost none at all so dull and slow as to be incapable of forming sound opinions or indeed of grasping all the most advanced sciences, provided they receive proper guidance. And this may be proved by reason. For since the principles in question [namely, those of the *Principia*] are clear, and nothing is permitted to be deduced from them except by very evident reasoning, everyone has enough intelligence to understand the things that depend upon them. (Descartes, 1974–86, vol. 9, p. 12)

But if the aim is to develop and refine natural-philosophical skills as one progresses, then we require something different:

> As for the individual, it is not only beneficial to live with those who apply themselves to [the study of philosophy]; it is incomparably better to undertake it oneself. For by the same token it is undoubtedly much better to use one's eyes to get about, but also to enjoy the colours of beauty and light, than to close one's eyes and be led around by someone else. Yet even the latter is much better than keeping one's eyes closed and having no guide but oneself. (Descartes, 1974–86, vol. 9B, p. 3)

'Using one's eyes to get about' is not something that everyone finds equally easy, however. What Descartes is seeking are those who can develop his system to completion:

> the majority of truths remaining to be discovered depend on various particular observations/experiments which we can never happen upon by chance but which must be sought out with care and expense by very intelligent people. It will not easily come about that the same people who have the capacity to make good use of these observations will have the means to make them. What is more, the majority of the best minds have formed such a bad opinion of the whole of philosophy that has been current up until now, that they certainly will not apply themselves to look for a better one. (Descartes, 1974–86, vol. 9B, p. 20)

We must recognise that some are more fitted than others to follow the path of instruction/enlightenment in natural philosophy. And in the *Recherche*, Descartes realises, practically, that people come to natural philosophy not with a *tabula rasa*, but with different sets of highly developed beliefs which are motivated in different ways and developed to different degrees. These rest upon various things, and this is what leads him, in the *Recherche*, to construct an image of the *honnête homme* as a model in which the moral sage and the natural philosopher meet, for, as he puts it in the prefatory letter to the *Principes*, 'the study of philosophy is more necessary for the regulation of our morals and our conduct in this life than is the use of our eyes to guide our steps'(Descartes, 1974–86, vol. 9B, pp. 3–4).

In the *Principia*, Descartes sets out to reform philosophy in its entirety, but he does not see the project as establishing the kind of stagnant system that Scholasticism had become, where what has caused the decline of the system was clearly in large part due, in his view, to the slavish adherence of its proponents to Aristotle. In this respect, Descartes is not in the slightest interested in winning over Scholastic philosophers to his system: they are simply not the kind of people who can develop it, and would only lead it to the kind of stagnation to which they have led Aristotelianism. *A fortiori*, they cannot act as paradigm philosophers, as sages whose wisdom can guide the rest. This role falls instead to those who, reflecting upon the current state of philosophy, have formed a low opinion of it, and have avoided taking it up. This low opinion, wholly merited, is what makes them *honnêtes hommes*, and it is precisely these whom Descartes sees as being potentially the new paradigm philosophers, marked by an intellectual honesty which rescues philosophy from the intellectual disgrace into which it has fallen.

Conclusion

Both Bacon and Descartes saw philosophy as being in desperate need of radical reform, and both of them saw this reform as being carried out by a wholly new kind of person: a philosopher, but not a philosopher in the sense in which those called 'philosophers' at their time (they were thinking primarily of clerical Scholastics) were philosophers. These wholly new kinds of philosopher were not simply people who carried out investigations in a different way from their predecessors: they had, and needed to have, a wholly different persona. The techniques of self-examination and self-investigation opened up by the wholesale attempt to transfer monastic religious values to the population at large and by the sense that one was responsible for the minute details of one's daily life in the form of new norms of appropriate behaviour gave way to the possibility of a new understanding of one's psychology, motivation and sense of responsibility, and of shaping one's personal, moral and intellectual bearing. Bacon and Descartes used this – in rather different ways, but with the same broad aims – to transform our understanding of what qualities, including personal qualities, one needs to be a philosopher. These personal qualities do not define an individual, as in twentieth-century biography, however, but a genre, a type, manifested in those individuals who have aspired to become philosophers. The study of this type is not biography, but it is closer to biography than the history of philosophical or scientific discovery and doctrine.

Bibliography

Adkins, Arthur W.H. (1970), *From the Many to the One: A Study of Personality and Views of Human Nature in the Context of Ancient Greek Society, Values, and Beliefs*, Ithaca, NY: Cornell University Press.

Ariès, Pierre (1975), *Religion populaire et réforme liturgique*, Paris: Edition du Cerf.

Bacon (1857–74), *Works*, ed. J. Spedding, R.L. Ellis and D.D. Heath, 14 vols, London: Teubner.

Barker, Sir Ernest (1948), *Traditions of Civility*, Cambridge: Cambridge University Press.

Boethius of Dacia (1987), *On the Supreme Good*, ed. and trans. J.F. Wippel, Toronto: Pontifical Institute of Mediaeval Studies.

Carrithers, Michael, Steven Collins and Steven Lukes (eds) (1985), *The Category of the Person*, Cambridge: Cambridge University Press.

Delumeau, Jean (1978), *La peur en Occident, XIVe–XVIIIe siècles: une cité assiégée*, Paris: Fayard.

Delumeau, Jean (1983), *Le péché et la peur: la culpabilisation en Occident, XIIIe–XVIIIe siècles*, Paris: Fayard.

Delumeau, Jean (1988), 'Prescription and reality', in E. Leites (ed.), *Conscience and Casuistry in Early Modern Europe*, Cambridge: Cambridge University Press.

Descartes, René (1974–86), *Oeuvres de Descartes*, ed. Charles Adam and Paul Tannery, 2nd edn, Paris: Vrin.

Detienne, Marcel (1967), *Les Maîtres de la verité dans la grèce archaïque*, Paris: Maspero.

Elias, Norbert (1982), *State Formation and Civilization*, Oxford: Basil Blackwell.

Elias, Norbert (1983), *The Court Society*, New York: Pantheon Books.

Gaukroger, Stephen (1995), *Descartes: An Intellectual Biography*, Oxford: Clarendon Press.

Gaukroger, Stephen (2001), *Francis Bacon and the Transformation of Early Modern Philosophy*, Cambridge and New York: Cambridge University Press.

Gaukroger, Stephen (2002), *Descartes' System of Natural Philosophy*, Cambridge: Cambridge University Press.

Hadot, Pierre (1993), *Plotinus or the Simplicity of Vision*, Chicago, IL: University of Chicago Press.

Hadot, Pierre (1995), *Philosophy as a Way of Life*, Oxford: Blackwell.

Mauss, Manuel (1938), 'Une catégorie de l'esprit humain: la notion de personne, celle de "moi"', *Journal of the Royal Anthropological Institute*, **68**, 263–81

Onians, Richard Broxton (1954), *The Origins of European Thought: About the Body, the Mind, the Soul, the World, Time, and Fate*, Cambridge: Cambridge University Press.

Rabbow, Paul (1954), *Seelenführung: Methodik der Exerzitien in der Antike*, Munich: Kösel-Verlag.

Ranea, Alberto Guillermo (2000), 'A "science for *honnêtes hommes*": *La Recherche de la vérité* and the deconstruction of experimental knowledge', in Stephen Gaukroger, John Schuster and John Sutton (eds), *Descartes' Natural Philosophy*, London: Routledge, pp. 313–29.

Rummel, Erika (ed.) (1990), *The Erasmus Reader*, Toronto: University of Toronto Press

Snell, Bruno (1960), *The Discovery of the Mind: The Greek Origins of European Thought*, New York: Harper and Row.

Vernant, Jean-Pierre (1965), *Myth et pensée chez les Grecs*, Paris: Maspero.

Vernant, Jean-Pierre and Pierre Vidal-Naquet (1972), *Myth et Tragédie en Grèce ancienne*, Paris: Maspero.

Wieland, Georg (1982), 'The Reception and Interpretation of Aristotle's *Ethics*', in Norman Kretzmann, Anthony Kenny and Jan Pinborg (eds), *The Cambridge History of Later Medieval Philosophy*, Cambridge: Cambridge University Press, pp. 657–72.

Chapter 3

Neither Genius nor Context Incarnate: Norman Lockyer, Jules Janssen and the Astrophysical Self

David Aubin and Charlotte Bigg

Parallel Profiles and Parallel Lives

In December 1872, the French Ministry of Public Instruction issued a medal commemorating the parallel discovery by Jules Janssen (1824–1907) and J. Norman Lockyer (1836–1920) of a spectroscopic method for studying solar prominences (Figure 3.1).[1] Four years earlier, on 26 October 1868, the two men's letters, mailed from India and England respectively, reached the French Academy of Sciences at about the same time, and were read out at its following meeting. Lockyer claimed he had anticipated the discovery by two years, as a published note on the subject testified. Lacking the means and time actually to try out the method, he had applied to the Royal Society for a government grant to subsidise the construction of an appropriate spectroscope. Learning from reports of the 18 August 1868 eclipse that prominences emitted certain bright lines, Lockyer had immediately set out to find them using his proposed method, and succeeded (Meadows, 1972, pp. 52–5).

But Janssen had beaten him to it. Sent to India for the specific purpose of applying spectrum analysis to the observation of the eclipse, he was struck by the brightness of prominence emission lines. In an oft-quoted statement, he confidently predicted right after sunlight had blinded him: 'Je reverrai ces lignes-là!' ('I will see these lines again!'). He thought it over during the night, and the next day he had perfected and tested his method. As a result, the spectroscope was turned into an instrument allowing the investigation of not only the chemical constitution of the sun, but also of its otherwise inaccessible physical attributes. In October 1868, Janssen wrote to his wife: 'I was sent to India to observe the eclipse for 5 minutes, and I am bringing back the perpetual eclipse.'[2]

1 See *Comptes-rendus de l'Académie des sciences*, vol. 75 (1872), p. 1491. See also Sarton (1927–30), vol. 13, pp. 353–5.

2 Jules Janssen to Henriette Janssen, 1 October 1868, Janssen Papers, Bibliothèque de l'Institut, Paris, Ms. 4,133, fol. 106, quoted in Aubin (1999), p. 81. For more on Janssen, see Chapin (1984) and Aubin (2002).

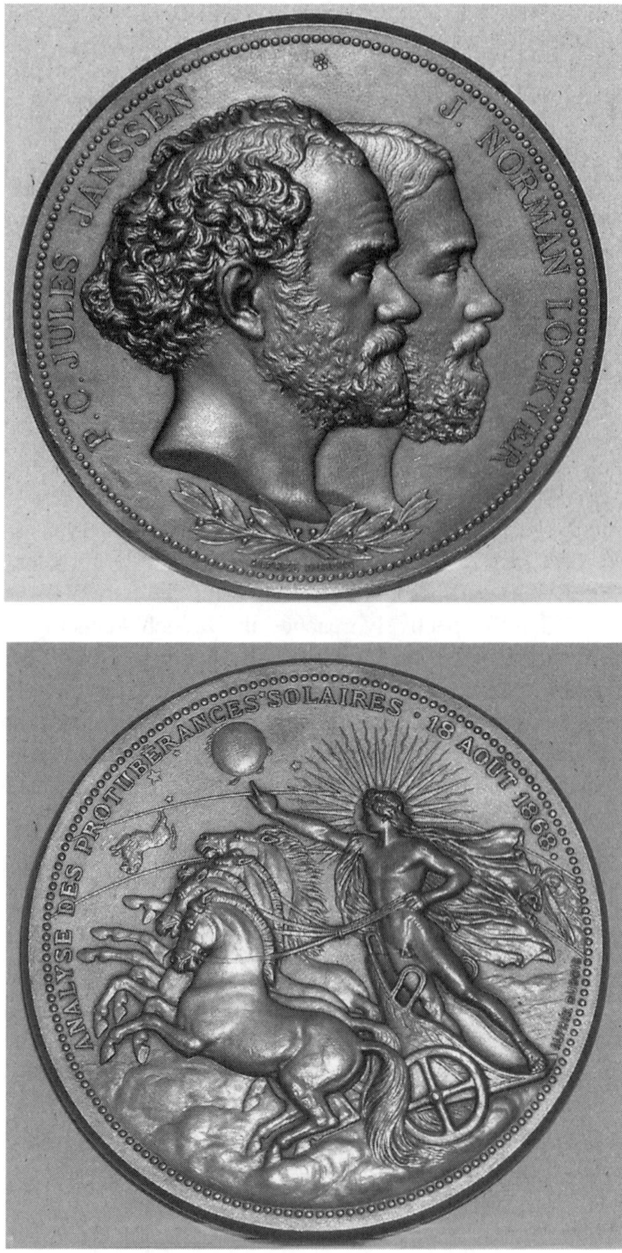

Figure 3.1 Alphée Dubois' medal issued by the Paris Academy of Sciences in 1872 in honour of Jules Janssen's and J. Norman Lockyer's parallel discovery of a spectrocopic method for seeing the Sun's prominences. The physical resemblance between the two scientists portrayed here may not be entirely coincidental. © Académie des sciences de l'Institut de France; photo: David Aubin.

Putting nationalist vindication aside, the British and French scientific establishments agreed to view Lockyer and Janssen's achievements as a case of simultaneous discovery. The medal struck in 1872 reflected this sentiment. Designed by the renowned medallist Alphée Dubois, it pictured Phoebus on a horse-drawn chariot pointing at solar flares and, in a subtle tribute to Janssen's priority, recalled the date of the eclipse. On the flipside, the profiles of both men were engraved imperial-style, Janssen to the fore. At first glance, the parallelism of the profiles is striking, the features of both men enhancing the aesthetic appeal of Dubois' work of art. Lockyer and Janssen cast their resolute gaze towards the right, somewhat above the horizon. Their noble, assertive foreheads, their well-defined eyebrows, their prominent but thin noses and their tight, grave lips all emphasise the scientific discoverers' stern outlook.

Janssen and Lockyer's resemblance on this medal is not fortuitous. Clearly it was the product of Dubois' mastery of a well-established artistic genre. In the second half of the nineteenth century, the medal enjoyed a renaissance as a form of art independent of the coin and the bas-relief. Artists such as Dubois could make profitable careers as medal engravers.[3] Medal-making, like other art forms, allowed artists to be creative within the limits of the genre. What distinguished the artist from the mere artisan was the ability to reproduce accurately not only the subject's physical features, but also character and expression. Medallic portraiture, according to the *Encyclopædia Britannica*, ought to be 'a thing of pure flesh and blood, suave and graceful in composition, and as pleasing in its purely decorative design as imagination can inspire or example suggest'.[4]

But Lockyer and Janssen's resemblance went beyond medallic convention. By the Victorian age one had come to expect active, innovative scientists to take the appearance of dignified, middle-aged European males with beards. One could have anticipated the medal's portrayal of the determination that had helped them rise from the low-level clerkships they once occupied to the position of officially acknowledged scientists. From then on, the two men, who remained on friendly terms for the remainder of their lives, were united in a common enterprise: the founding of a 'new astronomy' in general, and of astrophysical institutions in particular. But while Janssen was quickly successful in his bid for an independent, government-sponsored observatory, which was built for him in the Paris suburbs of Meudon in the 1870s, Lockyer struggled for years at the margins of the scientific establishment, his theories often greeted with scepticism and his professional status remaining uncertain. Historians now view these events as episodes in the establishment of a new discipline, astrophysics. Outsiders to traditional astronomy, Janssen and Lockyer had rapidly assimilated the new technique of spectrum analysis and made a fundamental discovery. Subsequently, they would lead numerous solar eclipse expeditions overseas.[5] Both became vocal public promoters of the 'new astronomy', and strove to establish it as an autonomous field.

---

3 On Dubois, see Mazerolle (1906) and *Revue numismatique* (1905), pp. 518–19.

4 S.v. 'Medal', *Encyclopædia Britannica*, 11th edn, vol. 18, pp. 1–2.

5 On solar expeditions, see Pang (2002).

On further examination of the medal, however, features individualising the two scientists become apparent. The unruly curly hair of the Frenchman stands in contrast to the British scientist's; under the beards, the chins seem to have different shapes. Despite their similar trajectories, despite the conventions of medal engraving, Janssen and Lockyer were different persons. While both operated in a broadly similar environment (late nineteenth-century Western Europe, the rise of precision sciences and scientific disciplines, colonial capitalist culture), they lived and worked in different local environments. Victorian London and Third Republic Paris were homes to differing national conceptions of science's utility and its relationship to the state.

Our work on the early history of astrophysics, and in particular on the figures of Norman Lockyer (by Bigg) and Jules Janssen (by Aubin), has led us to methodological and historiographic reflections on the treatment of the individual in historical writing. Neither conventional biographical approaches nor social history's methods seemed to us appropriate for accounting for these pioneers, founders and discoverers. In this chapter, we tease out some of the underlying assumptions of social history of science, technology and medicine, and of the biographical genre which has given rise to what we describe the 'genius versus context' dichotomy. We examine some of the negative effects this polarisation has caused, in particular the premature obliteration of the problem of individual agency. Picking up on the growing literature on the 'self', we explore ways of going beyond this binary opposition, which might be of use to social historians and to biographers alike.

Genius versus Context

The social history of science, technology and medicine has been founded in part on a rejection of traditional biographical writing: 'The bad old history of science of the early twentieth century', Thomas Hankins wrote in 1979, 'which we have all been taught to abhor, was largely biographical' (p. 2). The suspicion harboured by practitioners of social history against biography as a genre stems from an array of methodological, ideological and political objections. Biography appeared tainted by its historical debt to hagiographic writing. Social historians condemned it for being over-sympathetic to its subjects (and their narratives), and largely concerned with great minds and their discoveries. Further, since Robert Merton's incitement to conceptualise science as a collective enterprise, and the emergence of the sociology of scientific knowledge, the very focus on individual lives as the central unit of historical analysis has been judged wrong-headed. While scientific biographers themselves sought to redefine the genre and make it less politically conservative by including lesser-known figures, women and representatives of the non-ruling classes, most social historians remained adamant that the significant agent was the collective, not the individual.

Thus while they shifted the focus from abstractions to society and social groups, constructivist accounts of the sciences have paradoxically failed to move closer to the biographical genre. They have tended to push aside the person as an incarnated locus of intentions, desires and decision-making ability. When persons feature in these accounts, they appear often as ideal types of wider social entities – the invisible technician, the scientist-entrepreneur, the theoretical physicist. To put it bluntly: the task of exhibiting the social construction of knowledge has been carried out at the expense of the individual.

But the individual would not go away. As was clearly pointed out by the editorial board of the *Dictionary of Scientific Biography*: 'history of science … is made by men [*sic*] and not by themes or abstraction' (vol. 1, p. x). Some social historians have recently sought to make social constructivism more compatible with the individual.[6] They have in the process redefined the biographical genre as the study of individual trajectories through richly textured social spaces. As Yves Gingras, inspired by Pierre Bourdieu, put it: 'biography is *embodied* social history'.[7] One example emblematic of this renewal is Crosbie Smith and Norton Wise's celebrated volume on Lord Kelvin, *Energy and Empire* (1989). In these biographies, contextualist elements none the less remain most apparent. Thus, in the case of Wise and Smith, it is the public, the social Kelvin which is presented. As the authors significantly point out, a 'problem of constant concern' to them was the fact that very few letters had survived revealing Thomson's 'innermost thoughts or emotional response to major crises, such as the death of his father'. Refusing to engage in speculation, they acknowledged that scattered 'letters … unlike anything in the "official" archives' provided nothing less than 'a distorted sample'. Despite the comprehensiveness of their 800-page biography, they conceded: '[t]he truly private Thomson *must* remain a veiled figure' (Smith and Wise, 1989, p. xxii). Even in these works, social historians' ambivalence towards the biographical genre remains perceptible. In the introduction to his masterful book on Volta, Giuliano Pancaldi admits in convoluted terms that 'Insofar as it is a biography, this is a biography in context' (2003, p. 2).

For social historians and sociologists of science, 'biography in context' has been one strategy in their reassessment of the role of the individual. Parallel biography or prosopography are alternative ways of circumventing the genius versus context binary opposition. Just as Mario Biagioli examined a large panel of secondary mathematicians to identify what was exceptional about Galileo, a focus on two figures, such as Lockyer and Janssen, could help specify the general and local assumptions about the scientist's persona or his role in society and better apprehend the nature of the resources available for, and constrains limiting, the elaboration of astrophysicists in this period (see, for example, Biagioli, 1989; Pancaldi, 2003). Parallel biography might be seen as an experiment where the subjects of the comparison are probes immersed in different settings, revealing the resources offered to them in their particular conditions, and the constraints on their range of options. A comparative study of both scientists' trajectories tells us much about the general and specific, the

6 The complex and changing attitude of the history of science towards biography is not specific to the field: it is a recurrent subject of debate in the historical profession; for debates in the French context, see Marian (1986), Arnaud (1989) and Lévi (1989). Recent contextualist scientific biographies may be profitably compared to the biographies produced by the *Annales* historians, for example Duby (1984) and Le Goff (1996).

7 Gingras (2001), p. 125. See Bourdieu (2001) for his critique of science studies in relation to biography.

global and local character of the emergence of astrophysics as a distinct field, and of the configuration of state, science and society in which both men functioned. More broadly, the comparative dimension yields insights into the range of possibilities available to the enterprising individual seeking to invent a new practice, a new professional status and a new conception of the universe. It throws light on the tension between the universal, international and national conceptions of science, and how such a tension was managed by particular individuals. It helps us assess the singularity of historical actors against a shared background. One obvious benefit of studying Janssen and Lockyer in parallel has been a more nuanced understanding to 'context' and its determinisms.[8]

Very recently, isolated attempts have been made within sociology and social history of science, technology and medicine to confront the problem of individual actors head-on, to give this historiographical trend a theoretical backbone, a framework to conceptualise the 'singularity' of individuals. Taking the case of Robert Oppenheimer, Charles Thorpe and Steven Shapin (2000) have investigated 'charisma' using the methods of social studies of science. They define charisma as an 'interactional accomplishment' which serves the purpose of normative stabilisation of large technological organisations, such as Los Alamos in the 1930s and 1940s. Hélène Mialet (1999) has similarly explored 'singular subjectivity' through a study of the figure of the individual creator, taking Stephen Hawking as one example. She finds creativity to be 'distributed in specific tools, practices, and social networks', and thus proposes to understand the 'knowing subject as a distributed-centred subject'.

Whether these leads will be taken up in historical and sociological practice remains to be seen. But they are in any case significant expressions of a *malaise* in sociology and social history and recognitions that the issue of individual agency cannot indefinitely be brushed under the carpet. If social history of science, technology and medicine has been in part built on a rejection of the convenient and enduring straw man of biography, this has until recently been without engaging much with the genre itself or its representatives – and most damagingly, without addressing the central issues it raises for social history.

The Public Speech

How might we go beyond the 'false dichotomies in which historians are too often locked up' (Gingras, 2001, p. 125), for example between the individual and the collective? Specific arenas can be identified in which this binary opposition begins to dissolve, destabilising some of the underlying assumptions of both social history and biography. In our own work on the early history of astrophysics, and in dealing with Lockyer and Janssen in particular, we found that the figures – or types – of the *discoverer*, the *pioneer* of a new field and the *founder* of an institution were essential,

8 Several existing parallel biographies display this dynamic in its different variants. A study of John von Neuman and Norbert Wiener underscores the differences between both trajectories in the same context (Heims, 1980), while books on Hitler and Stalin (Bullock, 1991) and Einstein and Picasso (Miller, 2001) insist instead on the similarities of two trajectories in contexts conventionally perceived to be different or opposite.

yet hard to account for using the tools of social history – while, of course, being standard tropes of the biographical genre. While these types accurately describe one and/or the other character's activity, they are historiographically laden archetypes which require cautious treatment. Clearly, they are also contemporary roles our two protagonists played with. These types are central to the emergence and shaping of scientific disciplines in the late nineteenth century, as science took on the form of an increasingly collective endeavour, with groups of researchers identifying with well-defined procedures, conceptual frameworks, methods, approaches and instruments. Should we consider the individualistic types of the founder/explorer/discoverer, paradoxically endowed with all the more individuality as collective disciplines were subduing individual characteristics, as necessary counterparts, holding communities together by shared beliefs in them?[9] How to deal with these types without taking them for granted, as biographies so often have?

Lockyer and Janssen adroitly toyed with such figures of self-representation in the many public speeches they delivered in the course of their scientific lives. There are sound reasons underlying historians' suspicious attitude towards such sources. Speeches are careful reconstructions of scientific research crafted to fulfil the expectations of specific constituencies. As such, they are often little reliable factually, but can be precious in other ways. Speeches surely deserve more attention from social historians because they are central to the self-construction of science and scientists. Speeches are revealing of the fragile negotiations at play in the elaboration of new sciences and new scientists, and of the arenas in which a personality can assert his/her individuality. This can be shown taking the example of a speech given by Lockyer, which exhibits the various resources and limitations of this type of forum for someone seeking to found a new astronomy.

On Tuesday 16 December 1873, Lockyer introduced the subject of 'celestial chemistry' to his audience at the Quebec Institute, London. This lecture participated in the broader economy of public science, alongside the British Association for the Advancement of Science's annual meetings, the Royal Society *conversaziones* as well as the expeditions by pioneering scientists, tourists and mountaineers whose accounts continuously poured into the metropolis, together with a stream of exotic objects, images and stories to feed the growing popular press. This culture must be taken into account for a proper understanding of the establishment of the new astronomy in Victorian Britain and of Lockyer's place within it.

As described in this lecture, Lockyer's conception of astrophysics was based on the notion of the advancement of knowledge, of progress, with its religious connotations in a society which considered self-improvement and the personal quest for knowledge to be moral duties. In further tying the advancement of knowledge with the history of the human race, Lockyer situated his talk in a frame not only tinged with imperial overtones but also one shaped by the ideas of Darwin, whose *Origin of Species* was published in the same year that Bunsen and Kirchhoff elaborated the principles of spectral analysis. Indeed, a few decades later, Lockyer himself developed a theory of inorganic evolution to account for the functioning

9 On the connection between self-fashioning and historiography in the case of 'discovery', see Schaffer (1986).

of the astronomical ecosystem, a theory in which Darwin's evolution became 'an appendix, as it were, to the work of inorganic evolution' (Lockyer, 1900, p. 168). Another major theme in Lockyer's work was the chemical unity of the universe as revealed by the spectroscope. Here Lockyer made another analogy, this time with notions recently elaborated by physicists: the conservation of energy and the uniformity of a universe composed of matter in motion was extended to include the chemical composition of all bodies: 'the Spectroscope show[s] us that like matter is acting in like manner everywhere' (Lockyer, 1874, p. 411). Taking this principle quite literally, Lockyer did not hesitate to extrapolate from his laboratory experiments about solar phenomena, just as he claimed that the sun was an ideal laboratory, giving him indications about the behaviour of gases at temperatures and pressures impossible to achieve in the laboratory. Lockyer's astrophysics was finally in tune with a Humboldtian conception of knowledge, seeking as it was to uncover the connections between all phenomena, to 'knit the universe together' by means of travel, observation and speculation. It took a literary and aesthetic rather than a mathematical approach to phenomena. It operated largely in the field, using robust, portable devices rather than precise apparatus in custom-made, padded laboratories. These early astrophysicists disliked the obsessively precise positional astronomy (see Bigg, forthcoming), preferring to follow William Herschel, the 'natural historian of the heavens' (Schaffer, 1980).

Putting forward an astrophysics integrating recent natural history, physics and the field sciences (as well as, on occasion, archaeology, meteorology and anthropology) enabled Lockyer to navigate between different social and scientific milieus and reach a wide public. Representation permeated his activities, taking the form of a constant presence on the multiple stages of London's cultural and scientific life, but it was also central to his scientific approach. When in 1871 Lockyer obtained funding from the British Association for the Advancement of Science to lead an expedition, he implicitly agreed to represent both the association and the Empire on his travels. Once in India, not only the Sun but also the eclipse camp were photographed, as were the travellers and their local helpers. Returning to London to tour the most prestigious and fashionable venues, Lockyer re-employed the same visual technologies to project pictures of himself against different backgrounds, multiplying himself on the spot, asserting his presence and the importance of his work, and blurring the distinction between science and its popularisation, opening up a space in which the astrophysicist could exist. But the mitigated reception among scientists of many of his theories reveals the challenges inherent in putting forward a different kind of science and practitioner. Here probably lies the origin of Lockyer's reputation for eccentricity, which he in turn enrolled when taking on the figure of the (misunderstood) proponent of a new approach and founder of a new field.[10]

The lecture hall, together with other similarly public spaces such as the newspaper column, thus constituted an essential site for the elaboration and representation of Lockyer's astrophysical self and science. That Lockyer's self-fashioned image took this character owed as much to his own taste for public presence, to his lack of

---

10 On the significance of optical and spectroscopic practices for astrophysics, and on alternative strategies of establishing the field, see Bigg (2002) and (2003).

professional situation and his conception of astrophysics as to broader social and cultural transformations of late-nineteenth century Britain which brought into existence a new kind of audience and market for him and his science, consisting of urban, educated, bourgeois middle classes.

Discourse Analysis

In what sense can we say that Lockyer was the author of his speech at the Quebec Institute? The postmodernist project provides some useful insights for analysing such a speech and for undermining the genius versus context dichotomy. In his seminal 1969 piece 'What is an Author?', Michel Foucault asserts that the author has disappeared: 'a key-moment of individualization in the history of ideas, knowledge, literature ... philosophy and science', the author has given way to indifference towards it, an indifference which constitutes 'one of the fundamental ethical principles of contemporary writing'.[11] This claim exposes him in the ensuing discussion to the modernist criticism (made by structuralist sociologist Lucien Goldmann) that the group of philosophers to which Foucault belongs 'is characterized by the negation of man in general, and, from there, of the subject in all its aspects, and the author as well' (Foucault, 1994, vol. 1, p. 812).[12] To which Foucault answers, clarifying his position: whether the subject has an ontological existence or not is not the point.[13] The point is rather that, in examining a text, a particular method is applied, which supposes that the author has no significant existence. Foucault introduces the notion of 'author-function', as distinct from the actual author of the text, in order to draw attention to this other, long-ignored author: the authorial voice, which is expressed in the text itself, through a combination of style, pronouns and forms of language. Neither denying nor accepting a 'real' author, he simply dismisses the question of the author's existence, and this very dismissal creates interesting effects, such as the appearance of the author-function: 'The question I asked myself was the following: what does this rule by which the writer or author disappears allows us to discover? It allows us to discover the author-function at play' (Foucault, 1994, vol. 1, p. 817). And further: 'The author-function is thus characteristic of the mode of existence, circulation and working of certain discourses within a society' (Foucault, 1994, vol. 1, p. 798). This reasoning is applicable, Foucault insists, not only to the author of writings, but also more generally to the author of works of art, knowledge, including scientific, or even actions. In this wider understanding of 'discourse' which exceeds literal texts, the author-function appears as a fundamentally hybrid entity, resembling

11 Foucault (1969), p. 792 (our translation); cf. the incomplete translation in Foucault (1977).

12 On Goldmann's structuralist perspective, see Aubin (1997).

13 This and similar writings have often been charged with denying the existence of the subject and prompted violent debates among historians about the 'subject' with reciprocal misrepresentation of opposing positions: while the postmodernists often exaggerate the uncritical realism and essentialism of traditional historians (for example, Nye, 1983), modernists assert that their opponents deny the possibility of the subjects' existence (for example, Söderqvist, 1996, and Zagorin, 1999).

the actual author and his/her intention in some ways, but shaped by the (authorial and representational) conventions of discourse. But since any authorial expression necessarily operates within a discourse, Foucault implies that the author is as much the creator of the text as created by it. In this reading, it is meaningless to consider the author and the discourse separately.

Let us now turn to Janssen's speeches, where the construction of several types of author-functions appears clearly. Like Lockyer, Janssen thrived on public performances. On 17 February 1870, he gave a well-attended public lecture at the Sorbonne, which can be analysed usefully from a Foucaldian perspective. The picturesque side of his scientific travels, the imaging technologies of contemporary science and the widespread interest in questions of origins (of the universe and of man) were all drafted in this seduction operation. Somewhat candidly, he explained: 'I always thought that science is not widespread enough in our French society, and that this matter of fact is responsible for the slow development of our scientific institutions. … Today, nothing considerable can exist, nor develop without the support of public opinion. Science must therefore communicate its significance and usefulness to the new society'. This 'propaganda', as he called it, would attract public support for state-sponsored institutions (Janssen, 1903, pp. 300–301). From 1869 to 1874, Janssen consistently campaigned for the establishment of a government-sponsored astrophysics observatory where he would apply his new method on a daily basis. In the last years of the Second Empire, there was much talk about institutional reform in astronomy and its centralisation under Paris observatory director Le Verrier (Aubin, 2003). In democratising France, the establishment of scientific institutions had to be supported by public opinion, however.

To succeed in his bid, Janssen not only needed recognition from his peers, and public and government support, but also to fashion himself (his self) to become a credible answer to contemporary concerns. Another lecture, delivered at l'Académie des sciences in 1871, is highly revealing in this respect. Returning from an unsuccessful expedition to Algeria, where clouds had hindered the observation of the eclipse, Janssen chose to entertain savants and auditors with a detailed description of his escape from besieged Paris in a hot-air balloon (Janssen, 1871a; 1871b). This text is highly revealing of the various 'author-functions' in tune with Third Republic scientific discourses. For the sake of brevity, let us here focus on four aspects exhibited by this text.

First, Janssen underscored his 'scientific persona'. In Lorraine Daston and Otto Sibum's analysis, 'scientific persona' is an 'intermediate between the individual biography and the social institution: … a cultural identity that simultaneously shapes the individual in body and mind and creates a collective with a shared and recognizable physiognomy' (Daston and Sibum, 2003, p. 2).[14] In 1871, Janssen was concerned with the problem of solar constitution, and applying spectroscopy to the study of the corona, he explained, was his mission. Emphasised at various junctures in the adventurous account of his flight, this dimension showed how his scientific attitude crucially shaped his approach to technology and the world: he took regular

14 Daston and Sibum (2003) re-examines and develops the notion of persona as defined by Marcel Mauss (1938).

measurements of atmospheric pressure and temperature at duly recorded times, and gave scientific explanations for various phenomena he came across. Several footnotes show him constantly on the look-out for ways to improve air navigation. But this scientist persona sometimes gave way to other dimensions.

Janssen also emphasised another aspect of his constructed persona, namely his self-portrayal as an experienced traveller. He had never flown an aerostat, he said. But he was 'convinced that theoretical knowledge maturely acquired and experience of traveling would suffice in giving me the cool-bloodedness and the inspirations needed' (Janssen, 1903, p. 4). In no contradiction with his scientific persona, this aspect of his selfhood was emphasised at great length as the canvas of his literary technology. Squarely situated within the genre of the travel report, his account was complete with preparations, excitements, changes of perspective, surmounted difficulties, picturesque anecdotes, and final success.

Third, Janssen regularly insisted on the importance of this voyage for France and of himself as its devoted servant; in a time of war, this was an eminently nationalistic enterprise. Commenting on the topographic value of aerial photography, he stated that: 'It belongs to France, which has created aerostation, to endow science with this new branch' (Janssen, 1903, p. 13). Throughout the text, 'France' was identified with its institutions, the Academy, the Bureau des Longitudes, and especially its government. His visit to Adolphe Thiers, the new head of post-war France, in Bordeaux is depicted as the starting point of 'agreeable and precious' reports. But 'France' was also its people, peasants and local elites from the countryside. Reporting the patriotic sentiment of those among whom he had landed, Janssen saw it as bearing testimony of 'all that one could have received from France, had one known how to speak to it, to train it, and above all to organize it' (Janssen, 1903, p. 20).

There is a final aspect of Janssen's self-representation that needs to be noted: the poetic outlook that his condition afforded him:

Below us, in a heavy, obscure atmosphere, the nightly apparatus of a big city whose reddish, volcanic fires gave rise to the idea of an inferior world with its appetites, its passions, its violence, its misery. And what a coincidence! Was not Paris wrestling at this very moment with the ardent embrace of our enemies.

But if, breaking away from these ideas, one gazed toward our pure, diaphanous regions, already inundated by the morning lights of the breaking day, what a contrast and what a relief! One felt relieved and penetrated by a feeling of indefinable purity that gently carried one's thought up to the level of extra-terrestrial ideas. (Janssen, 1903, pp. 7–8).

As far as travel reports go, the narrator's ability to be moved by the extraordinary things he experiences is hardly exceptional. In this lecture to the Academy about a failed mission, Janssen's recourse to the lyric served as a reminder of the beauty of nature, the sun especially, whose investigation he had made his life's mission. This was an indication of the hopes that could be invested in science to create a new world, more serene, more beautiful, and geared toward truer concerns. For a defeated, humiliated France, science was both technical solution and poetic escape. In short, Janssen put forward a deliberate, global depiction of himself, as a complex individual who, having adequate knowledge and experience, could address some of the pressing problems of science and the nation, while retaining the ability to be exalted.

The Self as an Actor's Category

Searching for ways of reconciling essentialist and constructivist approaches to history and biography, we have been drawn to the self not only as a historiographical entity, but also as a historically situated one. Foucault suggests that the individual and the social are intimately linked. The person develops his/her identity in interaction with others, individuals and social groups, and the social landscape is constantly reinvented by individual initiatives. The way in which biological categories have been construed can provide pertinent guidelines. Since Simone de Beauvoir ('On ne naît pas femme, on le devient', 1949), the social roots of sexual categories have been one of the great lessons of gender studies. In discussing the problem of using personal experience, Joan Scott has further suggested that, counter to common perceptions, 'it is not subjects who have experience, but subjects who are constituted through experience. Experience in this definition then becomes not the origin of our explanation … but rather that which we seek to explain, that about which this knowledge is produced' (Scott, 1991, pp. 779–80). Categories such as 'gay', 'Black' or 'male', which she calls 'foundationalist concepts', need to be analysed rather than naturalised by historians. 'All categories of analysis', she adds, should be taken as 'contextual, contested and contingent', so that historians should turn to 'the history of foundationalist concepts themselves' (Scott, 1991, p. 796; for a critique, see Smith, 2001).

In Western societies, one especially important foundationalist concept has been 'the self'. The extant literature on the self is vast and highly heterogeneous.[15] For our purposes, one can distinguish three main threads: (1) philosophical, psychological and biological essays often rooted in cognitive science, that attempt to delineate the universal foundations of self-identity or consciousness; (2) mostly linear historical reconstructions of the emergence of the modern self, often seen to be intimately connected to Western capitalistic democracy, and (3) more recent historical analyses of various types 'ego-documents' (Schultze, 1996).[16] These three sets of investigations point to three conceptions of the self: as biological universal (in the same way as, for example, the eye is), as historically constructed entity (like science or nations),

15 On the history of the 'self', the literature is enormous. See an extensive conventional account by Taylor (1989) and revisionists essays in Porter (1997). See also Mascuch (1996) and, on the emergence of contemporary positive meaning in the mid-nineteenth century, Earman (2001).

16 We avoid discussing psychoanalysis, which would lead us far astray. Let us, however, recall that Michel de Certeau's analysis of Lacan's 'mirror stage' (de Certeau, 1986) underscored self-construction in relation to others: 'In a "flutter of jubilant activity" he [the child] discovers that he is *one* (a primordial form of the *self*), but this discovery occurs through that alienation which identifies him with what is *other* than him (a speculary image). The experience can be put into the formula, *I* am *that*. The self takes shape only in self-alienation' (de Certeau, 1986, p. 56).

and as historically active category (like ambition or love), playing specific roles for various people in different circumstances. Though distinct, these conceptions of the self all attempt to bring together different aspects of personhood.

It is not our intention to posit a stable, trans-historical conception of the self (compare, for example, Seigel, 1999). Nor do we assume that there is, for each given culture, a dominant conception that determines people's self-conception. As a historical category, the self becomes truly interesting when it is taken as an actors' category. It seems obvious that people have always had a sense of self. But as recent studies now show in a wealth of cases, this sense also is rooted in place, time and society. Historical writing that deals with individuals needs to be historically sensitive to the subjects' conception of themselves or of their selves. The social history of science has supplied an insightful critique of central though little theorised tenets of biographical writing: first, the presumption that a life, the *Lebenslauf* linearly unfolding from the birth to the death of the subject, exhibits continuity and unity, and second, the fact that persons can be taken as meaningful, unproblematic units in historical writing. It sometimes goes as far as altogether denying this unity of the life/person in the history of science. Taking its cue from postmodern scholarship, in particular the historical work of Michel Foucault, it questions the opportunity of adopting a view of individuality rooted in Enlightenment rationality, seeing in this (illusory) unity an ideological tool fostered by states to bring minds and bodies under control; the rational agents dear to modern economic theories being one of its later avatars. By affirming the fragmented self, it wishes to make a political statement against such means of buttressing conservative (neo-)liberalism, as well as to portray the self more truthfully. A wealth of theoretical and historical studies on selfhood has been produced in the wake of these reflections, which caution against the fundamental *Einheit* of an individual human being and explore alternative conceptions of personal identity (Taylor, 1989; Mascuch, 1996; Porter, 1997). The history of *mentalités*, Foucault's analysis of sexuality,[17] Scott's history of foundationalist concepts, Daston and Sibum's categorising of personae, Daston and Galison's discussion of the 'scientific self' or *microstoria* (critiqued, for example, in LaCapra, 1985) are, in this view, so many attempts at studying the self in this light. We want to suggest that the growing body of literature devoted to the self might be one of the most exciting and most promising recent developments for providing a better treatment of individuals in social history of science and a more reflexive attitude in biography-writing. Let us turn once more to Janssen's public performances to illustrate this point.

Public lecturers need to come to terms with specific aspects of their selves, such as, for example, the physical engagement with their audience.[18] After a public lecture in Lyons in 1873, Janssen put his impressions down on paper:

---

17 For the later Foucault, this notion was precisely the locus where the three dimensions discussed by Seigel intersected. See 'Subjectivité et vérité', and 'Les techniques du soi', in Foucault (1994), vol. 4, pp. 213–18 and 783–813, and, of course, Foucault (1978).

18 Another approach which we do not treat here is the study of material culture and instruments and their relation with the self, for example as extensions of the body (Schaffer, 1992).

At 8:10, I arrived. The Palace was full of people who climbed up the stairs to get into the amphitheatre. Walking in front of a door, one could see that the inside was packed, and a queue where people were pushing each other trying to get in. … Some had got in the area reserved for technicians and the police had to be called to force them to leave their spot. Behind me the *Association* [*française*] members had taken place, so that I was literally besieged with almost no freedom to move. But, above all, the heat was oppressive. … So many people, low ceilings, lights everywhere – that was too much. … As I progressed, my strength dwindled, my forehead sweated copiously in this sauna. We had been forced to open all windows. Strong efforts were thus required to be heard beyond the first ranks. … I judged that this should not be prolonged. … I shortened some parts, reached my conclusion and ended up by calling attention on the necessity of founding an observatory of physical astronomy. (Janssen to his wife, 25 August 1873, Janssen Papers, Bibliothèque de l'Institut, Paris, Ms. 4133, fol. 168)

Having been dropped on the floor by his wet nurse as an infant, Janssen limped for the rest of his life. His handicap was little hindrance, so it seems, to his climbing mountains and criss-crossing the globe, but this circumstance, as well as the description above, reinforces the impression that he had to surmount great physical strains and pains to deliver his scientific results to an eager public. In other words, Janssen saw himself, and wanted others to see him, as a servant of science and country. The analysis above should also be read in the same way. In the 1870s, the dedicated service of a daring, competent man was precisely what French society expected from its elite. Janssen's self-fashioning was a crucial component for establishing an institutional home-base for astrophysics in France. If Janssen could come across as a convincing astrophysicist, he might well become one; and of course, not only Janssen's performance was important, but how his proposed self fitted with current conceptions and expectations, from the most immediate (the persons he spoke to) to the most general (forms of politeness, jargon, adequacy of self-presentation with the accepted comportment of a serious scientist).

The self needs not always to conform. But as we saw with Lockyer, even non-conformity obeys certain rules. Unlike the situation in France, the new astronomy in Britain did not become institutionalised. This was in part due to the effective resistance of positional astronomers to this competing new branch of astronomy, and to the British government's continued disinterest in subsidising science. The ongoing assumption in Britain that scientific investigation was an activity for men of independent means made it difficult for the rising generations of middle-class scientists such as Lockyer who had to live from their work. To be a respected scientist in Victorian Britain did not necessarily improve career prospects. Thus in the 1870s, Lockyer, while a Fellow of the Royal Society, was employed as a third-class clerk in the War Office. While he advocated for many decades state support of science generally, and the establishment of astrophysical observatories in particular, Lockyer developed alternative strategies to put himself and astrophysics forward (and to feed himself, his wife and their eight children) when professional status proved unforthcoming. Lockyer invested in other personae: as editor of a scientific journal, as public lecturer, and as expedition organiser. In the absence of permanent and secure institutional identities, lecture halls, newspaper columns and eclipse camps were appropriated by Lockyer as so many stages where he could assert, even temporarily, his existence. He therefore practised science on the move, with portable instruments which were continuously packed and unpacked. His instruments, photographs and experiments always travelled with him, and were used and reused in a diversity of settings, in borrowed laboratories, on Indian mountains, for impromptu lectures on steamships, at soirees and in journals. Even the solar physics observatory which Lockyer founded and directed from 1879 was a temporary structure made of wood and sheets, and was always threatened, subsisting on the edge of the new South Kensington technical education complex until it was moved to Cambridge in the early twentieth century.

Conventionally, this has been interpreted as a failure. Helped by this 'failure', A.J. Meadows has, without resorting to heavy theoretical apparatus, produced an unconventional biography of the character (Meadows, 1972). Because they mostly deal with dead people, biographies often adopt a retrospective outlook on their subjects, one in which the inherent promise of success and 'genius' can be traced back to a range of personal and/or external factors. In Lockyer's case, the biographer has felt compelled to explain his subject's failure to institutionalise astrophysics by insisting on controversies. The biography therefore highlights Lockyer's constant struggles to construct his scientific self. But successful individuals could profitably be examined in this light, too, with a more systematic focus on the self-in-becoming.

Conclusion: The Self-in-becoming and a Fresh Look at the Sources

We have set out to explore the ways in which people self-fashioned themselves in relation to personae such as scientific pioneer, explorer and discoverer in nineteenth-century Europe, all endowed with strong connotations in the history of science, and which require closer investigation. The author of a recent biography on Galileo has described his work as a 'study of … identity in all its sociocultural dimensions, as well as a scrutiny of the processes through which such an identity is shaped' (Biagioli, 1993, p. 14). The self's ability to act, even transcend these processes and dimensions, is in no way denied; only the focus has shifted. To quote Scott again: 'Subjects do have agency. They are not unified autonomous individuals exercising free will, but rather subjects whose agency is created through situations and statuses conferred on them' (Scott, 1991, p. 793). What emerges from ideas of self-fashioning is a dynamic process perpetually recommenced in relation to others, and especially to socio-cultural representations of self made available by individuals' society and body. Nothing in the process remains stable, neither individuals, nor bodies, nor societies. The self is elaborated, perceived and reflected in a constant process of negotiation. In the cases we study, various identities (or foundational concepts) – such as scientist, professional or 'amateur', husband, father or son, explorer and pioneer, discoverer, *français* under the Third Republic or British under Queen Victoria, none of which is stable, but squarely situated in culture – are combined in an original way to form Janssen's and Lockyer's selves.

As our discussion of Lockyer's and Janssen's speeches has suggested, our focus on the dynamic processes of self-fashioning has led us to look at our historical materials in new ways. Commemorative medals, for instance, have rarely attracted much attention from historians of science. In a nine-part collection on 'medallic illustrations of the history of science', George Sarton (1927–30) seems to have considered them as little more than artistic illustrations of past scientists' greatness. The medals themselves (the objects) have mostly been deemed uninformative as sources for writing history of science. Without insisting on the *post facto* celebratory aspects of the commemorative medal – *éloges* and formal speeches can, after all, be counted among historians' standard resources – one may attribute this neglect mainly to the perception that the design and making of medals are constrained by too many social conventions. But rewards, medals, or rather the processes by which they were instituted and awarded, supply insights, for instance, into the early history of science policy (MacLeod, 1971; Bektas and Crosland, 1992; Grattan-Guinness, 1993).

Other forms of commemorations could be investigated, too. How are great savants celebrated in different times and places (for example, see Fara, 2001, or Rebekah Higgit's Chapter 9 in this volume)? Biographies are just one aspect of this elaboration, besides statues, banquets, funerals and *Oeuvres complètes*. Pierre Nora identifies these cultural manifestations as *lieux de mémoire*. Sources such as these, as well as their negative counterpoints (caricatures, satires), can be made to bear on the study of self-perception in societies and the way in which they shape individual struggles with the self. Recent cultural history has also seized upon autobiographies, which provide crucial insights into self-construction. One thinks here of *histoires de poilus*, accounts of First World War experiences (Audoin-Rouzeau, 2001; Congar, 1997), but also of Foucault's interest in texts such as *Moi, Pierre Rivière* (Rivière, 1973). Literary history – and the novel especially – could also be re-examined. How are biographies related to novels written in the same period?

Our focus on medals and public lectures is therefore not accidental. A shift of focus in the sources considered relevant, as well as a different approach to them, is entailed by our attempt to write self-based history. Taking the self into account makes us turn to hitherto dismissed sources in which the authorial voice is particularly present, such as public speeches, applications for funding or positions, representations of scientists' own selves and their science. To look at such sources, one has to consider the actors' stated intentions (see Cabrera, 2001). Generally, there has been a tendency to dismiss them as power-seeking strategies in disguise. This suspicion was a response to the perceived over-sympathetic attitude of conventional biographers and historians, who were seen to accept unproblematically the (often retrospective) rationalisations proposed by the actors themselves. Social historians have as a result tended to choose their sources guided by a concern to avoid any such accounts by the protagonists. We see 'propaganda' as a privileged locus for analysing the self-in-becoming: since the self constantly re-creates itself while it seeks to re-create its environment, it may be analysed in such programmatic statements which most clearly display this fiction of the self and its attempts at convincing, at being realised. Such utterings are taken here not so much as simple power-seeking exercises, but as revealing the construction of the self, the negotiation between the public's expectations and the scientists' beliefs and strategies; they are tests of credibility for the self-representation of the speaker; they are simultaneously efforts to shape the conceptions of others to make such a self-representation credible.

We would like to conclude with a plea for reflexivity. The point here is not so much that biographers should be more upfront about themselves and their intentions when writing, but rather that biographical writing needs to situate historically the construction of personhood and self.[19]

Bibliography

Arnaud, Claude (1989), 'Le retour de la biographie: d'un tabou à l'autre', *Le Débat*, **54**, March–April, 40–47.

Aubin, David (1997), 'The withering immortality of Nicolas Bourbaki: A cultural connector at the confluence of mathematics, structuralism, and the Oulipo in France', *Science in Context*, **10**, 297–342.

Aubin, David (1999), 'La métamorphose des éclipses de Soleil. En 1868, les éclipses deviennent objets de sciences et de politique', *La Recherche*, **321**, June, 78–83.

Aubin, David (2002), 'Orchestrating observatory, laboratory, and field: Jules Janssen, the spectroscope, and travel', *Nuncius*, **17**, 143–62.

Aubin, David (2003), 'The fading star of the Paris Observatory in the nineteenth century: Astronomers' urban culture of circulation and observation', *Osiris*, **18**, 79–100.

Aubin, David, Charlotte Bigg and H. Otto Sibum (eds) (forthcoming), *The Heavens on Earth: Observatory Techniques in the Nineteenth Century*, Durham, NC: Duke University Press..

Audoin-Rouzeau, Stéphane (ed.) (2001), *Cinq deuils de guerre, 1914–1918*, Paris: Noesis.

Bektas, M. Yakup and Maurice Crosland (1992), 'The Copley Medal: The establishment of a reward system in the Royal Society, 1731–1839', *Notes and Records of the Royal Society of London*, **46**, 43–76.

Biagioli, Mario (1989), 'Social status of Italian mathematicians, 1450–1600', *History of Science*, **27**, 41–95.

Biagioli, Mario (1993), *Galileo Courtier*, Chicago, IL: Chicago University Press.

Bigg, Charlotte (2002), *Behind the Lines: Spectroscopic Enterprises in Early Twentieth Century Europe*, PhD dissertation, University of Cambridge.

Bigg, Charlotte (2003), 'Spectroscopic metrologies', *Nuncius*, **18**, 665–77.

Bigg, Charlotte (forthcoming), 'Staging the Heavens: Popular Observatory Science in the Nineteenth Century', in D. Aubin, C. Bigg and H. Otto Sibum (eds), *The Heavens on Earth: Observatory Techniques in the Nineteenth Century*, Durham, NC: Duke University Press.

Bourdieu, Pierre (2001), *Science de la science et réflexivité*, Paris: Raisons d'agir.

Bullock, Alan (1991), *Hitler and Stalin: Parallel Lives*, London: HarperCollins.

Cabrera, Miguel A. (2001), 'On language, culture and social action', *History and Theory*, **40**, 82–100.

---

19 Grafton's 1999 study of Cardano's autobiography seems to us an interesting step in that direction.

Chapin, Seymour L. (1984), 'P. J. C. Janssen and the advent of the spectroscope into astronomical prominence', *Griffith Observer*, **48**, July, 2–15.

Congar, Yves (1997), *Journal de guerre, 1914–1918*, ed. S. Audoin-Rouzeau and Dominique Congar, Paris: Cerf.

Daston, Lorraine and H. Otto Sibum (2003), 'Scientific personae', *Science in Context*, **16** (1), 1–8.

de Beauvoir, Simone (1949), *Le Deuxième Sexe*, Paris: Gallimard.

de Certeau, Michel (1986), 'Lacan: An ethics of speech', trans. Marie-Rose Logan, in M. de Certeau, *Heterologies: Discourse on the Other*, Manchester: Manchester University Press, pp. 47–64.

Duby, Georges (1984), *Guillaume le Maréchal ou le meilleur chevalier du monde*, Paris: Fayard.

Earman, Elizabeth Deeds (2001), 'Agency in the discursive condition', *History and Theory*, **40**, 34–58.

Fara, Patricia (2001), *Newton, the Making of Genius*, New York: Columbia University Press.

Foucault, Michel (1969), 'Qu'est-ce qu'un auteur?', *Bulletin de la société française de philosophie*, **63** (3), 73–104.

Foucault, Michel (1977), 'What is an author?', trans. D.F. Bouchard and S. Simon, in D.F. Bouchard (ed.), *Language, Counter-memory, Practice*, Ithaca, NY: Cornell University Press, pp. 124–7.

Foucault, Michel (1978), *History of Sexuality*, 3 vols, trans. R. Hurley, New York: Pantheon.

Foucault, Michel (1994), *Dits et écrits, 1954–1988*, 4 vols, ed. D. Defer and F. Ewald, Paris: Gallimard.

Gingras, Yves (2001), 'Pour une biographie sociologique', *Revue d'histoire de l'Amérique française*, **45** (1), 123–31.

Grafton, Anthony (1999), *Cardano's Cosmos: The Worlds and Works of a Renaissance Astrologer*, Cambridge, MA.: Harvard University Press.

Grattan-Guinness, Ivor (1993), 'The Sylvester Medal: Origins and recipients 1901–1949', *Notes and Records of the Royal Society of London*, **47**, 105–8.

Hankins, Thomas L. (1979), 'In defense of biography: The use of biography in the history of science', *History of Science*, **17**, 1–16.

Heims, Steve J. (1980), *John von Neuman and Norbert Wiener: From Mathematics to the Technologies of Life and Death*, Cambridge, MA: MIT Press.

Janssen, Jules (1871a), 'Lettre à M. le secrétaire perpétuel, sur les résultats du voyage entrepris pour observer, en Algérie, l'éclipse de Soleil du 22 décembre dernier', *Comptes-rendus de l'Académie des sciences*, **72**, 218–22.

Janssen, Jules (1871b), 'Voyage aéronautique du *Volta* entrepris le 2 décembre 1870 en vertu d'une mission du gouvernement et de l'Académie des sciences', *Comptes-rendus de l'Académie des sciences*, **73**, 546–59.

Janssen, Jules (1903), *Lectures académiques, discours*, Paris: Hachette.

LaCapra, Dominick (1985), 'The cheese and the worms: the cosmos of a twentieth-century historian', in D. LaCapra, *History and Theory*, Ithaca, NY: Cornell University Press, pp. 45–69.

Le Goff, Jacques (1996), *Saint Louis*, Paris: Gallimard.

Lévi, Giovanni (1989), 'Les usages de la biographie', *Annales ESC*, **54**, November–December, 1,333.

Lockyer, J. Norman (1874), 'Celestial chemistry', *Nature*, **9**, 411–14

Lockyer, Sir Norman (1900), *Inorganic Evolution as Studied by Spectrum Analysis*, London: Macmillan.

MacLeod, Roy M. (1971), 'Of medals and men: A reward system in Victorian science 1826–1914', *Notes and Records of the Royal Society of London*, **26**, 81–105.

Marian, Michel (1986), 'L'histoire saisie par la biographie', *Esprit*, **117–18**, August–September, 125–31.

Mascuch, Michael (1996), *Origins of the Individualist Self: Autobiography and Self-identity in England, 1591–1791*, Stanford, CA: Stanford University Press.

Mauss, Marcel (1938), 'Une catégorie de l'esprit humain: la notion de personne, celle de "moi"', *Journal of the Royal Anthropological Institute of Great Britain and Ireland*, **68**, 263–81.

Mazerolle, Fernand (1906), 'Alphée Dubois', *Gazette numismatique française*, **10**, 1–27.

Meadows, Arthur Jack (1972), *Science and Controversy: A Biography of Sir John Norman Lockyer*, Cambridge, MA: MIT Press.

Mialet, Hélène (1999), 'Do angels have bodies? Two stories about subjectivity in science: The cases of William X and Mister H', *Social Studies of Science*, **29** (4), August, 551–81.

Miller, Arthur I. (2001), *Einstein, Picasso: Space, Time, and the Beauty that Causes Havoc*, New York: Basic Books.

Nye, David E. (1983), *The Invented Self: An Anti-biography from Documents of Thomas A. Edison*, Odense: Odense University Press.

Pancaldi, Giuliano (2003), *Volta, Science and Culture in the Age of Enlightenment*, Princeton, NJ: Princeton University Press.

Pang, Alex Soojung-Kim (2002), *Empire and the Sun: Victorian Solar Eclipse Expedition*, Stanford, CA: Stanford University Press.

Porter, Roy (ed.) (1997), *Rewriting the Self: Histories from the Renaissance to the Present*, London: Routledge.

Rivière, Pierre (1973), *Moi, Pierre Rivière, ayant égorgé ma mère, ma sur et mon frère Un cas de parricide au XIXe siècle*, ed. Michel Foucault, Paris: Gallimard.

Sarton, George (1927–30), 'Medallic illustrations of the history of science', *Isis*, **8**, 333–5; **9**, 359–62, 420–3; **10**, 485–8; **12**, 146–8; **13**, 353–5; **14**, 215–8, 417–19.

Schaffer, Simon (1980), 'Herschel in Bedlam: Natural history and stellar astronomy', *British Journal for the History of Science*, **13**, 211–39.

Schaffer, Simon (1986), 'Scientific discoveries and the end of natural philosophy', *Social Studies of Science*, **16**, 387–420.

Schaffer, Simon (1992), 'Self-evidence', *Critical Enquiry*, **18**, 327–62.

Schultze, Winfried (ed.) (1996), *Ego-Dokumente: Annäherung an der Menschen in der Geschichte*, Berlin: Akademie-Verlag.

Scott, Joan (1991), 'Evidence of experience', *Critical Inquiry*, **17**, 773–97.

Seigel, Jerrold (1999), 'Problematizing the self', in Victoria E. Bonnel and Lynn Hunt (eds), *Beyond the Cultural Turn: New Directions in the Study of Society and Culture*, Berkeley, CA: University of California Press, pp. 281–314.

Smith, Crosbie and M. Norton Wise (1989), *Energy and Empire: A Biographical Study of Lord Kelvin*, Cambridge: Cambridge University Press.

Smith, Jay M. (2001), 'Between discourse and experience: Agency and ideas in the French Revolution', *History and Theory*, **40**, 116–42.

Söderqvist, Thomas (1996), 'Existential projects and existential choice in science: Science biography as an edifying genre', in Michael Shortland and Richard Yeo (eds), *Telling Lives in Science*, Cambridge: Cambridge University Press, pp. 45–84.

Taylor, Charles (1989), *Sources of the Self: The Making of the Modern Identity*, Cambridge: Cambridge University Press.

Thorpe, Charles R. and Steven Shapin (2000), 'Who was J. Robert Oppenheimer? Charisma and complex organisation', *Social Studies of Science*, **30,** 545–90.

Zagorin, Perez (1999), 'History, the referent, and narrative: Reflections on postmodernism now', *History and Theory*, **38**, 1–24.

Chapter 4

Framing the Evidence: Scientific Biography and Portraiture

Patricia Fara

'So that to sit for one's Picture, is to have an Abstract of one's Life written, and published, and ourselves thus consign'd over to Honour, or Infamy' (Wendorf, 1983, p. 103). Writing in 1719, Jonathan Richardson, England's leading art critic, participated in enduring discussions of the relationship between biography and portraiture that date back to classical times. Artists cited Horace's dictum, *ut pictura poesis* ('as is painting, so is poetry'), to justify their insistence that poetry and painting were sister arts. Other ancient sources asserting this close relationship include Aristotle's *Poetics* as well as the direct parallel drawn by Plutarch, who made artists' focus on faces parallel his own bid to recreate lives rather than histories: 'just as painters get the likenesses in their portraits from the face and the expression of their eyes,' he wrote, 'so I must be permitted to devote myself rather to the signs of the soul in men, and by means of these to portray the life of each' (Wendorf, 1983, p. 101; Silver, 1983).

Biographers generally reproduce portraits of their subject, even though physiognomy and phrenology are no longer considered legitimate sciences. In 2000, confounding all predictions, the exhibition *Seeing Salvation: The Image of Christ* at London's National Gallery, showing images of Christ, attracted enormous crowds, demonstrating that people attach great importance to appearance, even for a divine being whose significance lies in His spirituality (Gombrich, 1963, pp. 45–55; Cowling, 1989). The secular equivalent of Jesus Christ is Isaac Newton, whose *Principia* provided a non-denominational Bible to spread the faith of Western science throughout the world (Cabral, 1996). Newton, commented one biographer, is almost purely mind, a person to whom 'sensual and aesthetic experiences were denied' (Hall, 1992, p. xiv). Nevertheless, biographers and readers remain fascinated by his looks.

Godfrey Kneller's *Isaac Newton* of 1689 (Figure 4.1) has become science's most famous portrait. Some people comment that he here resembles Christ, but for many, this Newton epitomises scientific genius. Richard Westfall, Newton's major biographer, saw 'an arresting presence, instinct with intelligence … Without difficulty, we recognize the author of the *Principia*' (Westfall, 1980, p. 482). This claim of instant recognition is suspect. We cannot know what Newton looked like: Westfall can only 'recognise' Newton because this particular picture, which differs from other representations, has become so popular. Moreover, Westfall's remark assumes a constancy in the depiction of supreme intelligence, and takes no account of shifting cultural perspectives and repeated reinterpretations. When Kneller painted this picture, the social category of 'scientific genius' did not yet exist; it originated around the end of the eighteenth century, and because Newton was the first member, his reputation helped to forge the characteristics defining how a superlative scientist should look and behave. Kneller's portrait of a melancholy recluse, an esoteric Cambridge scholar, belongs to the same genre as Robert Walker's *memento mori* reminder of John Evelyn holding a skull (Fara, 2005). But because we have inherited Romantic stereotypes of genius, we attribute a thin, pale face, dishevelled hair and fine fingers not to melancholy, but to mental brilliance. For modern viewers, the closest relatives of this detached Newtonian intellect with a minimalist body are Sherlock Holmes and Stephen Hawking.

Figure 4.1 Godfrey Kneller, *Isaac Newton* (1689) (Thomas Oldham Barlow's mezzotint of 1868). © Wellcome Institute Library, London.

Images are physical objects that merit their own narratives and yield their own information: tracking a portrait's history can cast new light on its subject. Pictures are, however, often shown wrested out of their original context. A large coloured oil painting hanging in a proud owner's home delivers a very different impact from the same picture in a museum or printed as a mediocre black-and-white copy in a book. Furthermore, reproduction does not necessarily mean replication, since engravers may alter the sitter's appearance, and add or eliminate decorative, (Gaskell, 2000, pp. 99–115).

Writers value textual over visual evidence, and many either scarcely mention images or else refer to them uncritically. By reading portraits literally at face value, biographers miss the opportunity to learn more about their subjects and their cultural context. Despite being unable to offer the unmediated insights into character that some authors would like to derive, portraits can – like texts and other sources – yield unique information about a sitter. They are instructive not just because of what they *do* show, but also because they are imbued with assumptions of what *should* be shown or concealed.[1]

By considering portraits as material objects, this chapter explores how their production, reproduction and reception affect shifting interpretations and provide valuable biographical information. The first section focuses on sitters' strategies of pictorial self-fashioning, while the second concentrates on scientific role models. The final part reviews the complex relationships between time, biography and portraiture.

Portraits and Propaganda

The slogan 'a picture is worth a thousand words' was coined for a 1920s advertising campaign, although scientific practitioners appreciated its force centuries earlier. Portraits were originally designed to honour the dead, but the Renaissance focus on an individual self encouraged living subjects to be portrayed (Porter, 1997, and especially Smith, 1997, and Burke, 1997). Because sitters promoted themselves visually, their portraits give insights not only into their appearance, but also into how they thought they should look. Although oil originals were only seen by limited audiences, engravings enabled propagandists to spread their image world-wide.

Self-promotional campaigns are sometimes obvious: Franklin observed with smug satisfaction in 1779 that his 'popularity has occasioned so many Paintings, Busto's, Medals & Prints to be made of me … that my Face is now almost as well known as that of the Moon' (quoted in Fortune and Warner, 1999, p. 119) (Figure 4.2). When a friend asked for his picture, Franklin took 'the Liberty to trouble you with the Care of six of those Prints to be distributed agreable to the enclos'd List' (quoted in Fortune and Warner, 1999, p. 120). Over the next few years, Franklin and his son ordered well over a hundred copies for friends and potential political contacts. His own library mantelpiece was crammed with busts and medals, and like many

1 The extensive relevant literature includes: Brilliant (1991), Burke (2001), Fortune and Warner (1999), Jordanova (2000), Leppert (1996), esp. Ch. 7, Pointon (1993) and Simon (1987).

Enlightenment natural philosophers, he owned a large collection of portraits that had been sent to him. As prints became cheaper and more popular in the nineteenth century, this practice continued, and scientists such as Michael Faraday built up large collections.[2]

Figure 4.2 Mason Chamberlin, *Benjamin Franklin* (1762) (Charles Turner's mezzotint, post-1840). © National Portrait Gallery, London.

2 Fortune and Warner (1999) is my major source of information about Franklin's portraits; see also Craven (1993).

In contrast, Newton's biographers often maintain that Newton shunned fame and was uninterested in art. Although this seems appropriate behaviour for a reticent genius, the visual evidence contradicts this received view and demonstrates that Newton was, like Franklin, actively engaged in fashioning his public image. The sheer number – over twenty busts and portraits, several of them paid for by Newton himself – testifies to his concern. He displayed some of them in his own home, and donated a large portrait to the Royal Society for more public appreciation. To reach still wider audiences, Newton sent different versions of himself as inducements or rewards for a particular favour.[3]

Paradoxically, portraits do not necessarily provide incontrovertible evidence of a subject's appearance. Westfall claimed to recognise Newton, yet one hagiographer commented in 1782: 'Various are the effigies of Sir Isaac, both in frontispieces, medallions, busts, seals, and other engravings, but most of them are dissimilar from his monument and from each other' (Maude, 1782, p. ii of the Appendix). At least four representations of Newton claimed to be an authentic likeness because they were made from death masks, yet even they look very different from one another.

One problem is that artists deliberately flattered their clients. Many biographers have commented on the intelligence radiating out of Newton's pictures like an aura of genius, yet the Bishop of Rochester, who knew Newton for twenty years, remarked to a friend that 'in the whole air of his face and make, there was nothing of that penetrating sagacity which appears in his composures' (Atterbury, 1783–85, vol. 1, p. 180). Piercing eyes and a large, broad brow have come to characterise Newton because they were the features that physiognomic handbooks instructed artists to exaggerate in order to signal mental aptitude: keenness of vision was associated with sharpness of intellect (Jay, 1993, pp. 21–147). Signs of aging were often concealed. In a private letter, an American visitor was brutally honest about one of Newton's last portraits, the choice for the frontispiece of the *Principia* that Newton presumably approved: 'by all those who have seen him of late, as I did, bending so much under the Load of Years as that with some difficulty he mounted the Stairs of the Society's Room. That Youthfull Representation will I fear be considered rather as an object of Ridicule than Respect' (Wolf, 1974, p. 349).

Nevertheless, valuable information can be gleaned from portraits. Most obviously, they provide major evidence for changing fashions in clothes and hair. Without pictures, it would be hard to reconstruct either the formal dress people adopted for social functions or the more leisurely attire that natural philosophers wore at home. Portraits also show equipment and places of work. For instance, images of instrument-makers at work or displaying their most celebrated inventions are (unlike representations of alchemists) probably reasonably faithful.

Occasionally, sitters' comments on their portraits survive. Charles Darwin disliked his first picture, a 64th birthday present, because it made him look 'a very venerable, acute, melancholy old dog' (Fara, 2000, p. 143). Even without such direct remarks, examining which pictures were reproduced for public viewing is biographically valuable because they provide evidence of a sitter's preference. What

---

3 My discussions of Newton's portraits are based on Fara (2002a), which contains full references.

is now the familiar first portrait of Newton (Figure 4.1) was virtually unknown until the middle of the nineteenth century, since Newton hung it in his own home and never had it engraved: he chose to be remembered by other images.

Men exchanged portraits to solicit patronage, cement friendships and express power relationships. Analysing the recipients of such pictures yields information about scientific networks and hierarchies. Newton agreed to sit for a portrait paid for by a colleague who held twin aims – to express gratitude, and to influence Newton's vote in a forthcoming Cambridge election. Newton later himself commissioned an expensive oil painting to bribe a French mathematician during negotiations over a de luxe Parisian edition of the *Opticks*. However, he only judged the editor of the second edition of the *Principia* worthy of an engraving, and he refused to send anything to allies of his arch-enemy Gottfried Leibniz.

Examining sitters' choice of pictures reveals which versions of themselves they liked to disseminate in their self-promotional campaigns. For the French mathematician, Newton paid Kneller a high price to be presented as an elegant, youthful *philosophe* with folded gloves, a sword and an elaborate wig; for the editor, he preferred to be shown wearing a sumptuous banyan (the philosophical version of leisure wear), posing as a distinguished Roman against a classical background.

Other men of science shared Newton's concern to control their image. This monitoring is particularly obvious in the Victorian practice of issuing photographic *cartes de visite*. Unsurprisingly, few copies survive of a studio photograph that Darwin thought made him look 'atrociously wicked', but he did buy copies of a more successful attempt – fashionably dressed in 'Great Exhibition' check trousers – to send out to family and friends (Browne, 1998). Similarly, Joseph Banks gave detailed instructions for the composition of his impressive portrait as President of the Royal Society (Figure 4.3). By prominently holding up the lecture of his younger rival, Humphry Davy, Banks reinforced his authoritarian rule. Ever alert to the competition, he also tried to get the best engraver, telling the artist: 'I beg however to hint that much as I admire the Burin of Schiavonetti I should was I to direct Employ Sharp he has engraved Boulton & John Hunter admirably' (Carter, 1987, p. 306). Absence can also provide evidence of control: no prints were made of two other versions of this portrait that revealed his age and ill health. Unsolicited pictures were harder to monitor, but the extreme rarity of one Banks caricature implies that he suppressed its publication (Fara, 1997).

Figure 4.3 Thomas Phillips, *Joseph Banks* (1808). © The Royal Society.

Banks' enthusiasm for flattering engravings suggests that he viewed them as counter-propaganda against caricatures. He was not alone: Edward Jenner's portraits illustrate how he sought to establish a respectable identity for himself despite the savage caricatures about smallpox vaccination. Caricatures and cartoons carry rich information about contemporary attitudes towards a biographical subject, and can furnish unique evidence of such controversies. Although often denigrated as an inferior art form, caricatures share many of the characteristics of conventional portraits, which emphasise favourable features. They can reveal unsuspected aspects of a subject because they exaggerate unflattering features that may be omitted from other sources: an early caricature of Banks shows him with gout long before his crippling illness was formally acknowledged (Fara, 1998; Jordanova, 2000, pp. 87–95; Donald, 1986; Browne, 2001).

Knowing a subject's personal choice of picture can correct retrospective interpretations. Particularly in America, Franklin is now celebrated primarily as a Revolutionary politician, but he advertised himself as an electrical experimenter (Figure 4.2). Following English conventions for portraying Enlightenment men of letters, Franklin is shown working in his study, his quill and paper prominently displayed to advertise his intellectual solidity. Electric bells and mutually repellent cork balls hang behind his right shoulder; on his left, the traditional draped curtain is drawn back to show an artificial cityscape demonstrating the advantages of Franklin's most famous invention, the lightning rod. Yet many biographers prefer Benjamin West's scene of Franklin 'drawing Lightning from the Sky', an imaginary and unfinished sketch drawn many years later for a large picture that, although never painted, would have depicted a mythical moment of discovery, since Franklin was pipped to the post by some French experimenters (Fara, 2002b). (Compare how Cicero wrote that Socrates called down philosophy from the skies.)

Figure 4.4 Engraving after Benjamin West, *Joseph Banks* (1773). © The Royal Society.

Portraits can themselves reveal propaganda tactics and yield information not directly obtainable from written sources alone. West's portrait of Banks shows him wrapped in a New Zealand cloak, a trophy from his recent voyage to the South Seas (Figure 4.4). Similarly, military men who adorned themselves in North American Indian costumes were boasting that they had survived their foreign experiences without being overpowered by them (Tobin, 1999, pp. 81–109). Here Banks presents himself as a young imperial adventurer with a keen eye for trading opportunities. He has incorporated two advertisements for his discovery of a new source for flax, sorely needed by the British Navy for making sails: his New Zealand flaxen cloak (to which he is pointing for emphasis), and the expedition's book of botanical drawings (near his left foot) open to show the plant itself.

Some portraits are deliberately deceptive. Checking the date of a James Hutton portrait betrays that the large pile of completed manuscripts on his table was wishful thinking: by then, Hutton had published only one pamphlet (Jones, 1996). When Carl Linnaeus wanted to convince his financial backers that he was an Arctic explorer with intimate experience of Sami (Lapplander) life, he distributed pictures of himself wearing a chaotic assembly of souvenirs, an ignorant parody of true Sami dress (Koerner, 1999, pp. 64–8).

The use of portraits as scientific propaganda does not, of course, end with the subject's death. As just a few examples, Linnaeus' picture was later offered as a lottery prize by a publishing entrepreneur, William Blake's ambiguous image of Newton has been re-created outside the British Library as a monument to national technological achievement, and the Nazis' destruction of Antoine Lavoisier's statue in Paris symbolised the chemical conflict between French oxygen and the German phlogiston of Georg Stahl (Fara, 2003a; Beretta, 2001, pp. 63–6).

Scientific Role Models

Portraits are not simply the product of an artist: there is an interactive triangular relationship between the sitter, the artist and the viewer (and sometimes a picture's commissioner as well). Therefore portraits both reflect and create exemplary ideals, simultaneously depicting, establishing, consolidating and advertising ideological constructs such as national character, appropriate gender behaviour and class structures. This multiplicity of functions makes the biographical use of portraits more complicated, but ultimately more rewarding. Portraits show an individual, yet simultaneously reveal the face of science itself.

Painters do not necessarily intend to produce a faithful likeness of their sitter. Some artistic devices are immediately apparent: Maggi Hambling gave Dorothy Hodgkin four blurry arms to emphasise her constant activity and to draw attention to her severely arthritic hands (Figure 4.5). In contrast, portraits produced in the eighteenth century may appear realistic, but deliberately set out to improve – as distinct from flatter – their subject by showing an ideal type. Continuing a humanistic tradition stemming from Aristotle and Alberti, writers stressed that 'the fine arts have a double purpose; they are destined both to *please* and to *instruct*' (Richardson, [1779] 1979, vol. 1, last page of unpaginated Preface). Portraits were designed not only to summon up the appearance of a sitter, but also to recapture the qualities that had made him famous (there were substantial differences in how men and women were portrayed).

Figure 4.5 Maggi Hambling, *Dorothy Hodgkin* (1985). © National Portrait Gallery, London.

In his lectures, Joshua Reynolds – Europe's leading advocate of portraiture – articulated philosophical ideas that applied to contemporary scientific and social models as well as to painting (Wind, 1986; Barrell, 1986). He taught that:

> there is one general form, which … belongs to the human kind at large, so in each of these classes there is one common idea and central form, which is the abstract of the various individual forms belonging to that class … [the painter], like the philosopher, will consider nature in the abstract, and represent in every one of his figures the character of its species. (Reynolds, 1959, pp. 47 and 50)[4]

Botanists were painting perfected specimens, and anatomists were drawing skeletons that conformed with (and hence corroborated) ideal shapes for men and women. Similarly, Reynolds was creating idealised versions of men that served as role models to be emulated.

Figure 4.6 Joshua Reynolds, *Joseph Banks* (1772). © National Portrait Gallery, London.

4 See Wind (1986) and Barrell (1986).

Reynolds' portrait of Banks (Figure 4.6) was first exhibited at the same time as West's (Figure 4.4). It instinctively seems more natural than West's obviously contrived studio portrait, yet Reynolds's picture also fulfils several rhetorical functions. For one thing, it validated Banks' personal status at a time when he was being cruelly satirised for his exploits in the South Seas. To reassure conservative spectators, Reynolds has rooted Banks in a traditional past, placing him in a scholar's study and using conventional symbols – such as the inkpot and quill – to signal his intellectual interests. Drawing on older Dutch styles, Reynolds has seated Banks in a gentlemanly pose and meticulously painted his fine clothes.

At the same time, this portrait legitimised scientific field research. Adventurous entrepreneurs showed themselves engaged in outdoor activities, while architects and medical practitioners were commissioning busts of themselves to insist that even gentlemen could work with their hands (Baker, 1988). Banks was helping to construct a new role model, the scientific explorer. Learned gentlemen had – like Newton (Figure 4.1) – traditionally been shown inside dark, enclosed spaces, yet Banks' study has a prominent window, and a terrestrial globe has replaced a celestial one, the conventional attribute denoting a natural philosopher. Banks is wigless, with a stubbly chin, and wearing an exotic Italian travelling coat, yet he is – Reynolds declares – a scholarly gentleman.

This exemplary function of Banks' portrait emerges still more clearly when it is compared with contemporary portraits of people following other occupations – musicians, artists, poets, architects, surgeons. Middle-class people were demanding portraits that advertised their success in the new careers that they were establishing. By emphasising the identifying characteristics of these various professionals, portraits reflected important cultural changes but also helped to consolidate them. Reynolds's portrait of John Hunter was often reproduced during the nineteenth century because it elevated the standing of surgery by idealising Hunter as a contemplative professional. Anatomical apparatus and specimens belonged to lower status surgeons, but Reynolds juxtaposed them with gentlemanly attributes – an ink pot, a quill, the traditional posture of meditation (the hand on the chin) (Jordanova, 1997).

Looking back, it seems that Banks personified two new scientific role models: the explorer (Figures 4.4 and 4.6) and – several decades later – the administrator (Figure 4.3). These three portraits were engraved and marketed in competition against caricatures. The early caricatures were being produced at the same time as the new social category of scientific explorer was being constructed, yet they placed Banks in a different group – the Macaronis (in Figure 4.7, a caricature of Banks as a 'Fly-catching Macaroni' straddling two halves of the globe is displayed in the top left-hand corner). The Macaroni satires were directed against foppish, etiolated young men who acquired Italian manners during their Grand Tour, and Banks tried hard to dissociate himself from them. Reynolds's portrait (Figure 4.6) rescues Banks by depicting a well-fed young man, with a belly whose curve mirrors the globe's. This image of a healthy explorer meshes with another role model that was being created: the sporty hearty Englishman (Deuchar, 1988; Barrell, 1980).

Figure 4.7 Matthew Darly, *The Macaroni Print Shop* (1772). © The Trustees of The British Museum.

By displaying Banks in an unexpected context, the Macaroni shop caricature provides immediate visual evidence beyond that available textually. It also brings biographical questions into sharp focus. How do we escape from our retrospective knowledge of a subject's future significance in a society with different values? How do we balance our awareness that an individual is unique against our wish to reconstruct the characteristics of a period by telling one person's life? This problem is particularly acute for Banks, who has become an emblem of Enlightenment, yet occupied a singular position within it.

Banks himself contributed greatly to strengthening science's prestige in the early nineteenth century. Even after scientific identities became well recognised, portraits were still designed to consolidate scientists' reputations and to establish stereotypes. There was, however, a new emphasis on authenticity. This arose even before the advent of photography, which first became popular because of the middle-class demand for portraits following traditional artistic conventions. This search for authenticity paralleled biologists' insistence on illustrating actual specimens rather than ideal ones. When Newton's portrait (Figure 4.1) was rediscovered in 1857, its appeal lay in its depiction of a genius at the height of his powers, almost at the creative moment itself. Victorian versions of Newton also claimed to be authentic. The sculptor of Newton's statue at Grantham worked with a death mask to produce a supposedly realistic face; reporters repeatedly stressed that it was a true likeness, drawing on phrenology for their glowing descriptions of a wide brow that unmistakably indicated Newton's towering intellect. Imagined portraits of Newton started to appear, which purported to be realistic representations of him in action at a specific instant – experimenting with a prism or watching the falling apple. Like other posthumous images, they reveal more about the period than about their subject.

Rather than concealing signs of age or illness, Victorian artists and photographers seemed to be aiming for brutal honesty. The most famous portrait of Darwin shows him in his seventies (Figure 4.8), looking weary and disillusioned, his thick cloak and warm hat betraying the invalid's need for protection. But despite his humble posture, Darwin is presenting himself as British science's elderly statesman and most distinguished sage. The sombre tones contrast with and emphasise his head's luminescent quality, which imparts the saintly aura of inspiration. The intense eyes and domed forehead testify to Darwin's intellectual powers, while the meticulously painted white beard gives him the authority of a Moses figure empowered to communicate God's laws of nature. Similarly, Julia Cameron's close-up of John Herschel is deceptively frank. She converted Herschel into a purely intellectual yet masculine being by photographing only his head surrounded by a glowing corona of hair. Although the portrait of an individual, this Herschel so closely resembles photographs of Albert Einstein with a lined face and shock of snow-white hair that it paradoxically becomes a universal evocation of scientific genius, an ambiguous counterpart of Kneller's *Newton.*

Figure 4.8 John Collier, *Charles Darwin* (1881). © National Portrait Gallery, London.

The Poetics of Portraiture

The picture of the Macaroni print shop (Figure 4.7) is itself a commentary on the nature of representation. Only a pane of glass separates the Macaronis on display in the window from the Macaronis and other spectators in the street. This is a fantasy gallery in which two sets of people are on display. The artist is deliberately forcing the external viewer (us) to confront the artificiality of representation, implicitly asking about the relationship between us, the diverse cast of characters in the street, and the people being satirised in the window. This image, strongly influenced by physiognomic principles, recalls political debates in which public figures were almost arbitrarily allocated to two opposing sides. To emphasise the distance between the viewer and the street Macaronis, they are all shown in profile, and pose visually the question of what it means to capture an individual with lines on a flat surface (Pointon, 1993, pp. 94–104; Donald, 1996, pp. 75–108).

Like caricaturists and engravers, biographers encapsulate people with black-and-white marks on paper. Particularly during the eighteenth century, writers and artists drew parallels between biography and portraiture. James Boswell set out to describe Johnson verbally in the same way that Reynolds painted their mutual friend Banks: 'I draw him in the style of a Flemish painter. I am not satisfied with hitting the large features. I must be exact as to every hair, or even every spot on his countenance' (Wendorf, 1983, p. 117). Especially in England, portrait painters sought to ally themselves with biographers to justify the value of a genre regarded as inferior to history painting. By 1817, portraiture had – remarked one sceptic – become 'one of the staple manufactures of the empire. Wherever the British settle, wherever they colonize, they carry and will ever carry trial by jury, horse-racing, and portrait-painting' (Shawe-Taylor, 1990, p. 7).

Biographers and artists tackle the common problem of deciding where to draw the lines between tactful disguise and blatant deceit, between faithfulness to minute details and concern for general character, between accuracy and imagination. John Dryden wrote to Kneller, summarising their shared dilemma:

> Thou paint'st as we describe, improving still,
> When on wild Nature we ingraft our skill:
> But not creating Beauties at our Will. (quoted in Wendorf, 1983, p. 100)

Like portrait painters, biographers lie at one apex of a triangle. Just as Newton was not a passive sitter to Kneller, so too Johnson interacted with Boswell; moreover, all were aware of their contemporary audiences as well as future interpreters who would bring new meanings. Both genres are also governed by conventions. Kneller's *Newton* (Figure 4.1) is a traditional representation of a melancholic scholar, and during the eighteenth century, most artists followed the rules for gentlemanly portraiture –a drawn-back curtain, an inkpot and quill, an enclosed study (Figures 4.2 and 4.6). Similarly, many biographers feel constrained to adopt a chronological framework, or to squeeze the lives of women into male plots structured by career patterns rather than by relationships or internal emotions (Backscheider,1999; Heilbrun, 1988; Wagner-Martin, 1994).

Despite these similarities, important differences exist. Many of them hinge round the long-standing spatial-temporal distinction between visual and literary arts expressed particularly influentially by Gotthold Lessing, an intellectual discrimination reinforced materially by the separate printing sites for images and text

(Mitchell, 1985, pp. 95–115; Park, 1969). According to this view, painting combines its symbols only in space, while poetry operates over time. Thus portraits fix sitters at a particular instant of time – Banks in the act of rising from his chair (Figure 4.6) or Franklin temporarily distracted from letter-writing (Figure 4.2). Because portraits are often painted to commemorate a particular event, they show the sitter at a definite age or in a certain way – such as Kneller's Frenchified Newton, or Banks surrounded by his trophies (Figure 4.3). Biographies, on the other hand, cover a long period and are often written after a person has died. There is room to discuss transient events, to compare different stages in life, and to present a character who moves and converses.

Biographers, artists and academics have argued against this harsh spatial-temporal distinction (Bal, 1991; Mitchell, 1980 and 1994; Souriau, 1949; Wagner, 2001). Viewers must pass physically through the space of a museum or a cathedral nave, where they may confront representations of people from different eras of the past, perhaps arranged chronologically as an implicit model of progress. In her picture of Hodgkin (Figure 4.5), Hambling struggles against portraiture's static constraints by using double arms to signify movement, an artistic device that relies on familiarity with cinema. Reynolds encapsulated a life's activities in a single image by insisting that portraits should represent an idealised version of the sitter, a sort of abstract condensation of his qualities. He surrounded Hunter with long-dead specimens testifying to his current prestige, and in his portrait of Banks (Figure 4.6) he incorporated the past of the voyage from which the traveller had just returned into the continuous present of a scene that typifies his current daily life. He also depicted a promised future (although it was never actually achieved), since the Latin text beneath Banks' determinedly clenched left fist is a famous quotation from Horace: 'Tomorrow we will again cross the immense ocean.' Conversely, biographies may be ordered spatially: Newton's life is often divided into three locations – Lincolnshire, Cambridge and London.

Nineteenth-century scientific biographers separated intellectual histories from personal lives, and created heroic narratives to establish disciplinary forefathers (Yeo, 1996; Schaffer, 1986). Parallel processes can be seen in scientific portraits. The quasi-journalistic pictures of experimenters in action, captured at a specific instant – such as Newton creating a spectrum with his prism – focus on scientific achievements. They seem different in kind from the more reflective, intimate studies of Darwin (Figure 4.8) or Herschel, who are portrayed without any books or instruments in a dark, timeless space.

In contrast, modern biographers and artists often seek to integrate the private and public, the emotional and rational facets of a subject. Hambling chose to paint Hodgkin at work, yet shows her in a domestic setting (Figure 4.5). The moving arms and uneaten sandwich convey Hodgkin's intense concentration, while the arthritic hands, magnifying glass and untidy white hair are displayed without concealment. When Henry Moore was asked to portray Hodgkin, he drew only her hands. Like Hambling, he decided to ignore the prevalent sexist convention of complimenting Hodgkin on her attractive face, and to focus instead on the painful hands which had not prevented her from winning the Nobel Prize. Hambling's portrait of an individual is also a celebration of scientific progress, since – as in a *memento mori* – the sandwich will inevitably decay in the future, while the molecular model represents permanent knowledge (Jordanova, 2000, pp. 153–8; Fara, 2003b).

Reynolds pithily reminded his students that, unlike a biographer, 'A Painter ... has but one sentence to utter, but one moment to exhibit. He cannot, like the poet or historian, expatiate' (Reynolds, 1959, p. 60). Portraits and biographies are undeniably different, yet they provide complementary ways of approaching an individual's character. Nineteenth-century readers appreciated this twinned role. Among the first popular illustrated biographical dictionaries was Edmund Lodge's *Portraits of Illustrious Personages of Great Britain*. Concerned to counteract reliance on physiognomic interpretation, Lodge told his readers that it is 'from the combination of portraits and biography that we reap the utmost degree of utility and pleasure which can be derived from them' (Pointon, 1993, p. 94). This remains a useful prescription for biographers.

Acknowledgements

I am grateful for the comments on an earlier version of this paper contributed by the participants in the Poetics of Scientific Biography conference held in Copenhagen in May 2002.

Bibliography

Atterbury, Francis (1783–85), *The Epistolary Correspondence, Visitation Charges, Speeches, and Miscellanies*, vol. 1, London: J. Nichols.

Backscheider, Paula R. (1999), *Reflections on Biography*, Oxford: Oxford University Press.

Baker, Malcolm (1988), 'Portrait busts of architects in eighteenth century Britain', in Charles Hind (ed.), *New Light on English Palladianism*, London: The Georgian Group, pp. 14-30.

Bal, Mieke (1991), *Reading 'Rembrandt': Beyond the Word–image Opposition*, Cambridge: Cambridge University Press.

Barrell, John (1986). *The Political Theory of Painting from Reynolds to Hazlitt*, New Haven, CT and London: Yale University Press.

Barrell, John (1990), 'Sir Joshua Reynolds and the Englishness of English art', in Homi K. Bhabha (ed.), *Nation and Narration*, London and New York: Routledge, pp. 154–76.

Beretta, Marco (2001), *Imaging a Career in Science: The Iconography of Antoine Laurent Lavoisier*, Canton, MA: Science History Publications, pp. 63–6.

Brilliant, Richard (1991), *Portraiture*, London: Reaktion.

Browne, Janet (1998), 'I could have retched all night: Charles Darwin and his body', in Christopher Lawrence and Steven Shapin (eds), *Science Incarnate: Historical Embodiments of Natural Knowledge*, Chicago, IL: University of Chicago Press, pp. 240–87.

Browne, Janet (2001), 'Darwin in caricature', *Proceedings of the American Philosophical Society*, **145**, 496–509.

Burke, Peter (1997), 'Representations of the self from Petrarch to Descartes', in Roy Porter (ed.), *Rewriting the Self: Histories from the Renaissance to the Present*, London and New York: Routledge, pp. 17–28.

Burke, Peter (2001), *Eyewitnessing: The Uses of Images as Historical Evidence*, London: Reaktion.

Cabral, Regis (1996), 'Herbert Butterfield (1900–79) as a Christian historian of science', *Studies in the History and Philosophy of Science*, **27**, 547–64.

Carter, Harold (1987), *Sir Joseph Banks (1743–1820): A Guide to Biographical and Bibliographical Sources*, Winchester: St Pauls Bibliographies.

Cowling, Mary (1989), *The Artist as Anthropologist: The Representation of Type and Character in Victorian Art*, Cambridge: Cambridge University Press.

Craven, Wayne (1993), 'The American and British portraits of Benjamin Franklin', in J.A. Leo Lemay (ed.), *Reappraising Benjamin Franklin: A Bicentennial Perspective*, Newark, DE: University of Delaware Press, pp. 247–71.

Deuchar, Stephen (1988), *Sporting Art in Eighteenth-century England: A Social and Political History*, New Haven, CT and London: Yale University Press.

Donald, Diana (1986), '"Characters and caricatures": The satirical view', in Nicholas Penny (ed.), *Reynolds*, London: Weidenfeld and Nicolson, pp. 355–63.

Donald, Diana (1996), *The Age of Caricature: Satirical Prints in the Reign of George III*, New Haven, CT and London: Yale University Press, pp. 75–108.

Fara, Patricia (1997), 'The Royal Society's portrait of Joseph Banks', *Notes and Records of the Royal Society*, **51**, 199–210.

Fara, Patricia (1998), 'Images of a man of science', *History Today*, October, 42–9.

Fara, Patricia (2000), 'Images of Charles Darwin', *Endeavour*, **24**, 143.

Fara, Patricia (2002a), *Newton: The Making of Genius*, London: Macmillan.

Fara, Patricia (2002b), 'Pictures of Benjamin Franklin', *Endeavour*, **26**, 1–2.

Fara, Patricia (2003a), 'Carl Linnaeus: pictures and propaganda', *Endeavour*, **27**, 14–15.

Fara, Patricia (2003b), 'Pictures of Dorothy Hodgkin', *Endeavour*, **27**, 85–6.

Fara, Patricia (2005), 'The melancholy of anatomy,' *Endeavour*, **29** (2005), 20–21.

Fortune, Brandon Brame and Deborah J. Warner (1999), *Franklin and His Friends: Portraying the Man of Science in Eighteenth-century America*, Philadelphia, PA: University of Pennsylvania Press.

Gaskell, Ivan (2000), *Vermeer's Wager: Speculations on Art History, Theory and Art Museums*, London: Reaktion.

Gombrich, E.H. (1963), *Meditations on a Hobby Horse and other Essays on the Theory of Art*, London and New York: Phaidon.

Hall, Rupert (1992), *Isaac Newton: Adventurer in Thought*, Oxford and Cambridge, MA: Blackwell.

Heilbrun, Carolyn G. (1988), *Writing a Woman's Life*, London: The Women's Press.

Jay, Martin (1993), *Downcast Eyes: The Denigration of Vision in Twentieth-century French Thought*, Berkeley, CA and London: University of California Press.

Jones, Jean (1996), 'William Hutton', in David Daiches, Peter Jones and Jean Jones (eds), *The Scottish Enlightenment 1730–1790: A Hotbed of Genius*, Edinburgh: Saltire Society, pp. 115–36.

Jordanova, Ludmilla (1997), 'Medical men 1780–1820', in Joanna Woodall (ed.), *Portraiture: Facing the Subject*, Manchester: Manchester University Press, pp. 101–15.

Jordanova, Ludmilla (2000), *Defining Features: Scientific and Medical Portraits 1660–2000*, London: Reaktion.

Koerner, Lisbet (1999), *Linnaeus: Nature and Nation*, Cambridge, MA and London: Harvard University Press.

Leppert, Richard (1996), *Art and the Committed Eye: The Cultural Functions of Imagery*, Boulder, CO: Westview Press.

Maude, Thomas (1782), *Viator, a Poem: or, A Journey from London to Scarborough, By the Way of York*, London: B. White.

Miles, Ellen G. (1993), 'The French portraits of Benjamin Franklin', in J.A. Leo Lemay (ed.), *Reappraising Benjamin Franklin: A Bicentennial Perspective*, Newark, DE: University of Delaware Press, pp. 272–89.

Mitchell, W.J.T. (1980), 'Spatial form in literature: Toward a general theory', *Critical Inquiry*, **6**, 539–67.

Mitchell, W.J.T. (1985), *Iconology: Image, Text, Ideology*, Chicago, IL and London: University of Chicago Press.

Mitchell, W.J.T. (1994), *Picture Theory: Essays on Verbal and Visual Representation*, Chicago, IL and London: University of Chicago Press.

Park, Roy (1969), '"*Ut pictura poesis*": The nineteenth-century aftermath', *Journal of Aesthetics and Literary Criticism*, **28,** 155–64.

Pointon, Marcia (1993), *Hanging the Head: Portraiture and Social Formation in Eighteenth-century England*, New Haven, CT and London: Yale University Press.

Porter, Roy (1997), 'Introduction', in Roy Porter (ed.), *Rewriting the Self: Histories from the Renaissance to the Present*, London and New York: Routledge, pp. 1–14.

Prescott, Gertrude M. (1985), 'Faraday: Image of the man and the collector', in David Gooding and Frank A.J.L. James (eds), *Faraday Rediscovered: Essays on the Life and Work of Michael Faraday, 1791–1867*, Basingstoke: Macmillan, pp. 15–32.

Reay, Barry (2002), *Watching Hannah: Sexuality, Horror and Bodily De-formation in Victorian England*, London: Reaktion.

Reynolds, Joshua (1959), *Discourses on Art*, ed. Robert R. Wark, New Haven, CT and London: Yale University Press.

Richardson, George ([1779] 1979), *Iconology*, 2 vols, New York and London: Garland.

Schaffer, Simon (1986), 'Scientific discoveries and the end of natural philosophy', *Social Studies of Science*, **16**, 387–420.

Sellers, Charles Coleman (1962), *Benjamin Franklin in Portraiture*, New Haven, CT and London: Yale University Press.

Shawe-Taylor, Desmond (1990), *The Georgians: Eighteenth-century Portraiture and Society*, London: Barrie and Jenkins.

Silver, Larry (1983) 'Step-sister of the Muses: Painting as liberal art and sister art', in Richard Wendorf (ed.), *Articulate Images: The Sister Arts from Hogarth to Tennyson*, Minneapolis, MN: University of Minnesota Press, pp. 36–69.

Simon, Robin (1987), *The Portrait in Britain and America with a Biographical Dictionary of Portrait Painters 1680–1914*, Oxford: Phaidon.

Smith, Roger (1997), 'Self-reflection and the self', in Roy Porter (ed.), *Rewriting the Self: Histories from the Renaissance to the Present*, London and New York: Routledge, pp. 49–57.

Souriau, Etienne (1949), 'Time in the plastic arts', *Journal of Aesthetics and Art Criticism*, **7**, 294–307.

Tobin, Beth Fowkes (1999), *Picturing Imperial Power: Colonial Subjects in Eighteenth-century British Painting*, Durham, NC and London: Duke University Press.

Wagner, Peter (2001), 'Representations of time in Hogarth's paintings and engravings', in David Bindman, Frédéric Ogée and Peter Wagner (eds), *Hogarth: Representing Nature's Machines*, Manchester: Manchester University Press, pp. 102–22.

Wagner-Martin, Linda (1994), *Telling Women's Lives: The New Biography*, New Brunswick, NJ: Rutgers University Press.

Wendorf, Richard (1983), '"*Ut Pictura Biographia*": Biography and Portrait Painting as Sister Arts', in Richard Wendorf (ed.), *Articulate Images: The Sister Arts from Hogarth to Tennyson*, Minneapolis, MN: University of Minnesota Press, pp. 98–124.

Westfall, Richard (1980), *Never at Rest: A Biography of Isaac Newton*, Cambridge: Cambridge University Press.

Wind, Edgar (1986), *Hume and the Heroic Portrait: Studies in Eighteenth-century Imagery*, Oxford: Clarendon Press.

Wolf, Edwin (1974), *The Library of James Logan of Philadelphia 1674–1751*, Philadelphia, PA: Library Company of Philadelphia.

Yeo, Richard (1996), 'Alphabetical lives: Scientific biography in historical dictionaries and encyclopaedias', in Michael Shortland and Richard Yeo (eds), *Telling Lives in Science: Essays on Scientific Biography*, Cambridge: Cambridge University Press, pp. 139–69.

Chapter 5

Biography and the Reward System in Science

Thomas L. Hankins

We scientific biographers confront an irony. We try to create a contextualised biography that shows how the subject's science is related to other personal, cultural and social forces in his or her life – things like religion, education, politics, social status, health, race and ethnicity, and sex (marriage, sexual preference, gender issues). And yet these are precisely the things that scientists themselves are forbidden to consider when rewarding their colleagues with tenure and promotion, prestigious appointments, society offices, prizes and all the other forms of reward. It is ironical that we want to write biographies about just those things that scientists in real life are supposed to avoid in judging their peers. Fortunately, we write biographies about dead people, which reduces the problem of political correctness, but there still seems to be a disconnect between what we biographers consider important and what the scientists themselves consider important.

The problem may be more apparent than real, resting on a dubious distinction between a 'career' and a 'life'. A career is what goes into the *curriculum vitae* and is what brings rewards to the scientist. A life is all the rest. But of course we cannot separate the career from the life. In the first place, it is the career, not the life, that decides whether a scientist gets chosen for a biography. In the second place, as Larry Holmes has pointed out, you cannot have a context without a text, or, in other words, you cannot have a background without something for the background to be background to (Holmes, 1981, p. 60). Therefore, we cannot really divide a scientist into a life and a career and treat only the life while ignoring the science that establishes a career.

This paradox raises the question of the relationship between biography and the reward system of science – how the life and the career fit together. The most obvious question to ask about biography and reward is whether biography itself is part of the reward system of science. We can easily answer that question in the affirmative. A biography honours a scientist. Even if it may not be very flattering, at least the author has found the subject important or interesting enough to warrant writing the biography. Also the eulogy, which is a particular kind of biography, is a time-honoured way of rewarding a distinguished career. Therefore we can conclude that having a biography written about one is a reward just as is receiving a prize or having one's name attached to a 'law', 'effect' or 'unit'.

A more important question is how the reward system of science, by its very nature, affects the writing of scientific biography. In some sense, the reward system in science has already selected the materials available for a biography. It has decided what materials will be retained in the archives, who gets a eulogy, and who is included in biographical dictionaries. Furthermore the reward system establishes norms for good scientific behaviour. Articles are more likely to be published if they are in the approved form; careers progress more rapidly if a scientist follows the prescribed path for advancement; granting agencies favour certain scientific behaviours. Therefore the reward system is inextricably woven into the fabric of the biography even when the biographer does not recognise it. Of course scientists are also rewarded by the pure pleasure of learning, but it would be naive to assume that recognition from their peers is not a major incentive for their labours.

In this chapter, I will investigate the relationship between biography and the reward system of science by looking at patents. This may seem rather far-fetched, because patents are awarded for technological innovations, not for scientific discoveries. As technology and science become indistinguishable, however, the differences in their reward systems become less obvious. There is a close analogy between the patent process in technology and the process of establishing priority in science. Applying for a patent is like submitting a scientific article for publication. Both claim priority for an original discovery. Both the patent application and the scientific article are reviewed by recognised authorities for value and originality. Granting of the patent and publication of the scientific article both stake claims of priority. We can see more clearly from patents than from scientific articles how the reward system affects biography.

Modern biography and modern patent systems have several things in common. First, they are both parts of the reward system of science. Second, they both took their modern form at the same time in the early nineteenth century. Furthermore, they both assume that scientific progress is the work of individuals and that individual scientists are the units worthy of study, protection and reward. (We will see that at times these common themes are more than analogies and that debates over patents have directly affected the writing of biography.) And finally, in both science and technology the system of reward is based on a narrative of discovery. Patents are awarded only to individuals, not corporations or other organisations, and they confirm an achievement in historical time. In science, prizes, positions and eponyms are given to individuals for their careers, which again reflect discovery narratives. Although scientific biographies serve many purposes, their central theme, even in the case of social or psychoanalytical biography, is this narrative of achievement.

Scientific Biography

Biographies from antiquity through the seventeenth century usually appeared in collections. One thinks of Plutarch's *Lives of Illustrious Men*, Diogenes Laertius' *Lives of Eminent Philosophers*, the numerous medieval collections of saints' lives, Giorgio Vasari's *Lives of the Artists*, and John Aubrey's *Brief Lives*. In the seventeenth century, one began to see individual lives such as Thomas Birch's *Life and writings of the Hon. R. Boyle* (1741) and Pierre Gassendi's *Life of Peiresc* ([1641] 1657). The emphasis in all these collections was on the moral virtue of the subject. They all assumed that it was exemplary moral conduct that led to success in natural philosophy. These early biographies were often collected because they assign virtue to a class or group of individuals engaged in a common endeavour.

The character of scientific biography changed somewhat in the eighteenth century with the advent of the *éloge* at the Paris Academy of Sciences. Fontenelle's *éloges* drew from Plutarch and from the classical *eclogue* – the scientist was a disinterested seeker after truth who sought no personal gain. His virtue was no longer a virtue of religion, however, but a virtue of objectivity (Paul, 1980). This was necessary because the Enlightenment viewed science as standing apart from religious truths. The virtue of the scientist lay not in his religious beliefs and commitments, but in his objectivity and disinterestedness. This was a new kind of moral worth, and it existed in tension with the continuing enthusiasm for natural theology, but it still honoured the scientist as a participant in a common endeavour.

The scientist still drew his virtue from the activity of a special group. Scientific biography in the nineteenth century moved away from the group and focused instead on the individual. In doing so, it accepted a new kind of discovery narrative based on the concept of 'genius'.

The Concept of Genius

The introduction of the concept of genius by the German Romantics in the late eighteenth century changed all this. Martha Woodmansee argues that the concept of 'author' had an unstable meaning in the eighteenth century (Woodmansee, 1984, p. 427). Writing had traditionally been regarded as a craft, with the writer putting into words ideas that were common property. The Romantic poets, however, especially Fichte in Germany and Wordsworth in England, argued that poetry was a unique creation of the inspired mind and was therefore the property of the poet alone. The question of who owned writing stimulated a debate over copyright that extended well beyond literature. It was a debate that suggested ideas of genius in other fields besides poetry. In particular, it promoted the notion of the genius in science and the heroic inventor in technology (Schaffer, 1990). At its very beginning, modern biography raised this question: is science the work of individuals, or is it the inexorable march of logic and experiment, with the individual playing a minor role? Although this debate was not as intense in science as it was in literature, it still affected the writing of scientific biography. When applied to science and technology, the word 'genius' had originally meant a knack or skill, not a divine inspiration, as can be seen from its derivatives 'engine', 'engineer', 'ingenious' and so on. The *École de Génie* at Mézières was a military academy where students learned artillery and fortification, not a school for writing inspired poetry.

The concept of genius claimed by the Romantic poets changed the meaning of the word in science and technology as well as in literature. David Brewster, in his biography of Newton, picked up the genius theme and made Newton into an inspired seer and prophet, while Augustus De Morgan, in his biography, called Newton an exceptionally hard worker, but not a genius (Theerman, 1985). Those like Joseph Priestley, who did not believe that science was inspired, tended to avoid biography. Instead he wrote histories, which described scientific advances as the work of many hands (Yeo, 1988, p. 263). The concept of genius obviously favours a biographical approach to the history of science, because it assigns discoveries to single individuals. But it also limits what biography can do, because a creation of genius, by definition, cannot be explained. It is a sudden inspiration coming from inside the mind with no external source. It is what R. K. Merton called a 'singleton' – the work of a single creative mind (Merton, 1973). If we conclude with Merton that true singletons never occur in the history of science, then we have to reconsider what biography can contribute.

Patent and Copyright as a Reward System

Just as the genius theory supported copyright, it also supported the patent system. Royal letters patent issued in the sixteenth and seventeenth centuries were monopolies granted to individuals either as rewards for services or as means of promoting certain industries. They were seldom given for invention. Most of them were import franchises. Patents for invention originated in the Venetian Republic, and moved from there to England. The first English patent for invention was granted in 1559 by Queen Elizabeth I to Giacopo Acontio, a Venetian, for his various wheel machines and furnaces (Bugbee, 1967, p. 29).[1]

As patents for invention spread, they became established by law and were justified by a number of arguments (Dutton, 1984, Ch. 1). The French Constitutional Assembly established a patent law in 1791 to replace the complicated system of privileges that existed under the *ancien régime.* The ideology of the French Revolution opposed privilege but defended property as a right under natural law, so the French advocates of patents evoked the concept of intellectual property, although the change of words from 'privilege' to 'property' was little more than a ruse to retain the status quo. Carried to its logical conclusion, this 'natural law thesis' called for patents in perpetuity, although in actual practice the Patent Law of 1791 granted *brevets d'inventions* only for a limited time (Hilaire-Perez, 1991).

In England, supporters of patents favoured the 'reward-by-monopoly' thesis. The Statute of Monopolies in 1623 went through numerous changes until it was regularised by the Patent Amendment Act of 1852. As Adam Smith explained, the inventor should be rewarded according to the usefulness of his invention (Dutton, 1984, pp. 18–20). The Crown issued a monopoly to the inventor for a specified number of years as a reward. If the invention proved to be valuable, the inventor would profit and thus be rewarded richly; if the invention turned out to be worthless, he would receive nothing. Thus the patent would automatically reward the inventor according to his deserts.

In Colonial America, the separate states had separate patent laws, which were unified under the Federal Constitution of 1787. Article I, Section 8, of the US

1 Pamela Long argues that patents and copyrights go back much further than has usually been recognised, but she also recognises that the meaning of 'intellectual property' has changed substantially and been rewarded differently in modern times (Long, 1991, pp. 847–8, and Long, 2001).

Constitution states that 'The Congress shall have Power ... to promote the Progress of Science and useful Arts, by securing for limited Times to Authors and Inventors the exclusive Right to their respective Writings and Discoveries.' Patents were at first granted by the President, until James Madison instituted the Office of Patents in 1802. Note that according to the Constitution, the purpose of patents was 'to promote the Progress of Science and useful Arts' – not to reward the inventor or to recognise his ownership. This 'monopoly-profit-incentive' thesis required a patent specification describing the invention in detail so that others could manufacture it freely after the patent expired. French and British patent law did not require a specification, which meant that the inventor could keep his invention secret and also obtain a monopoly right.

A fourth argument that appeared in most patent controversies was the 'exchange for secrets' thesis. According to this argument, the patent is a contract in which the inventor gives up the secret to his invention in return for protection against competition. The inventor has the alternative of trying to keep his invention secret, or publishing it in order to obtain a reward. This argument also requires a patent specification.

Scientific discoveries *per se* are not subject to patent because they are laws of nature, and not human creations. Thus the scientist cannot claim a property right in his discovery. His only reward is recognition, and the honour that pertains to that recognition. Hence it is crucial that he establish priority by publishing his discovery. In this case, rewards in science are like the last two patent theses argued above – they both require that the inventor give up his discovery in order to profit from it.

It has long been recognised that all patent law contains a certain degree of arbitrariness, and it is by no means certain that patents are advantageous to society as a whole. The patent system is a paradox, because as it stimulates invention, it also retards diffusion of that invention. Also, the period of monopoly under the patent is arbitrary. In most cases it has been set in multiples of seven years, because that has been the traditional time for apprenticeship. Some countries have abolished patents or have delayed establishing patent laws. The Netherlands abolished patents in July 1869, and did not restore them until 1912. Switzerland had no patent law until 1887. Germany established a uniform patent law for the entire *Reich* in 1877, although Chancellor Bismarck had declared his opposition to patents in 1868. The greatest compulsion to establish patents has been pressure from trading partners who have patent laws and regard manufacture in countries without such laws as piracy.

Because inventions in science have no legal protection, their rewards might be expected to be even more arbitrary. But the scientific reward system is, in some sense, self-adjusting, because the value of one's work may not be immediately apparent. As time goes by, work that seemed insignificant or even bizarre can become very important. Also, recognition can come from different places – from one's closest colleagues, from one's scientific peers or from the general public. As a result, one can expect that eventually the most important scientific achievements will receive the greatest recognition, while a patent award is not so adjustable.

The agitation in England to repeal the Patent Amendment Act of 1852 is particularly revealing. Lord Granville, the Chairman of the Select Committee in 1851, argued that 'the whole system was unadvisable to the public, disadvantageous to the inventors, and wrong in principle' in spite of the fact that Adam Smith, Jeremy Bentham and John Stuart Mill had all favoured patents (MacLeod, 1996, p. 142).

Agitation for patent reform began around 1827, with inventors and patent agents arguing that the existing system was expensive, clumsy and uncertain. Select Committees of Parliament and Royal Commissions investigated the system in 1851–52, 1862–65 and 1869–72. The abolitionists launched a strong campaign against patents after the Act of 1852 went into effect, but by 1870 they had lost the battle. Their arguments were strikingly similar to those expressed by the opponents of the genius theory. The engineer Isambard Kingdom Brunel argued that there was no such thing as true invention. New products and processes are 'the result of an observing mind, brought to bear upon circumstances as they arise', the same argument that Joseph Priestley had made in the eighteenth century. John Coryton argued in 1874 that all real work was done by the scientists and that patentable inventions were merely applications of ideas already in the public domain, an argument repeated in the United States in 1883 by Henry Rowland (MacLeod, 1996, p. 148).[2]

The supporters of patents were more skilled than the abolitionists in politics and propaganda, especially Bennet Woodcroft, head of the British Patent Office. Woodcroft argued for the heroic view of the inventor, and obtained the support of Henry, Lord Brougham, in his campaign. He attempted to create a National Portrait Gallery of Inventors and a Museum of Patents. Woodcroft argued that inventions were works of genius and he corresponded with biographers, especially John Timbs, who wrote the first composite biography of inventors, *Stories of Inventors and Discoverers in Science and the Useful Arts* (1860), at the direct urging of Woodcroft. Samuel Smiles also corresponded with Woodcroft. There is no evidence that he was hired by Woodcroft to write his much more famous *Lives of the Engineers* (1862) and *Self Help: With Illustrations of Character and Conduct* (1864), but they came at just the right time to support the proponents of patents in the patent controversy (MacLeod, 1996, p. 143). Thus in the nineteenth century the arguments for patents directly supported the notion of genius in scientific biography.[3]

Those who have opposed patents have suggested other systems of reward for inventors that do not involve granting a monopoly. These include prizes, subsidies and various forms of licensing. If these encouragements were substituted for patents, the reward system in technology would correspond much more closely to the reward system in science. An inventor might still choose to keep his invention secret rather than depend on the government or the courts for awards. Some still do. But most venture capitalists have preferred patents over secrecy as a way to protect their investments. Although an inventor must reveal details about his invention in order to obtain a patent, thereby reducing secrecy, the *anticipation* of a patent actually

2 On Rowland, see Reich (1985), p. 21.

3 Although the genius concept has been the mainstay of those arguing in favour of patents, it has worked both for and against inventors. In *Cuno Engineering Corp.* v. *Automatic Devices Corp.* (1942), Justice William O. Douglas argued that an invention must display a 'flash of creative genius' in order to earn a patent. For most inventors, this was an excessive demand, and in 1952 the US Congress changed the law to require only that the invention be 'non-obvious' (Silverman, unpublished).

increases secrecy in the research and development stages. As academic scientists increasingly seek patents, their ideal of open communication is placed in jeopardy (Nelkin, 1984, pp. 15, 97–8, 102).

Biography and the Reward System in Science

Between 1957 and 1963, Robert K. Merton and his collaborators published a series of studies on the reward system in science (Merton, 1973). They concluded that 'property rights in science become whittled down to just this one: the recognition by others of the scientist's distinctive part in having brought the result into being', and this recognition is based on only one thing: priority (Merton, 1973, p. 294). Establishing one's priority is important because so much hinges on it. It is, in theory, a winner-take-all mechanism for reward. However, Merton also recognised that there is seldom, if ever, a single discoverer in science. There are always others snapping at one's heels, or possibly secretly forging ahead. Also, one cannot establish when a discovery has occurred until well after it is supposed to have happened (Schaffer, 1986, esp. pp. 391–7). One might assume that science in our day is more competitive than it was in the past, and therefore more prone to priority disputes, but Merton argues that this is not the case. There have always been priority disputes, because they are a direct result of the institutional structure that rewards scientists. Merton says this explains, in part, why even the most generous and affable of scientists have worried about priority.

As a sociologist, Merton was interested in reaching generalisations that would be valid for all science at all times. As historians, we question such generalisations. Fighting to get support from an aristocratic patron in the seventeenth century is not the same thing as fighting for support from the National Science Foundation in the twenty-first. The rise of democratic institutions in the West has changed who grants rewards in science and technology. It is certainly no coincidence that new patent and copyright laws in France and the United States were instituted by revolutionary governments, and that those laws operated differently from the systems of royal letters patent and *privilèges* that they replaced.

Merton's argument that all scientific discoveries are multiples would seem to contradict the theory of genius, which has been the basis for much biography. Put simply, biography assumes singletons; Merton claims multiples. But his famous 1957 article 'Priorities in Scientific Discovery' is based almost entirely on biography (Merton, 1973). Merton found that even if a biographer claims priority for his subject, he is almost certain to discuss those who challenged that priority, and therefore to provide a useful source for Merton's research.

Patents and Biography

Since 1870, patent protection has been strengthened in most countries and extended to others, but the arguments of the opposition have never been fully answered. Current debates over the role of the author and attacks on the copyright system have revived the argument that creative acts are never the work of a single individual.[4] The question of authorship is especially complicated in scientific biography. If we reject the genius theory, then the subject's discoveries are not lone victories, but the work of many hands and minds. Yet the biographer needs to show why the subject is worthy of biographical treatment. The scientific community has recognized the reality of multiple discoveries to some extent by increasing the number of papers with multiple authors and awarding prizes to groups rather than to individuals. Has scientific biography followed this trend by recognising multiple discoverers, or has it followed the path of the patent system by limiting recognition to a single individual? It appears to have done neither. Scientific biography has responded by broadening its treatment of the social and cultural context of the subject while continuing to focus on the individual. If biography is still heroic, at least it has become heroic in a different way.

Biographers have not yet figured out how to deal with the collaborative research teams that characterised big science during the twentieth century. They have preferred to write about an earlier period, or, when they have ventured into the twentieth century, they have been attracted to theoretical physicists, who seem to fit better the notion of the lone genius.

The victory of the patent system in the nineteenth century may have stimulated biography, but it led it in an unsatisfactory direction. Heroic biography required that the author fudge the evidence in order to make his subject into a single discoverer. It denied the achievements of colleagues in order to establish a secure priority for the subject, and by making discovery an act of genius, it essentially made it inexplicable. The result was an ahistorical account of scientific achievement. It was this kind of biography that drew the scorn of prominent historians of science in the 1950s and 1960s.[5] An alternative would be group biography, but group biographies never seem to work. Instead, one writes the history of a research school, a scientific society, or the history of a particular scientific advance involving several contributors (Morrell and Thackray, 1981; Hunt, 1991). These are no longer biographies, although they may contain a great deal of biographical material. Another common alternative is a monograph describing an individual's contribution to an important area of emerging science. Often the author will disclaim in the preface that he is attempting a biography, and assert instead that he is describing only a particular aspect of the subject's life (Baker, 1975, p. vii; Kargon, 1982).

---

4 One is tempted to dismiss the criticisms of authorship by Roland Barthes, Michel Foucault, Martha Woodmansee, Peter Jaszi and others as post-modern excesses, but they warrant attention, particularly in non-fiction genres like biography and history. Biographers do not create the lives they describe; they can only find them.

5 George Sarton, the founder of history of science in the United States, praised scientific biography because it was the best way to understand 'genius', but Herbert Butterfield, in reaction to Sarton, called it 'an anecdotal kind of history which is really a rope of sand', and Henry Guerlac equated it with the worst kind of 'romance and antiquarian curiosity'. Guerlac was my thesis adviser at the time he published those comments, and it was only later, and with some trepidation, that I attempted to defend biography. See Sarton ([1936] 1957), Butterfield (1950, p. 51), Guerlac (1963, p. 798) and Hankins (1979). Thomas Söderqvist summarises all these quarrels in Chapter 15 of this volume.

As the character of science changes, we can expect our biographies to change. In particular, the trend toward increased patenting of scientific research, especially academic research, changes the assumptions of biography. In addition to seeking honour, the patentee now anticipates a pecuniary reward, and the biographer will have to adjust his interpretations accordingly.[6] In 1980, Chief Justice Warren E. Burger approved the first patent of a genetically engineered organism. In a five to four decision, the court concluded that 'a live human-made microorganism is patentable subject matter'. Soon afterwards, Herbert Boyer and Stanley Cohen patented a process for inserting foreign genetic material into a bacterial plasmid and assigned the patent rights to their institutions.[7] These first forays into genetic engineering touched off intense competition and numerous efforts to patent similar procedures and materials. The result has been both a bonanza and a curse for research in molecular biology. In the first place, it has blurred further the line between science and technology. The distinction between science and technology ceases to be philosophical, and becomes legal. The term 'genetic engineering' is, in itself, a statement that a process is patentable. In the second place, the materials left behind by a scientist for the benefit of future biographers have changed. Laboratory notebooks are different if they are prepared in anticipation of a patent. Industrial laboratories in the United States require that researchers keep daily records in hard-bound notebooks in such a way that all additions and deletions are clearly marked and no pages can be secretly removed or inserted. A superior signs each page stating that he or she has read and understood the page. The notebook then goes into a safe, and the researcher is not allowed to discuss the work without special permission. The reason for keeping this kind of record is to strengthen future patent claims. The narrative of discovery becomes a legal document that can be used in court to prove priority. Even as scientific research becomes more collaborative, it becomes more secretive. The ideal of open communication is sacrificed for financial benefit. Of course, scientists have known this particular kind of sin ever since the Second World War, when they accepted security restrictions in return for government support of their research, but the process has accelerated in the past quarter-century, with no indication that it will slacken in the future.

Conclusion

These changes are matters of great concern for many scientists, but should they worry historians and biographers of science and scientists? Probably not, because biographers have always had to take their subjects as they find them. As science changes, biographies will also change. But we must remember that history changes as well, and that the emergence of new forms of scientific biography has had more

---

6 One should not assume that the search by scientists for wealth is only a recent phenomenon. William Thompson, Lord Kelvin, was greatly honoured for his scientific labours, but it was his patented telegraph instruments that made him rich (Smith and Wise, 1989, pp. 698–710).

7 *Diamond* v. *Chakrabarty 447 U.S.* (16 June 1980), 303–22, quoted from Nelkin (1984), p. 13.

to do with changes in history than with changes in science. We need, therefore, to ask what kinds of biography contribute most to the history of science. There is little agreement on the subject. In their highly readable biography of Darwin, Desmond and Moore argue: 'Any new biography must take account of the recent upheaval in the history of science, and its new emphasis on the cultural conditioning of knowledge' (Desmond and Moore, 1991, p. xx). They place Darwin in the midst of radical English politics, and claim that because radical politics leads necessarily to radical science, we can skim over the science and understand Darwinian evolution as a product of politics. Larry Holmes takes just the opposite tack in his study of Lavoisier. Rather than get out of the laboratory and into the social whirl, he plunges ever more deeply into the laboratory: 'If we are to understand scientific activity at its core, we must immerse ourselves as fully as possible into these investigative operations, whether they be in the laboratory, the museum, the field, or the lecture hall, where scientists themselves spend the working days of their lives.' Holmes does not favour the older scientific biographies, but he says they at least held to certain standards of judgement. If we agree that science is merely the outcome of contingent social processes, 'the new standards risk being no standards at all' (Holmes, 1985, pp. xvi–xvii). As long as we have such disparate views of history, we will continue to disagree about how scientific biography should be written.

The current controversy between the advocates of 'social construction' and 'rational reconstruction' of science (to use only the most recent epithets in this quarrel) tends to polarise scientific biography into opposing camps. This is unfortunate, because the recent resurgence of scientific biography has shown that both perspectives have much to offer. As a possible cure for this rupture, I would like to point out that although the reward system of science is certainly a social construction, it has been created by scientists for scientists and rewards scientific, not social, achievements. It also determines who warrants a biography, and to some extent, what that biography will contain. It is therefore a middle ground that needs to be explored further. It also may not be fair. The rules for reward in science emphasise the individual, as do our biographies. This unfairness challenges us to find a way to write biography that criticises as well as exploits the fruits of this reward system. The history of patents shows how important a particular reward system has been in determining technological development. The history of science can benefit by raising the same issues in our biographies of scientists.

Bibliography

Aubrey, John (2004), *Brief Lives*, ed. Richard Barber, Woodbridge: Boydell Press.

Baker, Keith (1975), *Condorcet: From Natural Philosophy to Social Mathematics*, Chicago, IL: Chicago University Press.

Bugbee, Bruce W. (1967), *Genesis of American Patent and Copyright Law*, Washington, DC: Public Affairs Press.

Butterfield, Herbert (1950), 'The historian and the history of science', *Bulletin of the British Society for the History of Science*, **1**, 49–58.

Desmond, Adrian and James Moore (1991), *Darwin*, New York: Warner.

Diogenes Laertius (1942), *Lives of Eminent Philosophers*, Cambridge, MA: Harvard University Press.

Dutton, Harold I. (1984), *The Patent System and Inventive Activity during the Industrial Revolution 1750–1852*, Manchester: Manchester University Press.

Guerlac, Henry (1963), 'Some historical assumptions of the history of science', in Alastair Crombie (ed.), *Scientific Change: Historical Studies in the Intellectual, Social, and Technical Conditions for Scientific Discovery and Technical Invention, from Antiquity to the Present*, New York: Basic Books, pp. 797–812.

Hankins, Thomas L. (1979), 'In defence of biography: The use of biography in the history of science', *History of Science*, **17**, 1–16.

Hilaire-Perez, Liliane (1991) 'Invention and the state in 18th-century France', *Technology and Culture,* **32** (4), 911–31.

Holmes, Frederic L. (1981), 'The fine structure of scientific creativity', *History of Science*, **19,** 60–70.

Holmes, Frederic L. (1985), *Lavoisier and the Chemistry of Life: An Exploration of Scientific Creativity*, Madison, WI: University of Wisconsin Press.

Hunt, Bruce J. (1991), *The Maxwellians*, Ithaca, NY: Cornell University Press.

Kargon, Robert Hugh (1982), *The Rise of Robert Millikan: Portrait of a Life in American Science*, Ithaca, NY: Cornell University Press.

Long, Pamela O. (1991), 'Invention, authorship, "intellectual property," and the origin of patents: Notes toward a conceptual history', *Technology and Culture*, **32,** 846–84.

Long, Pamela O. (2001), *Openness, Secrecy, Authorship: Technical Arts and the Culture of Knowledge from Antiquity to the Renaissance*, Baltimore, MD: Johns Hopkins University Press.

MacLeod, Christine (1996), 'Concepts of invention and the patent controversy in Victorian Britain', in Robert Fox (ed.), *Technological Change: Methods and Themes in the History of Technology*, Amsterdam: Harwood Academic, pp. 137–53.

Merton, Robert K. (1973), *The Sociology of Science: Theoretical and Empirical Investigations*, ed. Norman W. Storer, Chicago, IL: University of Chicago Press.

Morrell, Jack and Arnold Thackray (1981), *Gentlemen of Science: Early Years of the British Association*, Oxford: Clarendon Press.

Nelkin, Dorothy (1984), *Science as Intellectual Property: Who Controls Research?, AAAS Series on Issues in Science and Technology*, New York: Macmillan.

Paul, Charles B. (1980), *Science and Immortality: The Eloges of the Paris Academy of Sciences (1699–1791)*, Berkeley, CA: University of California Press.

Plutarch (1876), *Plutarch's Lives of Illustrious Men*, trans. John and William Langhorne, London: Chatto and Windus.

Reich, Leonard S. (1985), *The Making of American Industrial Research: Science and Business at GE and Bell, 1876–1926*, Cambridge: Cambridge University Press.

Sarton, George ([1936] 1957), *The Study of the History of Mathematics*, reprinted in Sarton, George, *The Study of the History of Mathematics and the Study of the History of Science,* New York: Dover.

Schaffer, Simon (1986), 'Scientific discoveries and the end of natural philosophy', *Social Studies of Science*, **16**, 387–420.

Schaffer, Simon (1990), 'Genius in Romantic natural philosophy', in Andrew Cunningham and Nicholas Jardine (eds), *Romanticism and the Sciences*, Cambridge: Cambridge University Press, pp. 82–98.

Silverman, Robert J., 'Authors, Inventors, and Heroic Discoverers: Romantic Philosophy and the Reward Systems of Art, Science, and Technology', unpublished conference paper.

Smith, Crosbie and M. Norton Wise (1989), *Energy and Empire: A Biographical Study of Lord Kelvin,* Cambridge: Cambridge University Press.

Theerman, Paul (1985), 'Unaccustomed role: The scientist as historical biographer – two nineteenth-century portrayals of Newton', *Biography*, **8**, 145–62.

Vasari, Giorgio (1991), *Lives of the Artists*, trans. Julia Conaway Bondanella and Peter Bondanella, Oxford: Oxford University Press.

Woodmansee, Martha (1984), 'The genius and the copyright: Economic and legal conditions of the emergence of the "author"', *Eighteenth-Century Studies*, **17** (4), 425–48.

Yeo, Richard (1988), 'Genius, method, and morality: Images of Newton in Britain, 1760–1860', *Science in Context,* **2,** 257–84.

Chapter 6

The Tragedy of Comrade Hessen: Biography as Historical Discourse

Christopher A.J. Chilvers

Blood that Will Not Wash Away

The relationship between the history of science and 'scientific biography' is a far more complex configuration, counter-intuitive even, than is often recognised. The trivial proposition that biography enunciates a life history, thus 'scientific biography' offers the history of a life in science, neatly locates itself with the view of the history of science as a progressive enterprise dedicated to the revelation of human genius rather than the stupidities of politics and war (as historian of science Charles Singer contended, explaining his own philosophically driven passion at the Second International Congress for the History of Science and Technology in 1931).[1] Perhaps such a view has firm foundation when discussing Faraday, Einstein, Darwin, Mendeleyev or other distinguished scientists, though I fear not. The proposition loses all reflexivity when one begins to analyse the life of one of the more tortured souls who did not produce the powerful scientific theory or even progress in the chosen discipline with any particular speed.

If the subject did not display any such 'genius', or even the common characteristics, which would allegedly define such a concept, then much of Kuhn's normal science (and therefore much of modern science) finds itself ineligible subject matter. The obvious question to arise is whether a biographical subject is worthwhile without being able to tag the '… that changed the world'-suffix to their story. Without pre-empting the following discourse, an affirmative answer is indispensable to understanding science in its strength and weakness, in its relationship to other spheres in society, to understand it as a human activity bearing all the contradictions and fragilities of its actors. Further, when science is offered in an environmental frame, the restraints upon the concept of 'human genius' become more obvious. One may argue that the negative example does not confirm the positive.

If this admittedly simplistic formula cannot embrace the demanding nature of the biographical subject (Shortland and Yeo, 1996), what is the prospective biographer to turn to? One possible trajectory is the traditions of literary criticism and philosophy of literature. Reflecting upon Niels Jerne's life in these terms, Söderqvist declared that it 'epitomizes MacIntyre's claim that biography is truly "neither hagiography

1 'Place of Science in History: Conference on novel lines – no papers but full debates', *The Observer*, 28 June 1931, p. 11 (interview with Singer). See also Singer (1959), pp. 6–102. The argument was also outlined in stark form by Clark (1932).

nor saga, but tragedy." … I am convinced that this kind of tragedy is more widespread in today's scientific culture than we ordinarily wish to acknowledge' (Söderqvist, 2003, p. xxvi). If biographical writing is a literary form and an artistic discourse, it should be possible to develop aspects of that form and discourse in the service of the history of science and to translate literary genres into edifying biographical methodologies (Pachter, 1979). This presupposes the rejection of the crude dichotomy of scientific and artistic cultures codified in C.P. Snow's famous lecture (Snow, [1959] 1993; Edgerton, 2005). Specifically, consideration should be given to the genre of tragedy, replete with the spots of blood that will not wash away. The traditional and major obstacle to such translation is the tendency to consider tragedy only within the confines of 'the tradition', as literary drama. As Williams ([1966] 1979) noted, this will not necessarily suffice, particularly when recognising the distinct development of a varied meaning of tragedy in modern popular language. He contends that 'at particular stages of our own history, the revival of tragedy has been a strategy determined by this consciousness of a necessary tradition' (Williams, 1979, p. 16). The tradition of tragedy involves the interpretation of the past from the perspective of the present, and can thus be discontinuous as often as continuous, just as with historical interpretation (Evans, 1991). It is a genre firmly rooted in its social dimensions, whose *transitive* properties are partially reliant upon this social character. Its ability to convey epistemic messages across various epochs of the historical process relies upon certain contiguities between the periods in which it flourishes. If one can utilise tragedy in a manner that reveals its complex socially dynamic and historical character, its relevance to the scientific biography and the history of science will be manifest, though Williams cautioned against foolhardiness in such endeavour (Williams, 1979, pp. 13–16).

Aristotle's *De Poetica* defined tragedy as the perfect discourse (Aristotle, 1996, p. 10 [1449b, 23–5]), through which the cognitive qualities of human action in a given circumstance was dramatically portrayed for the gathered *polis*. A process of trial and error was developed in the public domain, normally in the character of a member of a ruling or wealthy family, as with Shakespeare (Williams, 1979, p. 18). The audience was not a peripheral element in Greek tragedy, but integral to the sense of comprehension and consciousness resulting from such a 'play'. This was an important aspect of the Greek political process, of Athenian democracy, with which the audience was actively engaged. It is hard to envisage such a participatory role for theatre in the modern *Zeitgeist* of popular alienation, though it was comparatively recently that audiences sought a sense of political and social engagement, or sedition even, from the theatre and the opera (Reiss, 1980, pp. 301–2; Patterson, 1989; Swingewood, 1975, pp. 190–222):

> What the form … embodies is not an isolated metaphysical stance, rooted in individual experience, but a shared and indeed collective experience, at once and indistinguishably metaphysical and social. (Williams, 1979, p. 18)

The dynamic quality of Greek tragedy with its epistemic role in the collective development of knowledge and the testing out of strategies for human action by reference to the past conveys the meaning of Aristotle's description of the perfect discourse. It was art in the service of the *polis*, a very political literary form (Reiss, 1980, pp. 300–302).

The transformations in tragedy through medieval times to the modern age are dealt with elsewhere and beyond the remit of this essay (Williams, 1979, pp. 15–45), though the enduring fascination, debate and industry derived from Shakespeare's mastery of the genre convey the continued epistemic relevance of tragedy as a discourse. Reiss, seeking to develop the philosophical ramifications of Aristotle's view of perfect tragedy, suggests a reason for the endurance of the genre. Tragedy has a historic character, surfacing in its genuine and most effective form during periods of great transition in social and political affairs. The precise form of the historical juncture may vary with the historical differences in the epoch as between the apex of Athenian democracy and the dilemmas of the rise of bourgeois society. However, in fundamental terms, these periods are decisive moments of great historical ambiguity when human action moves to the centre of the historical stage:

> The concept of history thus becomes that of a permanent crisis in human action. 'History' is the place of a fundamental ambiguity, which comes from the fact that such action must create its own meaningfulness. History and social action invent one another simultaneously and mutually. Once again theatre becomes creative of a discursive order, at length perhaps the invention of a new form of practice of what would previously have been called 'knowledge'. (Reiss, 1980, p. 301)

As such a period gathers and sections of society become aware of their prospective roles, the social and political implications of the paths before them require testing. Possible scenarios are discussed and developed before a newly gestating *polis* (Reiss, 1980, pp. 285–90).

Such periods in history acquire another characteristic, the growth in historical and social consciousness of large sections of the population. Thus, tragedy emerges as a forum for the collective investigation of possible courses of action, mainly through allegorical references to historical episodes that acquire the status of popular exemplars. The unity of Aristotle's perfect genre is partially that of the new *polis*, or collectively politicised audience involved in decision-making (though not necessarily in positions of established political power). Tragedy becomes a mechanism for the development of consciousness and an awareness of historical tasks faced. It is obvious, therefore, that, according to this view, tragedy is a distinct social phenomenon requiring a distinct set of social circumstances to blossom. Social environment is integral to tragedy as a genre.

Biography has the capacity to fulfil a similar function to this philosophical and epistemological conception of tragedy. Tragedy requires historical allegory as the forum for the development of social consciousness. It demands an agitated audience seeking to comprehend and question the limits of their historical role. Further, its foundation is a historical juncture when great change is in the air, 'an age of great aims' (Trotsky, [1925] 1991, p. 272). All of these facets of tragedy can be found in the realm of the 'scientific biography'. The historical allegory speaks for itself as the subject of the biography, while the requisite form of audience has mushroomed in the recent past as the print runs of popular science and popular history of science testify.

A crucial, if largely unnoticed, aspect of the development of the history of science is a relationship between the historian or 'history of science community' and the public consciousness of science. The platform of the great historical juncture is the very reason for the development of the history of science as a distinct public intellectual movement over the past century (Chilvers, 2005). Sarton's 'new humanism' (Sarton, 1924–25; Chilvers, 2005) and Singer's arguments for the role of the history of science at the 1931 Congress articulated the sense that humanity stood before the ultimate moment of historical ambiguity, when the development of science could lead to human liberation or to complete destruction.[2] Such a transitional juncture can give rise to the renewal of tragedy as a genre, alongside the history of science as a movement.

The careful documenting and analysis of the role of the individual in history can acquire a formative and instructive role for the audience. Biography's 'acquisition' of tragedy's historical epistemology can nourish the definitive features of the tragic genre (and arguably, of the social environment), raising such discursive issues as *hamartia*, *perepeteia*, even *katharsis.* Biography does not necessarily fit into particular Procrustean beds, but the philosophical issues underlying these Aristotelian concepts are common to reflective and critical biographical work. For instance, when the subject meets with ill fortune of some sort, whether physical, intellectual or in social standing, understanding the issues and actions that lay behind such 'poor luck' becomes an important part of the corpus. The analysis of human action within a given set of circumstances is an important task of the biographer.

The subject of this investigation, and the illustration of the proposed methodology, does not fit very easily within the conventions usually employed in defining the scientist. While being a historical figure of some note, he has appeared to be without biography. Far from an obscure laboratory technician, his name has echoed through the years and acquired a measure of fame (or infamy, depending on one's world view or even one's view of the current state of the history of science), if only in the dusty corridors of the academic history and philosophy of science department or the whispered words of the post-Soviet intellectual underground. Boris Hessen, the Professor of the Institute of Physics at the Moscow State University, was a convinced and intellectually able Marxist, who sought to develop both the historical treatment of science within the new regime of the USSR and to propagate the socialist revolutionary world view among scientists. He was thus something of a politician, a historian, a physicist and a revolutionary. Indeed, he organised teaching schedules and seminars at the Institute of Red Professors and can be characterised as an exemplar of the 'red professor' (Chilvers, 2003, pp. 422–35).

Noted within the history of science as the author of 'externalism', Hessen presented an audacious and famous paper to the 1931 Congress on 'The Social and Economic Roots of Newton's Principia' (Hessen, 1931). It should be noted that little of this paper was ever delivered to the Congress itself, due to the protracted skirmish between the Russian delegation and Charles Singer that prevented the delegation from reading their papers, even at the Special Session set to give the Russian

2 'Place of science in history: Conference on novel lines – no papers but full debates', *The Observer*, 28 June 1931, p. 11 (interview with Singer).

delegation a platform.[3] The paper was speedily published with the delegation's collection and handed out 'in brown envelopes' at this special session of the Congress.[4] Hessen's paper became the classic statement of the heuristic embedding the rise of modern science within the framework of the rise of capitalism, weaving its attendant technical and ideological concerns into the very motivations and science of the natural philosopher. Partially through the historical accident of the social environment of London in 1931 and the novelty of his approach, Hessen's concern for developing the Marxist history of science inspired many scientists, historians and even the odd sociologist to take up the articulation of this methodology, though the extent to which they understood the subtlety of his paper is open to question (Graham, 1985; Chilvers, 2005). The underlying strain of social, economic and political concerns helping to motivate the development of science set a new agenda for the scientific left, and along with the arguments of the Russian delegation to the 1931 Congress, the background of economic depression and the fear of war helped provide the platform for the launch of the 'social relations of science' movement of the 1930s and 1940s (see also McGucken, 1984; Wood, 1959; Crowther, 1941). The crystallographer J.D. Bernal, the Cambridge biochemist Needham and the scientific journalist for the *Manchester Guardian*, J.G. Crowther, were among those claiming to follow Hessen's work and method.

Two Cultural Camps

Much can be said for a 'two camps' philosophy, though not in the form espousing proletarian culture and science as separate from and in opposition to capitalism. Certainly, there are two histories of the influence of Marxism in the history of science, and it can plausibly be contended that there are two cultures of attitude to Marxism within the history of science community in the 'Western' world (Mayer, 2004; Christie, 1990; Young, 1990).

The first is the orthodoxy that dominates the texts of authors such as A.R. Hall or Floris Cohen, arguing that Marxism has little or no place within the history of science, despite the rather aggravating fact that the first academic faculty devoted to the subject was at Moscow University during the early years of Bolshevism (Cohen, 1974; Hall, 1954). Such orthodoxy is usually accompanied by dismissive attacks upon the papers delivered by the Russian delegation to the Second International Congress, especially Hessen's. Epithets such as 'simplistic', 'deterministic' and 'programmatic' are liberally sprinkled, though rarely followed by a detailed critique.[5] While such attacks form a certain socially necessary academic apparatus, there has been no attempt to analyse or criticise the other papers delivered at the Congress, especially the contribution of such a legendary forefather as Charles Singer, whose

3 See Chilvers (2004).

4 'Soviet delegates pulled up', *The Morning Post*, 6 July 1931, p. 14.

5 Floris Cohen develops a lengthy critique based upon a series of embarrassing misconceptions of Hessen and Marxism. His most notable abusive remark is 'this narrow minded piece of bigoted dogmatism at its Stalinist crudest' (Cohen, 1994, p. 332), which just about sums up his book.

paper has decomposed in a far more shocking manner than Hessen's (Chilvers, 2005).

Quietly, and without much fanfare, a rival history has been taking shape, consisting of a loose patchwork of papers, books and archival material that remain largely disparate and without the common theme of the reassessment of Hessen and the Marxist history of science (Josephson, 1991; Joravsky, 1961; Graham, 1985; Chilvers, 2003; Chilvers, Josephson and Schäfer, forthcoming). These have included limited biographical information and a concern for understanding Hessen's contribution to the history of science from the standpoint of his social environment. Graham referred to the importance of the outstanding problem in approaching Hessen's contribution:

> It is extremely ironic that radical historians of science who castigated old-fashioned historians of science for looking upon Newton's physics as if it 'dropped from the sky' (borrowing a phrase from Engels) accepted Hessen's paper exactly as if it had dropped from the Moscow sky. (Graham, 1985, p. 706)

A certain degree of reflexivity is required. Whether this lack of attention is because he was not one of the towering figures of the Bolshevik revolution, or even one of the many eminent figures forced into the humiliation of the Show Trials in the 1930s, it remains curious that the circumstances of the arrest and execution of one of the dominant figures in the history of science and physics have left so little mark. The difficulties posed by inaccessibility of the archives and the repressive character of the regime shaped by Stalin have been obvious contributory factors. However, the lack of attention paid to Hessen's history and contribution to the history of science mirrors the lack of attention given to his famous paper on Newton, and indeed, allows the convenient definition of his work as Stalinist Marxism, with most authors failing to draw attention to the dichotomy between classical Marxism and Stalinism. The problem with such analytically bland history is it is neither convincing in its denunciations of Marxism nor justifiable in recognising the magnitude of the issues involved in the reassessment of the Marxist history of science and the sociologically conscious methodology of 'externalism' (Enebakk, 2004; Chilvers, 2006).[6]

The 'Red Professor'

For many years, Hessen's fate was unknown even in Russia. As with Nikolai Vavilov, the respected geneticist who fell foul of the rise of Trofim Lysenko (Soyfer, 1994) and also attended the 1931 Congress (Vavilov, 1931, pp. 95–106), it was known he disappeared at some point in the 1930s, either to the *gulag* for hard labour or the execution cellars of the Lubyanka, headquarters of the NKVD. However, as the statues tumbled, light began to seep into the murky cellars of the *Yezhovshshina*. The fate of large numbers of dissidents, oppositional figures, the rebellious and the

6 Whether one can ascribe 'externalism' to Hessen's paper is questionable. A more convincing point of genesis is Merton ([1938] 1970), written to undercut the challenge of Hessen's method.

helpless begun to be unearthed, revealing the full extent of the terror that gripped the country (Getty and Naumov, 1999). Hessen is a case in point. There has been considerable uncertainty about when he died and the manner of his death – Graham (1985, p. 707) initially suggested 1938, while Josephson (1991, p. 242) contended 1937, and both thought he died in prison – though more recently released material confirms that his arrest took place on 21 August 1936, that he was tried before the notorious V.V. Ulrikh (Chair of the Moscow Show Trials) at a closed meeting of the Supreme Military Collegium Court on 20 December 1936 and was shot on the same day, along with A.O. Apirin. A.M. Reizen, tried alongside Hessen, was 'condemned to ten years hard labour and perished in detention' (Gorelik, 1995, p. 74). At the time of his arrest, he was the Director of the Institute of Physics, Professor of the Physics Faculty at Moscow State University (MGU) and a corresponding member of the Academy of Sciences, participating in the Commission on the History of Knowledge (with Bukharin), the forerunner to the history of science section (Josephson, 1991, p. 242).

At the time of the 1931 Congress, Hessen was already under suspicion in Russia, and was allegedly monitored by an OGPU agent (Ernst Kolman, a fellow delegate and Moscow-based Czech mathematician) throughout the visit (Graham, 1985, p. 713). Bukharin was accorded the honour of constant state surveillance from the British secret services, worried by the prospect of the spread of Bolshevik insurgency (Chilvers, 2005).[7] Graham contends Hessen's philosophically Deborinite views were the reason for the OGPU's interest, though his evidence is Kolman's attacks on Hessen's philosophy prior to the congress in the late 1920s and an interview with Kolman (Graham, 1985, pp. 713–15). Hessen's defence of the theories of relativity and of quantum mechanics from a Deborinite perspective (these physical theories would enhance the dialectical materialist world view) is therefore considered to be his *hamartia*. This is the dilemma of human action leading to tragic destiny, growing in magnitude as the tragedy unfolds, according to Aristotle. *Hamartia* is not necessarily a major event, and is often more poignant when its consequences far outweigh the original action (Aristotle, 1996, pp. 20–24 [1452b–1454b]). Given that Hessen was executed in 1936 and had figured prominently in the Deborinite–Mechanist debate as the interpreter of the 'new physics', the possibility exists that his philosophical views led to his later problems. However, further analysis of Hessen's background suggests alternative hypotheses, which help to explain his independent and distinctive position on the relationship between dialectics and science, his apparent *perepeteia,* or reversal, and the largely unrecognised sophistication of his paper on Newton.

Boris Mikhailovich Hessen was born in 1893 to a middle-class Jewish family in Elizavetgrad (known as Kirovograd during the Stalinist era) in the Ukraine. His father was a bank clerk, and the young Boris led a comfortable existence. His radical political awareness developed at his secondary school, the Gymnasium in Elizavetgrad where he was also noted as a talented mathematician. Because he was Jewish, he could not attend a university course in Russia, so he went to Edinburgh University in 1913–14 to study science and mathematics, joined by his friend I.E.

7 Bukharin Home Office file, National Archives, Kew, London, HO 45/14449, relates the surveillance.

Tamm, the future Nobel Prize-winning Russian physicist in 1958 with Cherenkov and Frank (for discovering 'the Cherenkov Effect'), who also attended the Elizavetgrad Gymnasium (as had the biologist B.M. Zavadovsky, also a childhood friend), and a whole community of Russian students unable to attend the Russian universities. Hessen's file at MGU notes that 'on current information, Tamm is a childhood friend of Gessen, they went to college together at Edinburgh University in Scotland' (Gorelik, 1995, p. 74). There were many other Russian students at Edinburgh and some at the University of Glasgow, such as Pyotr Kapitsa in the same year (Kedrov, 1984), most of whom were Jewish (some noted this next to their nationality on their matriculation forms). By 1917, Hessen had returned to Elizavetgrad (whether he did so before this monumental year is uncertain) and became secretary of 'an organization of internationalists. … a position that no doubt opened him to attack as a Trotskiite in the 1930s' (Josephson, 1991, p. 241). Hessen's ascension to the secretaryship of the Elizavetgrad Soviet in October 1917 suggests either an independent Marxist or a new Bolshevik joining during the period of 'dual power'. Josephson's reference also suggests the intriguing possibility that Hessen was involved with the small organisation around Trotsky that merged with the Bolsheviks in July 1917. Daniels' thesis claims the surge in Bolshevik membership and sympathy in 1917 was partly the result of the influx of many former Menshevik Internationalists, who opposed the war and threw their lot in with the Bolsheviks as the revolutionary crisis dragged on. Daniels further argues that this constituency became the political basis for the Left Opposition in the 1920s and 1930s (Daniels, 1960). Whether Hessen can be categorised thus is unknown, though he certainly moved in this milieu. Another possible *hamartia* has raised its head. Could he have actually been involved with Trotsky's grouping or Left Opposition politics?

While leading the revolution in Elizavetgrad, Hessen acquired a reputation for his revolutionary zeal. When the Bolsheviks nationalized the banking network, Hessen oversaw the seizure of his father's bank. Ever afterwards, he was known as the Elizavetgrad *Narkomfin*, or People's Commissar of Finance (Gorelik and Frenkel, 1990, p. 78). In the civil war, he fought in Trotsky's Red Army in the serious fighting across the Ukraine, and quickly became a political instructor for new troops and officers in Moscow in 1919. He continued lecturing (now in political economy rather than politics) at the Communist University until 1924, and enrolled as a student of natural science at the Institute of Red Professors (IKP), where he studied the 'structure of matter, the physics of heat and electricity, and theoretical physics' (Josephson, 1991, p. 241). In 1928, he became temporary director of the natural sciences division at IKP and, in 1931, Professor of Physics at Moscow State University (MGU) and a corresponding member of the Academy of Sciences in 1933, where he was assistant director of the Physical Institute (the director was S.I. Vavilov). Professionally, Gessen became increasingly renowned, publishing regularly in various journals such as *Science and Marxism*, *Under the Banner of Marxism* (PZM) and *Progress in Physical Science* (UFN). He was active in propaganda work at several factories, and lectured at the Institute of Philosophy, the Institute of History of Science and Technology and at MGU. At the time of his arrest, he was director of the Institute of Physics, Dean of the Physics faculty, both at MGU, and editor of UFN (Josephson, 1991, p. 241).

His association with an internationalist grouping suggests an independent route to revolutionary politics and a possible inclination towards 'Trotskyism'. If Hessen was inclined towards the Left Opposition, which developed in Russia in the 1920s and concentrated around Trotsky in Moscow, it raises the intriguing possibility of a connection between his philosophical outlook on relativity and quantum mechanics and his political outlook. It is interesting to note that Trotsky defended tolerance towards relativity and Freudian psychoanalysis in a speech in 1926 (Trotsky, [1926] 1973, pp. 227–49; Deutscher, 1959, pp. 178–80), shortly before Hessen entered the Deborinite–Mechanist debate in earnest. In the prevailing atmosphere of Moscow in 1936, 'the year of terror', such previous indiscretion would have been reason alone for the trip to Lubyanka.

The role played by Hessen in the deeply fractious and protracted controversy known as the Deborinite–Mechanist debate is also offered as the *hamartia* and *perepeteia* in this particular drama. Graham has not discussed Hessen's controversial philosophical arguments within the context of the Deborinite–Mechanist controversy at all, instead concentrating on the Stalinist attacks on Hessen, though he does suggest the source of Hessen's problems were his philosophical views (Graham, 1985, p. 707). Josephson contended that Hessen and the Deborinites were suppressed because of the 'revisionist' character of their view of dialectics, which emphasised the ever-changing nature of Marxist theory in an ever-changing world (Josephson, 1991, p. 246).

Josephson acknowledges that Hessen's politics could have created problems for him, though only in the form of malicious gossip. Generally, he takes a sympathetic view of the Deborinites, in stark contrast to Joravsky and Bakhurst, who both assail their 'formalism' and attempts to impose dialectics on science, while considering the Mechanists as defenders of science (Joravsky, 1961, pp. 182–95; Bakhurst, 1991, pp. 25–7). Interestingly, these authors merely regurgitate the criticism of Mitin and the other Stalinists who moved to suppress the Deborinites in 1930. Joravsky goes further, questioning Hessen's political and philosophical sincerity by alleging that neither his views on relativity nor quantum mechanics were original (derived partly from S.I. Semkovskii); he also alleges several incidents in which Gessen 'was both a harbinger and an ultimate casualty of the great break' by demanding the removal of editors of journals and publishing houses (Joravsky, 1961, p. 286). This is partially a reference to the 'ether incident' of late 1931 and early 1932, while Hessen was the editor of the physics section of the *Great Soviet Encyclopedia*. He wrote an article on 'Ether' for the 65th volume, defending its possible existence despite the predicted consequences of the general theory of relativity. His friends, Matvei Bronshtein and Lev Landau, along with several other young physicists at the Leningrad Physico-Technical Institute, sent Hessen mocking telegrams asking whether he would now write in defence of phlogiston. Whether Hessen initiated the response himself or not is unclear, but Bronshtein and Landau were promptly dismissed and sent for 're-education' 'for anti-social speeches against the article of comrade Gessen'. It is quite possible that party officials read the telegrams, a standard practice in this period (Gorelik and Frenkel, 1990, pp. 78–83).

Analysing Josephson, Bakhurst and Joravsky, it is noticeable that they all suggest an internal content-driven dynamic to the debate between the Deborinites, Mechanists and Stalinists. The risk is to account insufficiently for the extreme social context in which the reception of the 'new physics' occurred. In doing so, they are drawn into the terms of the debate based upon perceived content alone, and take sides accordingly. The desire to interpret Hessen's tragic fate from the perspective of his prominent position in the controversy is understandable and holds some merit. Before Hessen's intervention, the scientific community had largely supported the Mechanist argument that natural science, especially Newtonian mechanics, employed a sufficient form of logic from which to understand all material reality. The abstractness of Deborin's formulations on the need for a new 'social movement for dialectics' gave the impression that philosophers were extending their professional boundaries to become the final arbiters on policy relating to science. While philosophers and historians supported the Deborinite cause, few scientists concurred, viewing the intervention as a new form of pressure. The constituencies, themselves, are generalisations of a very complicated picture, as can be demonstrated by noting that both A.K. Timiryazev, who opposed the 'new physics' as bourgeois ideology, and S.I. Semkovskii, who wrote a book defending relativity, were supposedly Mechanists. The 'Gessen manoeuvre' of 1927–28 elaborated a dialectical defence of relativity and quantum mechanics, arguing that the scientific theories were compatible with a dialectical materialist world view. His sophisticated argument won over the sceptical Deborin, who previously resisted relativity, and the majority of the philosophers and scientists (Josephson, 1991, pp. 232-–6; Joravsky, 1961, pp. 170–95, 285–6). When, in April 1929, a conference of philosophers cemented the victory of the Deborinite faction, Hessen's rise was confirmed as one of the leaders of the Deborinites, along with the historian Nikolai Karev and others.

Within a year, the *perepeteia,* reversal, began with vicious articles from a new breed of philosopher appearing in publications such as *Pravda*. The new position questioned the right of philosophers to perceive themselves as interpreters of the Marxist philosophical position, arguing that this privilege belonged to the party, and to Stalin in particular. Stalin entered the arena himself shortly afterwards, describing the Deborinites as 'menshevizing idealists' (Josephson, 1991, pp. 247–8). Their suppression followed, along with the obligatory recantations and self-criticism (*samo-kritika*). Hessen gave way in early 1931, and was thus able to attend the Second International Congress.

Could Hessen's role as arch-protagonist in a philosophical debate have brought his fall? The intellectual content of the controversy and the defence of the new science offer only part of the process at work. The accusation that the Deborinites were the philosophical wing of the Left Opposition and 'Trotskyists' or 'wreckers' began to haunt the discussions in 1930. N.A. Karev, the historian and one of the Deborinite leaders, was forced to implicate Hessen in plotting against the regime when arrested in May 1936 (Gorelik, 1995, p. 74). Deeply political factors could not be divorced from the seemingly most abstruse discussions, and may well have contributed to Hessen's *hamartia*.

The Alliance with Bukharin

Hessen's wife, Yakovleva, was a close and childhood friend of Bukharin throughout this period, and had supported the Left Opposition when it congealed in 1923, but recanted her opposition when the resistance collapsed (Gorelik, 1995, p. 74). Consequently, she was considered both useful and potentially dangerous when the terror began in 1936. After her husband's arrest, Yakovleva was embroiled in the carefully constructed show trial of Bukharin, eventually becoming one of the major witnesses. Unlike Maria Spiridonova, who refused to give evidence, Yakovleva produced a horrifically virtuoso performance against her friend. Despite her pleasing testimony, she was shot, alongside Maria Spiridonova and 153 others, in Orlov Prison on 11 September 1941 (Conquest, 1968, pp. 537–40).

Hessen had a past of his own, as all the Old Bolsheviks did, and led a tenuous and blemished existence in the eyes of the increasingly paranoid rulers of Russia. Recognising his association with one of the more fearsome and oppositional figures in the history of the Russian Revolution tightens the threads around him quite substantially. It is well recorded that Stalin had a habit of leaving intact the women that he wanted to destroy, but scything through their loved ones instead. In Yakovleva's case, she was to prove useful in the crescendo of the terror. It is possible that the reasoning behind Hessen's arrest and disappearance was to apply pressure upon a crucial figure close to Bukharin. Hessen's own past and association with Bukharin, with whom he formed an alliance in defence of science after the 1931 London Congress, could only have helped to seal his fate. Thus, we face the dreadful possibility that Hessen's *hamartia* may actually have been his marriage, which acquired political significance along with his alliance with Bukharin, as much as the other threads of his life and ideas.

The Arrest of Comrade Hessen

Hessen's apparent eminence in Soviet academia continued up until his disappearance, and British Communist historians continued to lionise his paper long after his death. The unfortunate fact that he had been 'purged' along with millions of others was lost to the world for many years. However, Hessen was locked into an increasingly hopeless situation after December 1934, when Sergei Kirov's murder prompted the beginning of the Great Terror. Between 1932 and 1934, moderates rose to challenge and restrain Stalin's civil war in the cities and countryside. Kirov, Ordzhonikidze and others sought to roll back the police state and moderate the Five Year Plan. In this period, which Cohen (1974, p. 345) calls the 'Moscow Spring', Bukharin was partially welcomed back into the party leadership and Hessen's position became more secure, hence his election as corresponding member of the Academy of Sciences in 1933. He utilised this atmosphere to strengthen his particular niche acting as the political umbrella of Russian physics. While nominally the political and philosophical watchdog, Hessen inverted the role, taking up the political issues with the apparatus and philosophers and allowing the physicists to work unchecked. Ioffe was profoundly concerned to retain Hessen's services and support, looking dimly on any attempt to upset the balance (Gorelik and Frenkel, 1990, pp. 80–81). After 1934, the balance was destroyed, with the more Stalinist elements around Kolman, Lysenko, Prezent and Mitin gaining ground rapidly.

Hessen was arrested on 21 August 1936, two days after the beginning of the August 1936 trial of Zinoviev, Kamenev and other Old Bolsheviks accused, for the first time, of preparing a joint Trotskyist-Zinovievist centre for the purpose of terrorism, including the assassination of S.M. Kirov in December 1934. The preceding day, M. Tomsky committed suicide after allegations began to encircle him at the trial (Conquest, 1968, pp. 141–66). Hessen's charge was that he 'until very recently, personally supported and associated with those arrested for constituting a Trotskyist-Zinovievist terrorist center with N. A. Karev, N. Lurie and known counter-revolutionary Trotskiist workers' (Gorelik, 1995, p. 74). Karev, who had been deputy President of the Planning Committee at the Academy of Sciences, was arrested on 16 May 1936, though his trial with Zinoviev and Kamenev was set aside, for unknown reasons. He was secretly shot on 11 October 1936 (Gorelik, 1995, p. 74).

Karev's testimony of 5 June 1936 was the only evidence presented against Hessen, and it alleged that 'at the center of Zinovievist organization, from time to time, besides Bakayev and Yevdokimov, entered Kamenev, Zinoviev and Hessen. … Hessen knew organizational workers amongst the youth.' Given that the Zinovievist opposition was only active in Leningrad, the repression was more likely to be directed at the remnants of the Left Opposition in Moscow. Implicated with Hessen were Arkadii Osipovich Apirin (arrested 9 June 1936) and Arkadii Mikhailovich Reizen (arrested 29 October 1936), who were jointly accused of forming 'a terrorist group for the sake of organizing terrorist action against the leadership of the party and Comrade Stalin'. The interrogation records include 15 statements from Hessen and two 'confrontations' dating from 21 August to 18 November:

> On 20th December 1936 the case was closed at a meeting of the Supreme Military Collegium Court under the chairmanship of V. V. Ul'rikh. The verdict of the 'court' stated:- 'Gessen and Apirin are members of a counter-revolutionary Trotskiist-Zino'vievist terrorist organization to bring about the villainous assassination of Comrade S. M. Kirov and prepared from 1934–36, with help from fascist agents of the Gestapo, a number of terrorist actions against the leadership of BKP(b) and Soviet Government'. B. M. Gessen pleaded not guilty. A. O. Apirin did not plead. Both were shot on the same day – 20th December 1936. A. M. Reizen was condemned to ten years hard labour and perished in detention. (Gorelik, 1995, p. 74)

The discussion on the structure of this tragedy has concentrated upon the *hamartia* and the *perepeteia* of Hessen's life and demise. Here, I suggest that we have located a *katharsis*, as Hessen both refused to admit any guilt, despite obvious provocation, and refused to implicate others in order to save his life. Had he followed such a course, it is quite possible that he would have become one of the defendants in the Pyatakov or Bukharin show trials. However, those refusing to implicate themselves or others were generally shot in secret and their families led to believe they were in the *gulag*. If one views his position on science as contributory to his predicament, the *katharsis* could be viewed as the legacy his 1931 paper left mainly in the international arena.

He was certainly entirely innocent of the charge of terrorist activity, as were all those who had found themselves implicated with such preparations. It is probable Hessen was innocent of the charge of being involved with Trotskyists. Many who had opposed the regime at any point for a variety of reasons were labelled Trotskyists by a state seeking to criminalise all opposition. Most of the former Left Opposition had recanted their views and were thoroughly demoralised by 1936. However, the argument that Hessen's arrest was arbitrary or that it was due to some other misdemeanour is rendered questionable by three facets of the situation in August 1936.

Firstly, Hessen was in trouble with the state security apparatus from 1930, when he was publicly admonished for his views on science. His problems continued despite self-criticism on his part through to 1936. This could have been the product of professional jealousy or his views on science, though this is unlikely given the orchestrated and enduring character of the criticism. Secondly, the number of executions in 1936 for 'counter-revolutionary' activities was far lower than has previously been suspected (1,118). Most of those arrested at this time were sent to prison camps (219,418) or exiled (23,719). The number of executions that occurred in 1936 was also lower than any year since 1928, suggesting a highly targeted punishment, rather than one that was arbitrarily administered. The same cannot be said of the following year, when 353,074 were shot, with 429,311 being sent to the *gulag* (Getty and Naumov, 1999, p. 588). Thirdly, the timing of Hessen's arrest in late August coincides with the break-up of the allegedly Trotskyist network apparently taking shape in Moscow. These reasons for questioning the view of an arbitrary arrest are circumstantial and cannot be deemed historically rigorous, but they do suggest he was considered a significant figure as the arrests were carefully targeted at this moment.

Conclusion

If 'scientific biography' has the remit to study a scientific life, the life in question often exhibits complexities and dynamics beyond the immediate scientific or intellectual context. In Hessen's case, these complexities are extremely significant, in proportion to a highly complex character and social environment. If a society imposes an extremely repressive political environment on science, the complexities of the political environment are going to form an important part of the biography. The structure of the 'scientific biography' requires adaptability and a subtlety that can often be overlooked, in order to adequately convey the contexts of the modern scientific life.

Tragedy as a historical discourse offers some methodological markers for the potential biographer to grapple with the role of an individual in social history. The potentially tragic quality of human action can often be overlooked, especially when the subject is considered an exemplar of human genius. Understanding an exemplar of the more 'negative' aspects of the relationship between science and society offers an alternative possibility for the conduct of biography in the history of science, linking it to the historically epistemic project that many find themselves embarking upon in the epoch of the 'big science' of biotechnology, nuclear weapons and information superhighways. In portraying the case of Boris Hessen as historically tragic, three contentious assumptions have been made:

1. Tragedy is a historic discourse that fulfils a socially epistemic function at particular junctures in the historical process.
2. The Russian Revolution and Soviet society in the 1920s marked a historic juncture for human civilisation.
3. The significance of science to humanity is such that the consequences of mistakes in science are historically magnified.

The significance of science is the very founding block of the history of science as an intellectual movement and the foundation of tragedy as a relevant genre in the history of science. The social environment has been critical to their development, and is critical to the success of the 'scientific biography'. One of Boris Hessen's enduring achievements was his spirited attempt to analyse the relationship between science and society. Whether we wish it or not, this task now stands before us all, and tragedy as a genre seems an unfortunate but cognitively important integral component.

Bibliography

Aristotle (1996), *Poetics*, trans. M. Heath, London: Penguin.

Bakhurst, D. (1991), *Consciousness and Revolution in Soviet Philosophy: From the Bolsheviks to Evald Ilyenkov*, Cambridge: Cambridge University Press.

Chilvers, C.A.J. (2003), 'The dilemmas of seditious men: The Crowther–Hessen correspondence in the 1930s', *British Journal for the History of Science*, **36** (4), 417–35.

Chilvers, C.A.J. (2004), 'Debacle and diffusion: The myth of the Special Session at the 1931 congress and the left-wing scientists', unpublished paper presented at the *REHSEIS Biologists Engaged Colloquium*, Paris, 4 June 2004.

Chilvers, C.A.J. (2005), 'The Second International Congress of the History of Science and Techology (London, 1931): Themes and Contexts', unpublished DPhil. thesis, Linacre College, Oxford.

Chilvers, C.A.J. (2006), 'La signification historique de Boris Hessen', in S. Guerout (ed.), *Les Racines sociales et économiques de* Principia *de Newton*, Paris: Vuibert.

Chilvers, C.A.J., P.R. Josephson and W. Schäfer (eds) (forthcoming), *Boris Hessen Revisited* (provisional title), Boston, MA: MIT Press.

Christie, J.R.R. (1990), 'The development of the historiography of science', in R.C. Olby, G.N. Cantor, J.R.R. Christie and M.J.S. Hodge (eds), *Companion to the History of Modern Science*, London: Routledge.

Clark, G.N. (1932), 'Opening paper by G. N. Clark, Oriel College, Oxford', *Archeion: Archivio storia della scienza*, **13**, 272–3.

Cohen, S.F. (1974), *Bukharin and the Bolshevik Revolution 1888–1938*, London: Wildwood House.

Conquest, R. (1968), *The Great Terror: Stalin's Purges of the Thirties*, Harmondsworth: Penguin.

Crowther, J.G. (1941), *The Social Relations of Science*, London: Macmillan.

Daniels, R.V. (1960), *The Conscience of the Revolution: Communist Opposition in Soviet Russia*, Cambridge, MA: Harvard University Press.

Deutscher, I. (1959), *The Prophet Unarmed: Trotsky 1921–1929*, Oxford: Oxford University Press, pp. 178–80.

Edgerton, D. (2005), *The Warfare State*, Cambridge: Cambridge University Press.

Enebakk, V. (2004), 'Mellom de to kulturer: oppkomsten av vitenskapsstudier og etableringen av Edinburgh-skolen 1966–1976', unpublished PhD thesis, University of Oslo.

Evans, R.J. (1991), *In Defence of History*, London: Granta.

Floris Cohen, H. (1994), *The Scientific Revolution: A Historiographical Inquiry*, Chicago, IL: University of Chicago Press.

Getty, J.A. and O.V. Naumov (1999), *The Road to Terror: Stalin and the Self-destruction of the Bolsheviks 1932–1939*, trans. B. Sher, New Haven, CT: Yale University Press.

Gorelik, G.Y. (1995), 'Moskva fizika 1937 god', in V.A. Kumanev (ed.), *Tragicheskie sud'by: Repressirovannye uchenye akademii nauk SSSR*, Moscow: Akademii Nauk, pp. 54–75.

Gorelik, G.Y. and V.Y. Frenkel (1990), *Matvei Petrovich Bronshtein 1906–1938*, Moscow: Akademii Nauk.

Graham, L.R. (1985), 'The socio-political roots of Boris Hessen: Soviet Marxism and the history of science', *Social Studies of Science*, **15**, 705–22.

Hall, A.R. (1954), *The Scientific Revolution 1500–1800: The Formation of the Modern Scientific Attitude*, London: Longmans.

Hessen, B. (1931), 'The Social and economic Roots of Newton's Principia', in N.I. Bukharin et al. (eds), *Science at the Cross Roads*, London: Kniga, pp. 146–212.

Joravsky, D. (1961), *Soviet Marxism and Natural Science 1917–1931*, New York: Columbia University Press.

Josephson, P.R. (1991), *Physics and Politics in Revolutionary Russia*, Berkeley, CA: University of California Press.

Kedrov, F.B. (1984), *Kapitza: Life and Discoveries*, trans. M. Fradkin, Moscow: Akademii Nauk.

Mayer, A.-K. (2004), 'Setting up a discipline, II: British history of science and "the end of ideology", 1931–1948', *Studies in History and Philosophy of Science*, **35**, 41–72.

McGucken, W. (1984), *Scientists, Society and State: The Social Relations of Science Movement in Great Britain 1931–1947*, Columbus, OH: Ohio State University Press.

Merton, R.K. ([1938] 1970), *Science, Technology and Society in Seventeenth Century England*, New York: Harper and Row.

Pachter, M. (ed.) (1979), *Telling Lives: The Biographer's Art*, Washington, DC: New Republic.

Patterson, A. (1989), *Shakespeare and the Popular Voice*, Oxford: Basil Blackwell.

Reiss, T.J. (1980), *Tragedy and Truth: Studies in the Development of a Renaissance and Neoclassical Discourse*, New Haven, CT: Yale University Press.

Sarton, G. (1924–25), 'The new humanism', *Isis*, **6**, 9–42.

Shortland, M. and R. Yeo (eds) (1996), *Telling Lives in Science: Essays on Scientific Biography*, Cambridge: Cambridge University Press.

Singer, C. (1959), *A Short History of Scientific Ideas to 1900*, Oxford: Clarendon Press.

Snow, C.P. ([1959] 1993), *The Two Cultures and the Scientific Revolution*, Cambridge: Canto.

Söderqvist, T. (2003), *Science as Autobiography: The Troubled Life of Niels Jerne*, New Haven, CT: Yale University Press.

Soyfer, V.N. (1994), *Lysenko and the Tragedy of Soviet Science*, New Brunswick, NJ: Rutgers University Press.

Swingewood, A (1975), *The Novel and Revolution*, London: Macmillan.

Trotsky, L ([1925] 1991), *Literature and Revolution*, trans. R. Strunsky, London: Redwords.

Trotsky, L. ([1926] 1973), 'Culture and socialism', in *Problems of Everyday Life and Other Writings on Culture and Science*, New York: Monad Press, pp. 227–49.

Vavilov, N. (1931), 'The origin of the world's cultivated plants', in N.I. Bukharin et al., *Science at the Cross Roads*, London: Kniga.

Williams, R. ([1966] 1979), *Modern Tragedy*, London: Verso.

Wood, N. (1959), *Communism and British Intellectuals*, London: Victor Gollancz.

Young, R.M. (1990), 'Marxism and the history of science', in R.C. Olby, G.N. Cantor, J.R.R. Christie and M.J.S. Hodge (eds), *Companion to the History of Modern Science*, London: Routledge, pp. 77–86.

Chapter 7

Received Wisdom in Biography: Tycho Biographies from Gassendi to Christianson

Helge Kragh

Tycho Brahe is in no need of introduction. Described in the *Dictionary of Scientific Biography* with a few more lines than Niels Bohr, he may be the best-known Danish scientist ever. Tycho received international fame in his own lifetime, in part for his edition of his own correspondence, *Epistolae astronomicae*, printed at Uraniborg in 1598. He may even have been known to Shakespeare, in whose *Hamlet* the names of the figures Rosencrantz and Guildenstern are undoubtedly derived from two of Tycho's ancestors, Sophie Gyldenstierne and Erik Rosenkrantz. Indeed, it has been argued that *Hamlet* is an allegory for the competition between Tycho's cosmology and the infinite, Copernican world system of Thomas Digges (Usher, 1997; Usher, 1999). Whatever this speculation, already half a century after his death, Tycho Brahe became the subject of a major biography which further established his position as a pioneer and reformer of astronomy.

Biographies of great scientists often reflect the values of the author and the time he or she lived in. This elementary fact, valid for any kind of literary work, adds a time-specific element to biography and partly explains why biographies of a particular person, while based on the same sources, may none the less differ considerably. However, the variation that follows from different perspectives and periods is often counteracted by factors that tend to stabilise biographical accounts. Naturally, one such factor is the common foundation in a corpus of source material on which any serious biography must be based. I would like to suggest that another stabilising factor is of an inertial nature, namely that biographies often build on earlier exemplars. This may be a first work which, more or less consciously, comes to serve as a model (or in some cases, a counter-model) for later biographical portraits.

These general features are well illustrated in the bio-historiographical tradition concerning Tycho Brahe, one of the founding fathers of the new scientific enterprise that emerged in late Renaissance and early modern Europe. The first Tycho biography was written by the French natural philosopher Pierre Gassendi as early as 1654. The heritage of Gassendi's remarkable work can be followed in most later biographies, which not only took over Gassendi's plot but also much of his interpretation of Tycho's personality and activities. To trace this line of development is the primary aim of this chapter. It does not pretend to cover all or just most Tycho biographies, but focuses on a few select works of particular significance, most of them full-scale and scholarly biographies. For reasons of brevity, I do not consider the numerous works that are either of a popular nature or which focus on specific aspects of Tycho's life and work.

Gassendi's Tycho

Gassendi's *Tychonis Brahei ... vita* (1654) has a unique position in the history of scientific biography. Not only is it arguably the oldest full biography of a scientist – long before natural philosophers became scientists – it is also one of the most influential works within the genre.[1] Indeed, it cast its shadows for some three centuries. It was not Gassendi's first learned biography, though. In 1641 he had published a biography of the French bibliophile, naturalist and patron of science Nicolas-Claude Fabri de Peiresc, who was his personal friend and with whom he had carried out astronomical observations with the new telescope. *De vita N. Fabri Peirescii*, which was translated into English in 1657, described the life and career of a collector and patron who, in spite of working extensively with problems of science, never published his scientific work. For this reason he may not be counted as a 'scientist' and his biography may presumably be classified as dealing with a 'scholar' rather than a 'scientist'.

In November 1647, the Danish polyhistor and natural philosopher Ole Worm, Professor of Medicine at the University of Copenhagen, received a letter from the learned Frenchman Isaac Lapeyrère, an associate of Gassendi. 'Our friend, M. Gassendi', Lapeyrère wrote, was:

> an excellent man who values you highly and whose reputation I believe you are well aware of, has got the idea and wish to write the life of your famous Tycho Brahe. ... This will bring honour to your country, and I know for sure that you love it dearly. Add to this that this laudable man will mention you with honour and praise in this fine work. (Worm, 1968, p. 279)[2]

Lapeyrère requested Worm to assist the laudable man with his project and to send Gassendi, through Lapeyrère, as much information and as many sources he could lay hold on. Worm responded enthusiastically, adding that Tycho's long-time associate Longomontanus (Christian Sørensen) had unfortunately died only a few days ago. Instead, he promised to address Longomontanus' successor as Professor of Astronomy, Jørgen From, and ask him for help (Worm, 1968, pp. 284, 336). Apart from From, Worm also mobilised Stephan (or Steffen) Stephanius, royal historiographer and Professor of History in Sorø Academy. Stephanius' father, Johannes Stephanius, had been a *famulus* of Tycho at Hven in 1582–84, and his son had inherited his library

1 According to Christianson (2000, p. 245), the book was 'the first full-length biography of any scientist'. Of course, the claim depends on one's definition of 'scientist' and 'biography'; see further Liba Taub's contribution in Chapter 1 of this volume.

2 Norlind (in Gassendi, 1951, p. xiv) states twice that Worm was asked by Peiresc to supply material for Gassendi's fortcoming biography. This is obviously a slip of the pen, and should be Lapeyrère. Peiresc died in 1637. If he had correspondence with Worm, it is not included in Worm (1968).

as well as his loyalty to the legacy of the great astronomer. The younger Stephanius supplied important material from his father's notes and manuscripts about Tycho's life (Worm, 1968, pp. 355, 389). In 1648, Worm entered a correspondence with Gassendi, first in a long letter in which he gave detailed information about Tycho, including some charming anecdotes. In a later letter he provided information of Tycho's activity as a physician in the Paracelsian tradition and promised to send a transcript of Johannes Jessenius' funeral oration of 1601, *De vita et morte D. Tychonis Brahe* (Worm, 1968, pp. 317, 493). The oration included poems by Kepler and Paul Jensen Colding, both of whom had been present at Tycho's death. The material eventually became part of Gassendi's biography, where it was added in the form of an appendix.

Apart from the material he received from Denmark, Gassendi's biography mainly relied on Tycho's publications. Of particular value, from a biographical point of view, was the autobiographical sketch which Tycho had included in his *Astronomiae instauratae mechanica* of 1598. Tycho's account was, however, largely limited to an extensive review of his scientific work, and to astronomy in particular, whereas he gave little attention to personal matters. Several of the features of the autobiography can be found also in Gassendi's work. Thus, Tycho dealt in detail with his painstaking astronomical observations, whereas he referred only briefly to his world system, and then mainly to defend his priority against 'those who have not been ashamed to appropriate this hypothesis and present it as their own invention' (Brahe, 1946, p. 115).[3] He also dealt rather briefly with his work in astrology, yet without feeling any need to apologise that he had spent so much time on meteorological and judicial astrology. As to chemistry and alchemy – or 'chymistry', to use an appropriate term – Tycho clearly felt that his life-long work in this area was of great importance. However, he refrained from discussing his contributions to 'terrestrial astronomy' because of their secret nature, for 'it serves no useful purpose, and is unreasonable, to make such things generally known' (Brahe, 1946, p. 118).

Gassendi's detailed and carefully researched work was dedicated to Henri Louis de Montmort, the French nobleman and patron of science whose residence for a period served as a meeting place for Parisian natural philosophers. After some delay, the book appeared in 1654 – not long after the centenary of Tycho's birth – in part because Montmor encouraged Gassendi to complete it. *Tychonis Brahei vita* appeared together with much briefer biographies of Copernicus, Peurbach and Regiomontanus added on the request of the bookseller. Meanwhile, a brief account of Tycho's life had been published in *Almagestum novum* of 1651, the important work of the Jesuit astronomer Giambattista Riccioli, who was an advocate of the Tychonian world system, although in a slightly different version than the one Tycho had originally suggested. Riccioli admired Tycho as an astronomer, but his admiration was tempered by Tycho's Lutheran belief which made Riccioli assert that the Danish astronomer could not have achieved eternal felicity.

Gassendi organised his biography in six 'books' of which the first five were composed in a straightforward, chronological manner. Only in the final book did he

3 See also the facsimile reprint of *Astronomiae instauratae mechanica*, together with English and Czech translations (Brahe, 1996).

leave the chronological framework and offer a more general appreciation of Tycho's life and work, including a characterisation of his personality. Yet the major part of the book was a neutral, fact-oriented narrative in which the author simply listed, year by year, what happened in Tycho's life. Gassendi evidently found Tycho's scientific work, and his accomplishments in astronomy in particular, to be of greater interest than other aspects of his life and career. Even though an anachronism, one may characterise *Tychonis Brahei vita* as an excellent example of scientific biography in the style of positivist historiography.

In order to appreciate Gassendi's picture of Tycho, it is important to bear in mind the author's background and position as a key figure in the Scientific Revolution (Jones, 1981). Perhaps best known for his support of Epicurean atomism and the existence of void, Gassendi was among the leading natural philosophers within the Galilean empirical tradition who fought against the neo-Aristotelian tradition, and at the same time criticised what they considered the excessive rationalism of the new Cartesian science. A few years before he completed his biography of Tycho Brahe, he had published one of his main works, *Animadversiones in decimum librum Diogenes Laërtii*, a strong defence of experimental science and a long argument against Descartes' natural philosophy. Although not trained as an astronomer, he did important astronomical work and had a full grasp of the science of the heavens; he rejected astrology, and found nativity horoscopes to be absurd. Gassendi was known in particular for the first telescopic observation of a Mercury transit, which he described in *Mercurius in sole visus et Venus invisa* of 1632. Privately, he was in favour of Copernicanism, but as a Jesuit priest in Catholic France, he found it unwise to make his sympathy publicly known. He therefore supported the Tychonian world system as a compromise approved by the Church, if only as a substitute for the heliocentric system he believed in.

One might believe that Gassendi's biography was a piece of hagiography, but this is only the case to a very limited extent. He wanted primarily to describe Tycho's life and scientific work, not to praise it or to evaluate it morally. Contrary to Riccioli, a fellow Jesuit, he did not bemoan that Tycho was a Lutheran rather than a Catholic. Although he focused very much on Tycho as a pioneer astronomer, he did not eschew or systematically explain away other parts of his interests and works. Tycho, he wrote, had three primary interests, namely astronomy, alchemy and Latin poetry. As to his extensive occupation with chemistry in the spagyrical or Hermetic tradition, Gassendi described him as a true follower of Paracelsus, a subject he dealt with more objectively and in more detail than most later biographers (Gassendi, 1951, pp. 173–5).[4]

Gassendi firmly rejected astrology, yet he knew that Tycho took the subject very seriously, if far from uncritically. Tycho's attitude to astrology is a complex question, but what matters here is that he remained faithful to what he called the 'higher astrology' and the intimate correspondence between macrocosmos and

4 Unlike some of his friends, Tycho did not engage in *chrysopeia*, the art of gold-making. On Tycho as a Paracelsian chemist, see Figala (1972), Hannaway (1986) and Shackelford (1993). The links between heavenly and terrestrial astronomy are further examined in Shackelford (2002).

microcosmos. Gassendi dealt in considerable detail with Tycho's astrology, admitting that the great astronomer 'did not completely abandon this belief until the end of his life'. Eventually, Tycho did see the light, and in the end he turned into a resolute enemy of 'the frauds and vanities of astrology' (Gassendi, 1951, p. 170). It was important for Gassendi to present a picture of Tycho which did not openly conflict with the scientific values of the mid-seventeenth century. He therefore rationalised Tycho's advocacy of astrology as a means to attract interest to the true science of the heavens, observational astronomy. First and foremost a representative of the new empiricism, Gassendi found that Tycho's scientific life, in an appropriately modified version, might serve as powerful propaganda for the cause of empiricist science. It is no surprise, then, that his picture of Tycho was primarily that of an empirically minded scientist, a masterful instrument-builder and an untiring collector of accurate astronomical data. 'Even though Tycho's published works are worth their weight in gold, his observations are even more precious,' he wrote. And, with a hidden reference to Kepler: 'One must praise Tycho to the highest because he did not deny us the results of his observations, and he is even more worthy of still more fame because he made and wrote down so many excellent observations from which others could draw the same or even more important conclusions' (Gassendi, 1951, p. 192).

Gassendi's Tycho was a genius of observational astronomy, and it was this aspect which filled the pages of the biography. Although Gassendi was careful to mention also Tycho's theoretical contributions, he gave them low priority. Likewise, the geocentric Tychonic world system received only scant attention and was described in uncommitted language. After all, Gassendi was convinced that it was a wrong inference based on otherwise admirably accurate observations. He gave a fair account of Tycho's reasons for his non-Copernican alternative, but chose not to mention it in the summary of Tycho's life and work in Book 6. By over-emphasising observations and giving theory short shrift, Gassendi created what John Christianson has called an enduring myth of Tycho's island (Christianson, 2000, p. 244).

A Bio-historiographical Tradition

Gassendi's biography became a success. A second edition was printed in The Hague in 1655, and reprintings appeared in 1658 and 1717, as part of Gassendi's own *Opera omnia*. It took nearly three hundred years before it appeared in a language other than Latin, and then, of all languages, in Swedish (Gassendi, 1951). The translator, Wilhelm Norlind, Head Librarian at the University of Lund, Sweden, and a life-long student of Tycho Brahe, added very detailed commentaries which make up almost a separate book, a kind of preparation for his later biography. Readers of the book should notice that the translation is not complete, as it includes neither Gassendi's extensive preface nor the appendices (also it leaves out the biographies of Copernicus, Peurbach and Regiomontanus). Only recently has a full translation in French appeared, translated from the 1658 reprinting by Jean Peyroux (Gassendi, 1996).

Because of its wealth of data and the author's position as an eminent man of science, *Tychonis Brahei vita* came to function as the work on which later biographies were naturally based. In 1745, Oluf Bang, a Copenhagen schoolmaster and genealogist, published a book-length biography based on Gassendi, and the following year, on the occasion of the bicentenary of Tycho Brahe's birth, the historian and archivist Jacob Langebeck published a collection of articles and sources, several of which were not known to Gassendi (Langebeck, 1746). The two works received a broader audience when they, a decade later, were translated into German and published as two volumes by the Copenhagen bookseller Christian Gotlob Mengel (von der Weistritz, 1756).[5] Together with Gassendi's original work, Mengel's book – itself strongly relying on Gassendi – remained the chief biographical account for nearly a century and a half. For example, they formed the background for Jean-Baptiste Delambre's detailed 110-page chapter on Tycho in his influential *Histoire de l'astronomie moderne* of 1821.

The next full Tycho biography appeared in 1871, written by the Danish historian Frederik Friis and published in Danish only.[6] Friis was on his way to become the country's leading scholar of late Renaissance culture, and would later edit new collections of Tychonic sources and also publish biographical accounts of Sophie Brahe and some of Tycho's assistants, including Peder Jakobsen Flemløse and Elias Olsen Morsing.[7] However, his biography of Tycho scarcely superseded that of Gassendi, on which it relied to a considerable extent. The structure and style was largely the same, a somewhat dry factual narrative, but unlike the seventeenth-century Frenchman, Friis lacked scientific competence and therefore passed lightly over Tycho's contributions to astronomy. On the other hand, *Tyge Brahe* was generally more contextual, and it gave relatively high priority to Tycho's travels, family, economic conditions, political involvement and his relationship to King Frederik II and his son, Christian IV. Friis' description of Tycho's network and *familia* on Hven was only superseded by Christianson's book (2000).

In some respects, as when it came to alchemy and Paracelsianism, Friis' work had a decidedly whiggish flavour. As for alchemy, Friis suggested that young Tycho's interest in the subject was merely an attempt to make money to pursue his astronomical studies; as for Paracelsianism, Friis simply chose to ignore it. The Tycho we meet in *Tyge Brahe* is a Renaissance nobleman as well as a scientist who, following Gassendi's plot, was obsessed with accurate observations and with neither interest in nor capacity for theory. In fact (and contrary to Gassendi), Friis did not mention Tycho's lunar theory or, for that matter, other of his contributions to theoretical astronomy. Also, Friis dealt only rather superficially with Tycho's cosmological model. It was an error, of course, but Friis argued that the attempt was

5 Philander von der Weistritz was Mengel's pseudonym. From Dreyer's biography, one is led to assume that Weistritz and Mengel were two different persons. For information about the works of Bang, Langebeck and Mengel, see Ehrencrone-Müller (1924–35).

6 Johann T.B. Helfrecht, a German schoolteacher and organ player, published in 1798 *Tycho Brahe geschildert nach seinem Leben, Meynungen und Schriften*. I have been unable to procure a copy of this work.

7 See the biographical dictionary in Christianson (2000), and also Christianson (1998).

justified in so far that the Copernican system was not yet empirically superior to the Ptolemaic model:

> Whereas Copernicus was a philosophical investigator who mostly lived in his study chamber, Tycho Brahe was a practical astronomer for whom it was important to find a system which did not merely agree with experience, but also could be applied to the most important task of determining at any time the positions of the planets. For this purpose the Copernican system was not yet sufficiently developed, ... and it is not without reason that Tycho several times claimed that his own system fitted the observations much better than the Copernican system. (Friis, 1871, pp. 124–5)

Friis' work became the starting point of the first comprehensive and scholarly Tycho biography in a non-Scandinavian language since Gassendi, published in 1890 by the Danish-Irish astronomer Johann Dreyer.[8] His book quickly achieved an authoritative status, as indicated by its several reprintings, including a best-selling Dover edition of 1963 and an even later reprint of 1977. About 1910, Dreyer would embark on the herculean work of editing and publishing Tycho Brahe's *Opera omnia*, which appeared in Latin in 15 volumes between 1913 and 1929. However, in 1890 he did not yet have a full grasp of the sources, and as a result his biography bore the imprint of Gassendi's old work to no less a degree than that of Friis. We find in Dreyer's book many of the same elements as in earlier biographies, including an emphasis on instruments and observations, and a corresponding neglect of Tycho as a theoretical astronomer. Likewise, and now no longer surprisingly, Dreyer had little to say about astrology and alchemy, and nothing about Tycho's work within the Paracelsian tradition of a holistic conception of the earth and the heavens.

Tycho's legacy, according to Dreyer and the tradition on which he built, was the transformation of astronomy into a high-precision observational science in which data were all-important. Tycho was no theorist, and yet he unknowingly and crucially contributed to the new world picture that emerged in the seventeenth century: his accurate data served as the empirical foundation for Kepler's laws, and these were again the results from which Newton constructed his theory of gravitation, the law that revolutionised astronomy and symbolised the maturity of the scientific world view. Or, in Dreyer's words: 'His works became the foundation on which Kepler and Newton built their glorious edifice, and the star of Cassiopeia [of 1572] started astronomical science on the brilliant career it has pursued ever since, and swept away the mist that obscured the true system of the world' (Dreyer 1963, pp. 196–7).

Much the same message can be found in John Gade's biography, published for the American-Scandinavian Foundation in commemoration of the 400th anniversary of Tycho's birth. 'Knowing little of astronomy', as he admitted, Gade dealt in his readable biography only very superficially with Tycho's scientific work. Yet his

---

8 Johannes L.E. Dreyer (1852–1926) left for Ireland in 1874 after having graduated in mathematics and astronomy from the University of Copenhagen. He worked for most of his active life as Director of the Armagh Observatory, where he did important work on catalogues of stars and nebulae. On his life and career, see Knobel (1927) and Gingerich (1982). Apart from his scholarly work on Tycho Brahe, he also published a small, popular biography, which appeared in Danish only (Dreyer, 1901).

evaluation of Tycho's legacy did not differ from that of the astronomer Dreyer. According to Gade, Tycho's greatest achievement was that he formed 'a link in the great chain: Copernicus, Brahe, Kepler, Galileo, Newton'. Moreover, 'Tycho's findings … formed the basis of Kepler's discoveries and … Kepler supplied the groundwork for Newton's discoveries, the starting point of modern astronomy' (Gade, 1946, p. 189).

After many years of study, Norlind published in 1970 an impressively detailed and scholarly biography of Tycho, a work which corrected some earlier errors and made use of primary sources previously unknown to scholarship. Unfortunately, it was never translated from Swedish and therefore remained inaccessible to most historians. Norlind organised his book much in a librarian's style, with a focus on the works rather than the man. The result was erudite tabulation rather than narration, a compilation of materials for further research rather than a biography integrating life and science into a contextual whole. Norlind was primarily concerned with Tycho's works, not his astronomy nor his place in the history of science. Although he went considerably beyond Dreyer, added many details and filled out many lacunae, from a historiographic point of view he remained within the Gassendi tradition, only was his biography more complete and sophisticated.[9]

Norlind's general acceptance of Gassendi's plot did not imply that he uncritically followed either Gassendi or Dreyer. For example, earlier biographers standardly reported that while a student at the University of Copenhagen, 13-year-old Tycho observed a partial eclipse on 21 August 1560, and that it was this observation which turned him towards astronomy. A partial eclipse was indeed visible from Copenhagen on this date between 11.49 a.m. and 1.46 p.m. Yet, as Norlind pointed out, Tycho nowhere himself mentioned the 1560 eclipse. Norlind argued that the account given by Gassendi and taken over by later authors was 'an etiological myth' invented by Gassendi with the purpose of 'explaining what it was that motivated Tycho to devote himself to the science within which he should become so famous' (Norlind, 1970, p. 15).[10]

The striking similarity between the biographies of Gassendi and Norlind, separated in time by more than three hundred years, can be illustrated by an episode from Tycho's important correspondence with the Kassel astronomer Christopher Rothmann, an early advocate of Copernicanism. In a letter to Rothmann of 1589, Tycho argued against the diurnal rotation of the earth by means of the following argument: 'Some people think that if a missile were thrown upwards from the inside of a ship, it would fall in the same place whether or not the ship were moving. They offer these assumptions gratuitously, for things actually happen quite differently. In fact, the faster the motion of the ship, the more difference will be found' (Thoren, 1990, p. 278). It is the irony of history that in 1640 Gassendi was the first to demonstrate experimentally what Galileo had only argued theoretically, namely that a body falling from the vertical mast of a ship will share in its fall the motion of the

9 The turn from the old tradition in Tycho scholarship to a new one can be glimpsed from the reviews of Norlind's book by Christianson (1971) and Thoren (1971).

10 Gassendi's claim is repeated in Bang (1745, p. 94), Friis (1871, p. 7) and Dreyer (1963, pp. 13–14). Thoren (1990, p. 12) considers Norlind's objection to be convincing.

ship. Referring to Tycho's letter, he found it 'strange' that Tycho, the empiricist, had not investigated the matter experimentally: 'Had he made the experiment, he would have found his claim to be incorrect and realized that it rather proves the rotation of the earth' (Gassendi, 1951, p. 110). Norlind spoke in he same moralistic voice:

> Tycho's pondering shows that in some cases he did not, after all, have the critical attitude that one might expect him to have. ... How easy would it not have been for him to prove that his claim was wrong. There were enough opportunities on his numerous travels across the Øresund. But he never made such an experiment, it was merely a thought experiment. (Norlind, 1970, p. 112)[11]

Like earlier biographers, Norlind concluded that 'Tycho was not an accomplished theorist' (p. 325) and that his greatness lay in precise observations. In the very last lines of the biography, he followed Dreyer and Gade in associating Tycho's legacy with the successes obtained by his successors, that is, he adopted a teleological or whiggish point of view:

> The discovery of the three laws which are named after Kepler would never have occurred had he not had access to Tycho's observational material mainly collected during the years at Hven. ... Copernicus' world system had been perfected by Kepler; what was left was to show that the motion of the earth is physically possible and further to explain why the earth and the planets moved in the way Kepler had found. It was Newton who solved this problem. (Norlind, 1970, p. 329)

A Wider Picture: Some Recent Revisions

As we have seen, the biographies in the Gassendi-Dreyer-Norlind tradition became increasingly detailed and sophisticated without breaking with the basic framework of the original biography of 1654. Such a break only occurred in the late twentieth century, not so much by adding new sources as by presenting known material in a new way and by focusing attention on perspectives that until then had been more or less ignored. Since the late 1960s, specialists in late Renaissance astronomy, including Kristian P. Moesgaard, Yas Maeyama and Owen Gingerich, had studied Tychonian astronomy in great technical detail and come to recognise that Tycho was as much an accomplished theorist as he was a master observer and instrument-builder. For example, Tycho was the first to develop a sophisticated theory of observations (Thoren, 1988), and he devoted much effort to developing theories of the moon and the sun. Only now did the true nature of these works, and their intimate connections to Tycho's instruments, become fully understood and appreciated (Thoren, 1973).

The result was a substantially new picture of Tycho's scientific work, which was first presented in a coherent way in Victor Thoren's magisterial biography of 1990, *The Lord of Uraniborg*, generally acclaimed as the best biographer of the Danish astronomer.[12] 'It is in the description of Tycho's scientific work ... that previous

---

11 Øresund is the sound between Denmark and Sweden where Hven is located.

12 The biography was started as the project of John Christianson, who drafted some of the chapters and contributed in various other ways. Yet only Thoren's name appears on the

biographies are now truly obsolete,' Thoren wrote, referring to the traditional lack of appreciation of Tycho's theoretical astronomy (Thoren, 1990, p. ix). His own biography included detailed chapters on Tycho's solar and lunar theories. Thoren not only demolished the old picture of Tycho, the narrow empiricist and mediocre theorist, he also highlighted – as Friis had done earlier – the importance of his noble upbringing and what it meant to his possibility of a making a career. 'The implications and ramifications of *privilege*', he wrote, 'are infinitely more subtle than are the technicalities of Tycho's astronomy and every bit as significant for his career' (Thoren, 1990, p. ix).

This feature, Tycho Brahe's life as a distinguished nobleman and his manoeuvring in the patronage system of his time, was given even fuller attention by John Christianson in his remarkable book on Tycho as networker and scientific entrepreneur. *On Tycho's Island* (2000) is not just another Tycho biography, and perhaps it may not be classified as a biography at all, as it starts its account only around 1572 and has as its main topic the human and material contexts of Tycho's life and science. It is a work primarily on Uraniborg and Tycho's many assistants – his *familia* – rather than on Tycho himself. Yet precisely because of this unconventional focus, it succeeds in presenting Tycho in a new and fruitful light.

Apart from the emphasis on astronomical theory, two further topics loom large in the new generation of Tycho biographies and help to separate them from the older tradition. First, the religious context of Tycho's work and career is taken seriously; second, his intense and life-long occupation with Paracelsian chemistry and medicine is given much higher priority than in earlier works. These aspects have been examined in detail by Jole Shackelford, among others, and they form the central part of the biography written in 1994 by the Danish intellectual historian Alex Wittendorff. More recently, the themes have been investigated by Morten Fink-Jensen, a student of Wittendorff (Fink-Jensen, 2004). Experts disagree about how important Paracelsianism was for Tycho, more precisely, but there is no doubt at all that it was much more than merely a pastime amidst his serious astronomical work. The same holds for Tycho's obsession with Latin poetry, a theme which has been investigated in detail by Peter Zeeberg (1995) in particular (and which was also noted, if not elaborated, by Gassendi).[13] Zeeberg's work shows how Tycho successfully used his poetry to promote his science and self-image throughout Europe. And more than that, for poetry was not merely ornamental to Tycho, it was part and parcel of his quest to understand the secrets of the universe.

Finally, it is now evident that Tycho was in no way above the religious struggles of his time, when Philippism in Denmark came under pressure from an increasingly militant Gnesiolutheranism. Nor did he consider his astronomy as merely a scientific and secular project. Contrary to Galileo, Tycho believed in the literal meaning of the Bible, and in general he saw his astronomical research as anchored in a Hermetic-Christian context. Experts differ in their opinions of how important religion was to Tycho, what his religious attitude was, more exactly, and also about the consequences of his attitude. Was he an 'aristocratic heretic' who became a victim

---

title page.

13 For other titles, see the bibliography in Christianson (2002).

of the campaign of orthodox Lutheranism (Wittendorff, 1994, p. 233)? Or was he rather a liberal Philippist who merely wanted to stay outside the religious turmoil of the period? According to Christianson (1998, p. 477), Wittendorff went too far in his characterisation of Tycho as a Neoplatonic heretic and critic of the Church. I tend to agree, and also feel that Fink-Jensen has exaggerated the degree to which Tycho's work was embedded in a religious context. Whatever the truth of the matter, in any future biography of Tycho, it will be impossible to ignore the religious component.

Conclusion

We have seen that Gassendi's first biography of Tycho Brahe set the framework for later biographies during a surprisingly long period. Still in the twentieth century, the shadow of Gassendi was unmistaken in works such as Norlind's that in detail and erudition went far beyond the original. The problem with the works of Gassendi and his followers was not that they reflected values particular to their time – they did, but the very fact that the tradition can be followed through three centuries shows that it was not seriously constrained by such values. It was rather a lack of balance between the various components in Tycho's life and work. The different, richer, and undoubtedly more authentic picture of Tycho that emerged in late twentieth century can be ascribed basically to two reasons. First, a better, more fine-grained technical insight into Tycho's astronomical works revealed the central position theory had in his research programme. Second, and possibly even more important, the new generation of historians placed a much stronger emphasis on aspects of Tycho's life that had been known for long but rarely given high priority. By casting a wider net, it was shown how patronage, religion, Paracelsianism, architecture and poetry were deeply integrated in his vision of astronomy, and out of this contextualistic insight a new portrait of Tycho took shape.

Tycho Brahe was eager to leave a legacy. On his deathbed, the night before he died, he kept saying 'Ne frustra vixisse videar' – 'Let my life not be in vain' (Christianson, 2002, p. 231). He did leave a legacy, and later historians have discussed exactly what kind of legacy it was. If Tycho in his heaven – or where he might have ended – could read the biographical literature, he would probably rejoice, if also at time despair. Perhaps he would appreciate how excellent a description of his life Gassendi's work was, even compared with most later biographies.

Bibliography

Bang, Oluf (1745), 'Den store vidtberömte Danske Astronomus [...] Tyge Brahe', *Samling af adskillige nyttige og opbyggelige Materier*, vol. 2, Copenhagen: private printing.

Brahe, T. (1913–29), *Tychonis Brahe Dani Opera omnia*, ed. J. Dreyer and E. Nystrøm, Copenhagen: Gyldendal.

Brahe, T. (1946), *Tycho Brahe's Description of his Instruments and Scientific Work*, trans. and ed. H. Ræder, E. Strömgren and B. Strömgren, Copenhagen: Munksgaard.

Brahe, T. (1996), *Instruments of the Renewed Astronomy*, trans. and ed. A. Hadravová, P. Hadrava and J. Shackelford, Prague: Koniasch Latin Press.

Christianson, J.R. (1971), [Review of Norlind, *Tycho Brahe*, 1970], *Isis*, **62**, 545–6.

Christianson, J.R. (1998), 'Tycho Brahe in Scandinavian scholarship', *History of Science*, **36**, 467–84.

Christianson, J.R. (2000), *On Tycho's Island: Tycho Brahe and His Assistants, 1570–1601*, Cambridge: Cambridge University Press.

Christianson, J.R. (2002), 'The legacy of Tycho Brahe', *Centaurus*, **44**, 228–47.

Delambre, J.-B. (1821), *Histoire de l'astronomie moderne*, 2 vols, Paris: Libraire les Sciences.

Dreyer, J.L.E. (1901), *Tycho Brahes Fortjenester af Astronomien*, Copenhagen: Gad.

Dreyer, J.L.E. (1963), *Tycho Brahe: A Picture of Scientific Life and Work in Sixteenth Century*, New York: Dover Publications.

Ehrencrone-Müller, H. (1924–35), *Forfatterlexicon omfattende Danmark, Norge og Island indtil 1814*, 12 vols, Copenhagen: H. Aschehoug.

Figala, K. (1972), 'Tycho Brahes Elixier', *Annals of Science*, **28**, 139–76.

Fink-Jensen, M. (2004), *Fornuften under troens lydighed: Naturfilosofi, medicin og teologi i Danmark 1536–1636*, Copenhagen: Museum Tusculanum.

Friis, F.R. (1871), *Tyge Brahe: En Historisk Fremstilling efter Trykte og Utrykte Kilder*, Copenhagen: Gyldendal.

Gade, J.A. (1946), *The Life and Times of Tycho Brahe*, New York: Princeton University Press.

Gassendi, P. (1654), *Tychonis Brahei, Eqvitis Dani, Astronomorum Coryphæi, Vita*, Paris: Mathurini Dupuis.

Gassendi, P. (1951), *Tycho Brahe: Mannen och verket*, trans. W. Norlind, Lund: C.W.K. Gleerup.

Gassendi, P. (1996), *Vies de Tycho Brahe, Copernic, Peurbach et Régiomontanus*, trans. J. Peyroux, Paris: Librairie A. Blanchard.

Gingerich, Owen (1982), 'Dreyer and Tycho's world system', *Sky and Telescope*, **42**, 138–40.

Hannaway, Owen (1986), 'Laboratory design and the aim of science: Andrea Libavius versus Tycho Brahe', *Isis*, **77**, 585–610.

Jones, H. (1981), *Pierre Gassendi, 1592–1655: An Intellectual Biography*, Nieuwkoop: B. de Graaf.

Knobel, E.B. (1927), [Obituary notice of J. Dreyer], *Monthly Notices of the Royal Astronomical Society*, **87**, 251–7.

Langebeck, Jacob (1746), 'Rare og utrykte Efterretninger om Tyge Brahe', *Danske Magazin*, **2** (18–24), 161–372.

Norlind, W. (1970), *Tycho Brahe: En Levnadsteckning med Nya Bidrag Belysande hans Liv och Verk*, Lund: C.W.K. Gleerup.

Shackelford, Jole R. (1993), 'Tycho Brahe, laboratory design, and the aim of science', *Isis*, **84**, 211–30.

Shackelford, Jole R. (2002), 'Providence, power, and cosmic causality in early modern astronomy: The case of Tycho Brahe and Petrus Severinus', in John R. Christianson et. al. (eds), *Tycho Brahe and Prague: Crossroads of European Science*, Frankurt am Main: Harri Deutsch, pp. 46–69.

Thoren, Victor (1971), [Review of Norlind, *Tycho Brahe*, 1970], *Journal for the History of Astronomy*, **2**, 205–7.

Thoren, Victor (1973), 'New light on Tycho's instruments', *Journal for the History of Astronomy*, **4**, 25–45.

Thoren, Victor (1988), 'Prostaphairesis revisited', *Historia Mathematica*, **15**, 32–9.

Thoren, Victor (1990), *The Lord of Uraniborg: A Biography of Tycho Brahe*, Cambridge: Cambridge University Press.

Usher, Peter (1997), 'Shakespeare's cosmic world view', *Mercury*, **26**, January, 20–23.

Usher, Peter (1999), 'Hamlet's transformation', *Elizabethan Review*, **7**, 48–64.

von der Weistritz, P. (1756), *Lebensbeschreibung des berühmten und gelehrten dänischen Sternsehers Tycho v. Brahe*, 2 vols, Copenhagen and Leipzig: F.C. Pelt.

Wittendorff, A. (1994), *Tyge Brahe*, Copenhagen: Gad.

Worm, O. (1968), *Breve til og fra Ole Worm*, trans. H. D. Schepelern, vol. 3, Copenhagen: Munksgaard.

Zeeberg, Peter (1995), 'The inscriptions of Tycho Brahe's Uraniborg', in Minna Skafte Jensen (ed.), *A History of Nordic Neo-Latin Literature*, Odense: Odense University Press, pp. 251–66.

Chapter 8

The Programmatic Function of Biography: Readings of Nineteenth- and Twentieth-Century Biographies of Niels Stensen (Steno)

Signe Lindskov Hansen

When discussing the function of the biography – especially scientific biography – as a genre, we often find ourselves debating its merits as history, literature or didactics, or even as an edifying, ethical and existentially authentic genre (Söderqvist, 1996, 2006). Based on examples taken from Danish-language biographies of the Danish seventeenth-century anatomist, geologist and theologian Niels Stensen (a.k.a. Steno), I will suggest that biographies can also fulfil a *programmatic* function. The term 'programmatic' is broadly construed as an underlying agenda, as a conscious attempt to influence readers' attitudes, as a tactic, and as a means by which different groups of intellectuals protect their own interests. Several of Stensen's biographers use their texts to promote certain views or programmes concerning areas, such as religion, national policy, research policy and aesthetics. This programmatic (or argumentative) function manifests itself in certain storytelling strategies that are visible in a variety of rhetorical manoeuvres and methods aimed at seducing the reader.

In the following, I will demonstrate how these specific argumentative strategies – or rhetorical seduction strategies – are present in a number of Danish nineteenth- and twentieth-century Stensen biographies, and how these strategies form an integrated part of the narrative techniques and strategy of each particular biographical work.[1]

There are a few examples of cross-biographical, comparative investigations of biographies about particular scientists, such as Robert M. Young's (1987) and Frederick B. Churchill's (1982) studies of Darwin biographies, Paul Theerman's (1985) examination of Newton biographies, Henry Guerlac's (1954) review of Lavoisier biographies and Dorinda Outram's (1976) work on Cuvier biographies. Outram sceptically points out how biases and polemical elements in a biography control the way the subject is portrayed, thereby indicating the dangers inherent in to the programmatic function of the biography: the demagogical and myopic repetition of stereotypical portrayals. However, the other extreme – the uncontrolled and uncontrollable biography, which uncritically reproduces sources and factual information – is just as problematic as it is unreadable. The vast majority of

1 This chapter was translated from the Danish by Heidi Flegal; it draws on a chapter in my PhD dissertation, revised in Hansen (forthcoming).

biographies of scientists and scholars are located somewhere between these two extremes. Of course, no biography can avoid the influence of narrative-technique control, or avoid certain interpretations of the links between the research and the life lived, and thereby give the biography a certain direction. And, as will be evident from my reading of a number of Stensen biographies, the underlying interpretational directions of these biographies can be seen to promote highly varied aspects of the biographee and be used to argue in favour of specific opinions and points of view.

Profile of the Stensen Literature[2]

The first major Danish-language biography of Niels Stensen (1638–1686) was published in 1865, and has since been followed by some twenty biographical works in Danish.[3] Although a respected scientist both in Denmark and abroad, Stensen was hidden in his native country by a veil of biographical obscurity for almost two centuries after his death. In fact, for a long time he was better known abroad; during the first centuries after his death, biographies appeared in French, Italian and other languages. The earliest of these is Johannes von Rosen's account of the final years of Stensen's life, which von Rosen spent living with Stensen in Germany.[4] During the 1770s, two Italian Stensen biographies appeared, written by Fabroni (1779) and Manni (1775); along with von Rosen's account, these two books served as the basic sources for a number of the Danish biographies, in particular the early ones published towards the end of the nineteenth century.

From a present perspective, it may seem remarkable that no Stensen biography in Danish was written until the mid-nineteenth century. In a Danish context, however, Stensen was a scientific, religious and geographical exile, a status that he retained far into the nineteenth century. It is possible that Stensen's life in exile contributed to the lack of domestic interest in his person for almost two hundred years. In many ways, Stensen was an ill-adapted figure on the Danish intellectual scene; as I will discuss later at greater length, many of his biographers therefore emphasise those parts of his personal history that contribute to the understanding of his life as a man exiled, as a genius shunned, as an original and professionally uncompromising scientist, and as a man dedicated to his faith.[5]

The first Danish biography of Stensen was written by the historically interested parson and Member of the Danish Parliament Jørgen Wichfeld (1865). Two decades

---

2 This profile of the Stensen literature is limited to biographical works in Danish. One implication of this is that the works of the Austrian-born scholar Gustav Scherz (for example, 1956, 1987–88), whose extensive scientific productions concerning Stensen – published in Denmark from the 1950s and onwards, but in German and English – fall outside the scope of the study.

3 Including novelistic biographies, but not short biographical accounts, such as dictionary entries.

4 The exact publication year of von Rosen's (b. *c.* 1653) account remains unknown.

5 He had not been completely forgotten, however. For instance, he appears in passing in Worm's prominent work (1771–84) on learned men in Denmark, Norway and Iceland, which features a brief one-page biography followed by one page of bibliographic references.

later, in 1884, the historian A.D. Jørgensen presented his 231-page *Nils Stensen* (re-published in 1958 in a second edition annotated by Gustav Schertz). During the following decades, the biographical writings on Stensen mainly took the form of brief articles in journals and newspapers and entries in biographical dictionaries and other reference works. In 1887, the historian C.F. Bricka, Jørgensen's successor to the office of Keeper of the Public Records, began the project that would become the national biographical dictionary, *Dansk biografisk Lexikon* (Bricka, 1887–1905). When, in 1902, the editorial staff reached the letter 'S', Nils Stensen was featured in a 13-page biographical article, indeed quite lengthy according to the standards of the dictionary. The early inter-war years saw the publication of the handy three-volume biographical dictionary *Dansk biografisk Haandleksikon* (Dahl and Engelstoft, 1920–26), in which Stensen appears in a two-page biography. He is also prominently represented in Carl S. Petersen and Vilhelm Andersen's comprehensive illustrated history of Danish literature, the four-volume *Illustreret dansk Litteraturhistorie* (1924–34), which contains a 26-page biographical section on Stensen in a long chapter on the natural sciences in Denmark in the seventeenth century that also contains biographies of and references to figures such as C.S. Longomontanus, Thomas Fincke, the Bartholins, Ole Worm, Simon Paulii, Ole Borch and Ole Rømer.

Over the next twenty years, a number of Stensen biographies appeared in rapid succession; several were occasioned by the tercentenary anniversary of his birth in 1938. The first was a small Catholic-oriented biography by Johannes Metzler (1928), followed by a short scientifically oriented commemorative paper by a medical doctor (Meisen, 1932) on Stensen's life and scientific research. Next came a broadly oriented memorial tribute by the medical historian Anker Aggebo (1937), followed a year later by the 307-page poetically grandiloquent biography in epic verse written by the Danish lyricist Marinus Børup (1938), and an article in the Catholic journal *Nordisk Ugeblad for katholske Kristne* (Anon., 1938). It was followed by J.T. Suhr's succinctly biographical lecture on the theme 'Niels Steensen the Man' given to the Danish Medical Society and printed in 1942 (Suhr, 1942). The new edition of the Danish biographical dictionary, published in 1933–44, also contained a new, revised, version of Stensen's biography (Engelstoft and Dahl, 1933–44).

From 1950 to 1980, roughly two Stensen biographies were published in Danish each decade. In 1953 came Karen Plovgaard's strongly Catholic-oriented biography, and in 1958 A.D. Jørgensen's often-quoted biography from 1884 was re-published. The Stensen scholar Gustav Schertz wrote a minor biographical work in 1963, and 1966 saw the appearance of a thin volume by medical doctor Egill Snorrason under the auspices of the Swedish pharmaceutical corporation Mölnlycke.[6]

Up until the 1970s, the Stensen literature was mainly dominated by traditional, hagiographic works. But from 1970 onwards, new types of Stensen biographies emerged. In 1972, Peter Beck's small but richly illustrated Stensen portrait appeared in an educational biographical series; it had clearly didactic intentions, targeting high school students. Also in 1972, Danish fictional prose writer Karl Bjarnhof

6 This work took the guise of a 'New Year's greeting' of medical history 'on the occasion of the tercentennary of Niels Stensen's lecture in Paris on the anatomy of the brain in 1665', as the company noted in its preface (Snorrason, 1966, p. 9).

published an extensive monographic biography which more emphatically than any of is predecessors took a psychologically inquisitive look at Stensen. One need only get past the subtitle, *På spor af Niels Steensen* ('tracking Niels Steensen') – to perceive its sleuth-like, investigative approach. Incidentally, it is worth noting that the original subtitle in Danish uses the word *spor* as the indefinite plural form of 'track' or 'trail', whereas the Danish idiom calls for the definite singular form (*sporet*). This makes for an obvious intentional reference to several possible tracks that are initially unknown – thereby heralding a more open and pluralistic interpretive biographical strategy.

Throughout the 1980s and 1990s, it was particularly the Catholic-oriented biographies that gained ground, that is, books released by Catholic publishers, written by Catholic authors, and promoting or propagandising in favour of Catholicism. The efforts in Catholic circles to have Stensen beatified mark the Stensen literature during that period. In 1981, a Danish catholic press published a volume entitled *Niels Stensen og Danmark: Kirkens fattige tjener* ('Niels Stenden and Denmark: The Poor Servant of the Church') (Kongsted et al., 1981). The year of Stensen's eventual beatification (1988) also witnessed an unusual co-operation between the Catholic Stensen scholar Gustav Schertz, the Catholic Diocese in Denmark, and the Danish Ministry of Foreign Affairs – a minor biographical study (Schertz, 1988) that focused on the meeting between the Catholic and domestic Danish 'marketing interests' in connection with Stensen's beatification. Five years later, in 1993, another ambitious Stensen biography appeared, a thorough 340-page biography written by a Catholic nun, Sister Miriam Mortensen. Also in 1993, the genre profile of the Stensen literature was expanded to include the novelistic, with the appearance of author Jacques Berg's biographical novel *Himmelstigen: En roman om den lærde Niels Stensen, der blev Guds og troens lidenskabelige tjener* ('The Stairway to Heaven: A Novel of the Learned Niels Stensen, Who Became the Zealous Servant of God and Faith') (1993) – another title that clearly reveals the strong focus on Stensen's conversion,[7] his faith, and his unwavering religious fervour.

It is interesting to observe that so many of the biographies are mainly concerned with Stensen's religious devotion rather than his anatomical or geological works. Not that Stensen's scientific accomplishments have remained unchronicled (they have indeed), but these achievements are usually documented in non-biographical treatises. The late 1980s saw the release of two more biographies: Harald Moe's pictorial biography (1988), which was published to commemorate the 350th anniversary of Stensen's birth, and a minor study by Jens Jørgen Nygaard (1989).

The third, updated, edition of the Danish biographical dictionary (Bech, 1979–84) contained a new and more compact Stensen biography written by Egill Snorrason. Also, the new nine-volume material-historical work on literature history, *Dansk Litteraturhistorie* (1983–85), contained, like its predecessor written by Carl S. Petersen and Vilhelm Andersen in the 1920s, a lengthy chapter on eighteenth-century science; but the biographical section on Stensen was reduced to about one page, and focused on Stensen's scientific merits only; the author, Jens Hougaard, concluded that Stensen had 'become a bird under strange skies' in his native Denmark. The Danish encyclopaedia project *Den Store Danske Encyklopædi* (Lund, 1994–2001)

7 Niels Stensen converted to Catholicism on 2 November 1667.

also included a brief account of Niels Stensen's life and scientific works. Finally, a short didactic biography by the philosopher Carl Henrik Koch was published in 2003.

The Comprehensive Genre

The literature on Stensen is interesting for several reasons. It is first and foremost a good example of how heterogenous the genre of scientific biography is. Based on the same framework – same person, same period and more or less the same source material – the biographies of Niels Stensen unfold in a wide variety of diverse narrative accounts. The Stensen literature encompasses dramatic stories of decline and decay written by Protestant or non-religious biographers who regard Stensen's conversion to Catholicism as a great loss to science and to the nation. Conversely, Stensen's Catholic biographers portray his life as an account of the purification of a man whose talents only truly came to fruition in his theological studies. Last but not least, there are the fictionally tinged biographies in which Stensen's life and deeds are coloured by the biographer's own aesthetic project.

Several of these Stensen biographies show quite clearly that the biographical narrative is being used for a different, and higher, purpose than presenting the life and research of the biographee. As the following examples will show, the variety of the ideological or aesthetic projects of the biographers are woven into the biographical narrative, and it is not infrequent to see them exerting a decisive influence on the interpretations of the subject. Brief analyses like this one cannot cover all aspects of the multifaceted views of Stensen. They can, however, present a number of significant characteristics in the programmatic approach of the biographies to their subject.

Wichfeld: The Theatrical Protestant

The first case is the biography written by the historically interested parson and Member of Danish Parliament Jørgen Wichfeld.[8] As mentioned earlier, Wichfeld's 111-page account (1865) was the first Stensen biography published in the Danish language. It kindles an interest in modern readers too, not least because the author is fairly outspoken in his personal pronouncements on his subject; he does not shy away from giving explicit assessments of Stensen, his contemporaries, and a posterity that the author openly criticises for having consigned Stensen to oblivion.

Wichfeld unfolds a narrative that recounts a destiny of tragic dimensions, depicting the formation, greatness and decline of an eminent scientific persona. In this narrative, Stensen is subject to a series of events and fatal external influences which cause his life to veer drastically in a tragic direction. As in a Greek tragedy, these events remain just outside the scope of the main character's own free will:

8 Jørgen Wichfeld (1800–1888) played no prominent role in Denmark's political life, and in addition to his biography of Niels Stensen, his activities as an antiquarian and history buff resulted in a number of dissertations on ancient relics on the island of Lolland (compare Bricka, 1887–1905, vol. 18, p. 526).

his life lives him, and not vice versa. Although Stensen's life propels him around the amphitheatre, he retains – at least in Wichfeld's interpretation – a purity of the soul. As Wichfeld puts it: 'no purer soul walked the Earth. All things corporeal and earthly were to him foreign and inconsequential, while Science and the interests of the higher mental faculties were everything' (Wichfeld, 1865, p. 2).

The pivotal issue in Wichfeld's trisected biographical narrative is the course of events surrounding Stensen's conversion to Catholicism.[9] The religious aspects of Stensen's life dominate the biography to such a degree that, particularly at the interpretative level, they completely eclipse his scientific merits. Even though Wichfeld spends many pages on a description of Stensen's scientific work, most such passages are referential, often using direct quotations. In other words, in scientific matters the biographer remains on the outside of the events, whereas in religious matters, particularly in the story of Stensen's conversion, the author delves deeper, embarking upon analyses as well as independent interpretation.

Wichfeld initially describes Stensen's conversion as an intentional act. Stensen 'abjured the faith of his fathers', and 'devoted himself to the service of a foreign church' (Wichfeld, 1865, p. 2). Even so, Wichfeld's account of his subject's change of faith is a story of conversion in the original sense of the word. Stensen is described as falling prey to quick-witted Catholics who had the ability to convince and convert him. In Wichfeld's rendition, the course of events assumes conspiratorial proportions, as Stensen is seen to become the object of calculated rhetoric, impassioned Mediterranean female cunning and calculating clergymen who are fully aware of the prestige the Catholic Church will gain by bringing such a great mind under its protection.[10]

The struggle between Stensen's Protestant roots and the massive Catholic influences he encountered in Italy is the driving conflictual force in Wichfeld's account. The tensions between the two denominations, as evidenced in Stensen's person and in his development, are transformed into an integral part of the biography as a means of narrative progression.

The controversy between Protestantism and Catholicism permeates Wichfeld's text at several levels: both in the interpretation of Stensen's development and life story and in the perception of the contemporary spiritual, scientific and political climate, not to mention in the composition of the text, its style and its rhetoric. An antithetic way of reasoning colours Wichfeld's style, and the antithesis is used as a syntactic figure. Wichfeld describes Protestantism and Catholicism as 'two adversarial, principal conditions for all human beings', and the complementary adversaries are made tangible by drawing a parallel to the antithetic juxtaposition of the concepts 'intellectual freedom' and 'intellectual serfdom' (Wichfeld, 1865, p.

9 Wichfeld divides Stensen's life into three main phases centred around his 'intellectual propensities' (cf. Wichfeld, 1865, p. 4): first, the period 1660–66, when Stensen studied at universities abroad; second, Stensen's years in Italy in 1666–75; third, 1675–86, when Stensen worked as a Catholic priest until his death. These three chapters are preceded by a brief four-page introduction to Stensen's life and scientific importance, which also contains a panoramic overview covering the first 22 years of Stensen's life.

10 Compare, for instance, Wichfeld (1865), pp. 4f., 17, 27, 32, 53 and 73.

4).[11] Intellectual freedom becomes synonymous with Protestantism, while intellectual serfdom is equated with Catholicism. Within Wichfeld's interpretational framework, this judgement makes it inconceivable that Stensen should have entered the ranks of Catholicism of his own volition; it seems absurd that he could have made such a decision as a conscious, positive choice, for how could anyone, especially a man of science, choose intellectual serfdom over intellectual freedom? In Wichfeld's narrative, Stensen's conversion to Catholicism becomes a hidden temptation, a shadow that gradually sneaks up on him. The activities of the Jesuits are described as 'propaganda' (Wichfeld, 1865, p. 5), and Stensen's dealings in Paris with the great French orator and prelate Bossuet in the mid-1660s make the 'temptation to switch over to Catholicism' move closer, although the prelate's 'most ardent persuasions' were still in vain, as Wichfeld so portentously writes (Wichfeld, 1865, p. 15). The metaphors are saturated by the author's views, and remain so throughout the biography. For example, Wichfeld speaks of Catholicism as 'delicate nets that are positioned all around him', about the attentions of the Italian court being borne by a hidden intention, and about the actions of the Italians towards Stensen as 'peculiar machinations' set in motion to win Stensen over to Catholicism (Wichfeld, 1865, p. 17, p. 19, and p. 23).

Wichfeld's tendency to draw a sharp line of demarcation between good and evil, Protestantism and Catholicism, is also obvious in the way he depicts Stensen's character. Stensen meets two women in Italy: the noblewoman Sra Arnolfini and the nun Maria Flavia. Wichfeld's portrayal of the two women, particularly Arnolfini, is unmistakable tinged with mysticism and Mediterranean exoticism. She is portrayed as a sophisticated, attractive yet virtuous woman; the portrait fits surprisingly well with Wichfeld's overall interpretation of Stensen's captivation with the Catholic faith. As a colourful and highly emotional female figure, she is a sort of incarnation of the Catholic faith, which by virtue of its worship of the Virgin Mary features femininity as an integral part of its religious practices. As for the exotic aspects of Sra Arnolfini, Wichfeld's views are not inspired by fascination and admiration, but rather by a fear of the demonic, which exerts an irresistible attraction upon even the purest of hearts. Wichfeld's rhetorical construction of this woman as a figure in the biography seems to be concerned with mastering and taming her perilous nature, in order to retain the image of Stensen as an untainted soul.

These dramatic contrasts and the battle between good and evil tie in well with the classical tragedy as described by Aristotle in his *Poetics*: 'Again, since Tragedy is an imitation of persons who are above the common level, the example of good portrait painters should be followed. They, while reproducing the distinctive form of the original, make a likeness which is true to life and yet more beautiful' (Aristotle, 1948, Section 2, Part XV). If one accepts Wichfeld's interpretation of Stensen's external life as characterised by breaks and dramatically staged twists of fate, then one will find that the element that creates narrative uniformity is the very self-sacrifice and

11 'The antithesis as a syntactic figure seems almost indispensable to the dialectic thought process' ('Antitesen som syntaktisk Figur synes næsten nødvendigt at høre til den dialektiske Tankeproces'), Ulla Albeck (1961, p. 184) points out with reference to Søren Kierkegaard and Rasmus Nielsen, who were Wichfeld's contemporaries.

purity in Stensen's moral character. The purity of our protagonist remains untouched throughout the entire biography, while his close Catholic friends are the ones made responsible for the fateful compromising of character that his conversion comes to represent. Aristotle too points to deceit among people who harbour a mutual love for each other as the preferred theme of tragedy (Aristotle, 1948, Section 2, Part XIV).

The latter part of Stensen's life, so deeply ascetic and full of suffering – which Wichfeld mainly passes on to his readers through an eight-page direct quotation from a description written by Johannes von Rosen, Stensen's contemporary and companion during the last years of his life – serves to conclude Wichfeld's tragic account of the noble-minded Protestant's conversion and demise. Throughout the biography, Wichfeld stresses that Stensen's ascetic lifestyle was a consequence of the religious awakening that had taken place in him, occasioned by the Italian female catalysts of conversion. This consequential line of reasoning is in fact in line with the classical tragic form chosen by Wichfeld for his biography. This story of suffering (as it must indeed be called, since Stensen spent his last years growing increasingly ill, getting a minimum of food and sleep, and wandering around barefoot, impoverished and in rags) follows the goal of the Aristotelian tragedy of bringing catharsis – emotional purification – by means of a precise, pitiable description of the suffering inflicted upon the main character by the tragedy at hand. In his *Poetics*, Aristotle emphasises precisely 'the Scene of Suffering' (pathos) as an essential component in the tragic narrative, defining it as 'a destructive or painful action, such as death on the stage, bodily agony, wounds, and the like' (Aristotle, 1948, Section 1, Part XI). The audience's empathy with the suffering endured by the tragedy's main character completes the process of purification (catharsis).[12]

In Wichfeld's account, however, the horrible descriptions of the extreme lengths to which Stensen takes his ascetic lifestyle do not only serve the purpose of catharsis. They also seem to be intended to function as illustrations of the downright atrocities occasioned by the corruption of the Church, and to show how badly a man can fare if he gets involved with it. In other words, the biography has a morally instructive aim, and with its unequivocal message and deliberate narrative strategy, it borders on anti-Catholic propaganda. Wichfeld's text embodies a blatantly programmatic dimension, in the sense that is intentionally written to influence the attitudes of the readers.

A.D. Jørgensen: The Psychological Historian

Nils Stensen by the historian A.D. Jørgensen (1884) is regarded by the subsequent Stensen literature as the first scholarly biography of Stensen in Danish. It certainly takes a much different approach to its subject than Wichfeld's. Whereas Wichfeld used the breaks as the key element in his understanding of Stensen's character, and the discrepancies between Protestantism and Catholicism to describe and define Stensen's relations with the world around him, Jørgensen's general approach was to

12 Compare Aristotle (1948, Section 1, Part VI). See also Chris Chilvers' Chapter 6 in this volume.

regard Stensen as a coherent individual who interacted openly and reciprocally with contemporary society, and whose life and scientific activities were interwoven. In that sense, Jørgensen was much more interested in Stensen's psychological condition than Wichfeld was. In his epilogue, Jørgensen expressed a hope that his book would 'be able to make a contribution to the perception of an historical *character* who is so closely akin to the best in *our national character*' (Jørgensen, 1884, p. 213; my italics).

One might reasonably ask which of Stensen's many personal characteristics Jørgensen was attempting to depict as 'akin to the best in our national character' (he refers here to the great project of the 1800s, during the heyday of historical science, of building the Danish nation-state)? Basically, there are at least six: the sickly yet responsive boy, the young man who defends his country on the ramparts, the diligent student, the scientific genius and discoverer, the learned conversationalist and companion, and the humble Christian. The basic characteristic that links these multifarious roles together might perhaps best be described as an inquisitive approach to life, combined with a sense of self-sacrifice and an unselfish will to put the 'cause' ahead of oneself. In Stensen's case, the 'cause' turns out to have many faces: scholarship and learning, king and country, science, and last but not least, faith and religion.

As far as the author's approach to Stensen's religious life is concerned, it is interesting to note that Jørgensen is far more impartial than Wichfeld in the way he regards Stensen's conversion to Catholicism. Jørgensen's perception of religion seems more modern, because he considers faith to be an internal, psychological issue. For Jørgensen, faith is mainly a question of emotional sincerity, a question of a religious feeling from within that bears on the development of one's personality, and only to a lesser degree a matter of professing one's faith to this or that rival denomination.

Overall, Jørgensen takes a surprisingly psychological approach to Stensen, especially considering the fact that the biography appeared in 1884 – before Freud and psychoanalysis, but also long before the psychobiographical approach began to gain ground. Whereas Wichfeld freezes Stensen in the image of a pure, intellectual, scientific genius, Jørgensen's portrayal of his subject is so empathic and so finely nuanced that his understanding of the individual reaches far into the twentieth century. Jørgensen (1884, pp. 135, 204) explicitly acknowledges that elucidating Stensen's personal development is the main goal of his biographical project, rather than for instance explaining minute details in Stensen's scientific endeavours or describing his work for the Church during the last years of his life. All in all, Jørgensen's biography holds a special position in the early Stensen literature, precisely because he is able to contemplate and combine Stensen's personal, psychological, scientific and religious development as a fully formed continuum.

Jørgensen identifies Stensen as a seeking soul who, inspired by his early reading of Descartes and others, carried doubt as an integrated part of his scientific methodology, but also as an important element in his religious quest. Jørgensen (1884, p. 51) quotes Stensen as referring to the philosophy of Descartes and the ability to doubt as one of the most significant reasons for his converting to Catholicism.[13] What is more, Jørgensen does not view Stensen's conversion as an abrupt breaking off of his scientific career, but rather as a direct – and self-elected – consequence of the nature of his personality:

> It is certainly not, as has been claimed, the switchover to Catholicism that prompted Stensen to give up science, nor the circumstance that he later became a priest. Catholic sources have submitted, and quite rightly so, that neither one fact nor the other would in the least way preclude scientific pursuits. The reason lies deeper, and the time for his giving up is another. The sight of the free and fertile scientific life unfolding in Amsterdam in 1660 had suddenly transformed the young doctor into a scientist; in the same location a decade later, the sight of the intellectual plight of Christianity caused a break in his investigatory ingenuity. (Jørgensen, 1884, p. 152)

Jørgensen describes Stensen's scientific method as a unique preoccupation with the observation itself, a tireless empiricism. Leibniz's critique of Stensen for transforming from an excellent anatomist into a passable theologian is countered by Jørgensen with a scathing defence: as an anatomist, Stensen never placed his trust 'in the examinations and descriptions of others'. Jørgensen emphasises that Stensen:

> carried out his own examinations, and any theory that proved contradictory to his own findings lost its value to him, no matter who had presented it. He assumed the very same position as a theologian, with the sole exception that the object of his investigations here was the personal life, the relation he himself and others had with God, engendered in and nurtured by the church. (Jørgensen, 1884, pp. 180f)

Even though Jørgensen's assessments of Stensen's religious development are more clement and more nuanced than, for example, Wichfeld's, his assessment of Stensen's relation to the Danish academic community – or rather, its relation to Stensen – is much more callous. In various places throughout the biography, Jørgensen openly ridicules the fact that the Danish elite was not up to accommodating Stensen and giving him the position in Copenhagen's academic circles that Jørgensen believes his biographee deserved.[14] Jørgensen actually predicted that if Stensen had been living in Denmark at the time when his religious quest began, he would have remained a Protestant. He could have cultivated a deeper Christian life along the lines of the religious poet-bishop Thomas Kingo; in Jørgensen's enraptured phraseology, he could have become a 'Protestant Pascal'.

Jørgensen's biographical account is therefore not a tragic story of decay, inflated, as Wichfeld's is, to Greek proportions. On the contrary, it is a biographical narrative of an industrious, talented, ever-seeking scientist who, motivated by a feeling of inner necessity, chose to dedicate the final years of his life to theology instead of science. Consequently, the element of drama in Jørgensen's work does not lie in Stensen's conversion to Catholicism, but in his own native country's lacking ability to accommodate him. From this perspective, Jørgensen presents an important

13 Stensen did, however, express his scepticism of Descartes' philosophy, and he regarded the deification of human thought as dangerous (compare Jørgensen, 1884, p. 61).

14 See, for example, Jørgensen (1884), pp. 46f., 63, 120, 130, 211 and 213.

underlying argument for rehabilitating Stensen's reputation in his native Denmark and reinstating him as a national figure, and as 'an historical character who is so closely akin to the best in our national character' (Jørgensen, 1884, p. 214). Jørgensen's Stensen biography is thus part of an agenda for building a Danish national identity.

Børup: The Forced Poet

Marinus Børup's biography *Steno* from 1938 holds a special place not only in the Stensen literature, but in the entire Danish tradition for scientific biography. Marinus Børup (1891–1959) was a lesser-known author and lyricist,[15] and the second edition of the Danish biographical dictionary proclaims the Stensen biography to be his principal work. It is a long, 300-page prose poem, a literary genre which, to my knowledge, has never been used in the history of Danish scientific biography. It could be claimed that the term 'prose poem' is too restrictive to characterise the bursting, collage-like exposition in this biography. Børup himself applies the somewhat vague term 'images in verse' (as indicated by the subtitle of the book: *Hans Livs Historie fremstillet i Versbilleder*, 'The Story of His Life in Verse Images') to outline the form; and an unruly form it is, at least in terms of genre definition. The author's own description of his narrative technique and its underlying motivation runs as follows:

> The earliest attempts to write the Steno book – as a prose novel – or as a narrative poem – were made in 1932, but soon had to be aborted in the face of the vast subject-matter. When the aim was to include, as pregnantly as possible, all things significant to Steno – all his discoveries and the phases in his life – the form that proved to be applicable was a mosaic of scenes: a narrative conglomerate of direct dialogue – upon an underpinning of lyrical fugue motifs – and occasional monologues on certain ruminations and moods; and all of this in verse, not for ornamentation, but for concentration and originality, shifting with the nature of the narrative: triumph, ending in martyrdom. (Børup, 1938, p. 305)

It is evident from the finished text that Børup got carried away by the process, that the subject matter atomised, as it were, his work with the form, and that the coherent entity he was obliged to abandon – a revelation he saves for his epilogue – was perhaps too grand an ambition. Most of the biography consists of dialogues conducted between Stensen and the people who figured in his life and in the circles he frequented. The actual dialogue sequences make the text appear very much like a drama, a tendency that is reinforced by the structuring of the text into scenes and tableaux.

Even though Børup writes in the epilogue that 'the most significant content of the book [is] original material from the past, often reproduced verbatim', and that 'only a few portrayals of nature – with the exception of Holland – are based on self-experience' (1938, pp. 305–6), the text clearly shows that Børup has taken a number of dramatic and artistic liberties in relation to the subject matter. The very fact that he decides to arrange the narrative as direct dialogue implies that he is, quite

15 Børup is not mentioned in *Dansk Litteraturhistorie* (1983–85).

literally, putting the words into the mouths of his characters. The dialogue-based form and the scenic organisation mean that the pace of the narrative is slow, and that the dominating point of view is what in the French tradition (after Pouillon, 1946) is called *vision avec*, which means that the reader gains a temporal sense of following the events in real time, as though they 'were there'. This type of narrative normally reinforces the reader's participatory role in the course of events (as opposed to the panoramic portrayal, in which brief, summary passages can cover a far-reaching complex of events or span several centuries). And indeed, one's first impression of the characters appearing in Børup's text is stronger and more vibrant in the dramatised presentation. Yet, although in the above-mentioned quote Børup emphasises that the material has 'dictated' the form, and that the verses are not 'for ornamentation, but for concentration and originality', one can hardly disregard the way in which the distinct artistic reworking of the material creates a peculiar distance between the text and the reader.

In general, one must conclude that Børup's hefty dramatisation of Stensen's life contains elements that transcend the genre. His biography challenges what one might choose to call the epic naturalness of biography as a genre. It seems as though the author, with his artistic manoeuvrings, never ceases to wedge himself into the reader's direct relation to the text, as though one is never really allowed to 'be alone' while reading it. A more modern way of expressing it would be to say that the text draws attention to itself as a text. However, it does not do so by virtue of a meta-textual level in the text itself, nor by explicitly discussing its own textual status (typical examples of mechanisms referred to when describing a text as pointing back at itself – its self-referentiality). Børup's text draws attention to itself by means of its strong formal expression, for example in the passages with dialogue that appears as rhyming verse.

The fact that Børup primarily builds his biographical narrative in the form of dialogue has a significant impact on his approach to Stensen's life and research. For the most part, the point of view is that of Stensen and the people with whom he surrounds himself, and Børup therefore has limited opportunities for treating Stensen and his research *externally*. This means that his biography is mainly limited to providing a summary of Stensen's scientific work, while refraining from summarizing and analysing Stensen's work in broader terms. This non-critical and non-analytical position in respect to his biographee's scientific achievements permeates the biography, even though the author uses relatively long passages to describe Stensen's scientific work. This, too, mostly takes the form of direct speech, for instance when Stensen gives lectures and performs anatomical demonstrations,[16] engages in scientific discussions with colleagues,[17] or sits in his student's lodgings

16 For example, Stensen shows Paulii, Bartholin, Worm and others that the heart only is a muscle (Børup, 1938, pp. 63ff.). In another case (1938, pp. 96ff.), Stensen gives a lecture on the anatomy of the brain at the French scientific academy in Paris.

17 For example, Stensen discusses anatomical craftsmanship with Blasius, discovers the secretory duct of the parotid salivary gland while labouring at dissection (Børup, 1938, pp. 27ff.), discusses scientific theory with Spinoza and Swammerdam concerning empiricism

monologically 'philosophising'.[18] The dialogue-based presentation thus largely prevents Børup from regarding Stensen's life and works from without, from an analytical perspective, while simultaneously preventing the author penetrating into Stensen's inner spheres of thought and regarding his development from a more psychological or existential perspective. And it is actually characteristic that when describing his own perception of Stensen's development, Børup uses terms that denote external conditions, such as 'triumph, ending in martyrdom' (Børup, 1938, p. 305).

In Wichfeld's biography, the biographer's strongly judgemental perception of Stensen's religious conflict determines the course and strategies used in the narrative. In Børup's versified biography, on the other hand, the aesthetic and in literary terms formalistic project – that of the lyrical form – defines the framework within which the biographical narrative is able to unfold. And in this way, Børup's biography is a salient example of how aesthetic choices – on a par with the analytical and interpretational choices – can have a considerable impact on the way a biography portrays its subject. Judging by the perseverance with which Børup pursues his formal project, his aesthetics seems to become a programmatic project in its own right. The prominent position that poetry holds in Børup's narrative universe is underscored by a curious dialogue between the poet-dramatist Molière and his physician, who turns up rather unexpectedly and seems suspiciously like a ruse included to allow Børup to embed an alter ego in the words of Molière. The Frenchman declares:

> But if one is a poet, one may smile in the knowing
> that a poet will outweigh a doctor phil. in the showing.
> And knowing of so many a doctor in the grouping
> who tried his hand at poetry, then the critic's way went stooping
> makes doctoral designations seem even less desirous
> – unless they can endear one to one's patients-cum-admirers!
> Whatever stuff this offers to a lyrical concoctor,
> I still remain a good friend of one specific doctor,
> exempting him alone from physicians *in pleno* –
> *The Physician*: My thanks!
> *Molière*: That's quite all right – but the man's name is Steno. (Børup, 1938, p. 92; my italics)

Here, Børup makes his strategy quite clear: poetry takes precedence over medicine, although the author does grant access to Stensen, and to him alone.

---

versus theory (1938, pp. 39ff.), and debates his geological findings and observations of crystals at the academy in Florence (1938, p. 121ff.).

18 For instance, Stensen philosophises hypothetically about cells and microcosmos (Børup, 1938, p. 73), and philosophises about his own geological observations (1938, p. 133).

Plovgaard: The Fervent Catholic

Karen Plovgaard's 161-page biography, entitled *Niels Stensen. Anatom, geolog og biskop* ('Niels Stensen: Anatomist, Geologist, and Bishop') was published in 1953, a revision of a series of 26 articles that had appeared in the Danish Catholic journal *Katholsk Ugeblad* ('Catholic Weekly').

Plovgaard's biography is well researched and updated to reflect the newest Stensen studies. On top of this foundation of solid source work, however, there lies a educationally agitatorial coating. Plovgaard persistently includes the reader by using the pronoun 'we', which in an authoritative – some might even say dictatorial – way involves the reader in an involuntary interpretational agreement with the narrative element in the text. Although the effect of this type of targeting is partly contingent on personal preferences, the chosen form has an inherently authoritative inequality or imbalance, well known from didactic situations such as the classroom, agitatorial types of exposition like the political speech, and the typical caricature of the nurse speaking to a patient in the patronising 'we' mode.[19]

A more detailed investigation into other layers in Plovgaard's text reveals a religiously agitatorial track that runs parallel to the biographical narrative. Whereas Wichfeld strongly disapproves of the Catholic influence Stensen encounters during the travels of his youth, referring to it as 'propaganda' (Wichfeld, 1865, p. 5), Plovgaard speaks of 'the powerfully Catholic atmosphere' in Paris, claiming with reference to Stensen's conversion that there can hardly 'be any doubt that it served to prepare him for what would later befall him' (Plovgaard, 1953, p. 26).

Furthermore, whereas both Wichfeld and A. D. Jørgensen describe the period around Stensen's conversion as being marked by doubts and apprehension,[20] Plovgaard characterises the period following his conversion as 'a very happy time indeed' (1953, p. 49). Generally speaking, a number of interesting things concerning the author's narrative technique take place in her chapter on Stensen's conversion. The chapter begins with a description of the feeling of joy that suffuses those who are newly converted, including quotations from an anonymous woman about the peace and feeling of happiness she experienced after her conversion, and a passage from the English convert Arnold Lunn's book *Within that City* (1936) about the carefree feeling after a conversion. The author concludes that although the sources do *not* mention anything about how Stensen *himself* felt during his first months as a convert, 'his friends let us know'. She briefly quotes the Catholic Viviani, who expressed his enthusiasm at the conversion of Stensen, but gives no direct reports of any potential feeling of elation in Stensen himself (1953, pp. 49–50). This section of the book comes dangerously close to manipulating the source material, because in her narrative technique of embedding various accounts of the joy experienced by

19 As Albeck points out (1961, p. 64): '*we* is used to denote *you* (in the singular) to convey an amicably condescending tone, for instance in medical language: "What a healthy lad we are today" … or in the way mothers speak to their children, or similar protective tones'.

20 Jørgensen actually speaks of how Stensen later gives 'several different, apparently contradictory accounts of his conversion, all depending on how, at different times, he more strongly emphasizes this or that aspect' (1884, p. 119).

other converts, Plovgaard deduces that Stensen must have felt the same. The facts of the matter remain unknown, but the arguments serve the biographer's underlying errand of extolling the blessings of the Catholic faith.

Even the very composition of the biographical narrative as a whole indicates which elements Plovgaard considers important in Stensen's life. The section on his conversion appears as early as a third of the way into the account, whereas Wichfeld, Jørgensen and Børup relate the event halfway or more into theirs. In other words, Plovgaard dedicates a relatively larger proportion of her biography to the description of the time after her subject's conversion, and therefore to the time during which he actively served the Catholic Church. Plovgaard's principal interest lies with Stensen's religious development and his work for the Catholic cause. This is also why the last two-thirds of the biography are liberally strewn with direct quotes from Stensen's correspondence and writings in which he defends the Catholic faith and discusses issues relating to the profession of faith. This also means, however, that his scientific work slips into the background in Plovgaard's biography. Stensen's principal works are mentioned almost in passing, and since Plovgaard concentrates most of her biographical account on the years following his conversion – when his is endeavours in natural science are phased out and gradually cease – her portrayal of Stensen as a scientist is limited to a minimum.

Hence, in the final chapters of the biography, when Plovgaard presents an overall analysis of the course of Stensen's life, she also chooses to concentrate on his religious dedication:

> When seeking to trace the extraordinary course of Stensen's life, immersing oneself in his reactions to the phenomena of life unfolding in its various stages, one gradually begins to form a picture of a remarkable, richly faceted religious figure. Like an elaborately cut diamond, it shines with a rare clarity and purity. (Plovgaard, 1953, p. 147)

Plovgaard then proceeds to emphasise the number of beneficial traits in Stensen's personality, a virtual catalogue of positive qualities that supports the religious interpretation of Stensen as a person: he was '*receptive* to religious things', he felt 'in *awe* of the Creator', he had a 'pronounced *compassion* towards those who suffered, both humans and animals', he did not disappoint others thanks to 'the *unselfishness* in his character', his religious personality was characterised 'by his *honest search for truth*. The search is supported by his urge to carry out research' (and note here how the scientist's search for truth is made subservient to that of the religious seeker), he possessed a distinct sense of '*self-knowledge*' and personal '*humility*'. His '*sense of responsibility*' is stressed as being 'a very notable trait in Steno's religious personality', and is linked to a '*firmness*, with which he, as a man of the cloth, embraces and tends to his calling'. Plovgaard concludes her four-page review of Stensen's religious virtues by emphasising that the 'the most typical trait in Niels Stensen's life as a practicing Christian … is the *unremitting consistency* in the way he lives his faith' (1953, pp. 147–50).

This review of positive personal characteristics can serve as a model for the receptive reader, and this is where Plovgaard unambiguously reveals the underlying narrative strategy in her biography: the religious and morally didactic influence on the reader. As a consequence, Plovgaard's biography becomes inscribed in the centuries-old tradition of biography as an exemplary moral tale. The author underscores this function of her biography in the last chapter, where she presents the following conclusion on Stensen's works:

> Stensen has left behind learned works. He has also preached and laid out rules for others to live by. The results of his scientific research will be important for many years to come. His words on morals and faith have had their consequences and will always continue to do so. But he has mainly served through the most costly educational method of all: *by way of example*. (1953, p. 150)

It would be difficult, if not impossible, to express the goal of the morally didactic biography clearer than this. Plovgaard's biography reaches beyond the traditional edification intrinsic to the biographical genre, as her work explicitly pleads the cause of the Catholic faith. This is where biography becomes a rhetorically seductive genre, intended to promote the views of a certain religious group; where it becomes, in effect, a programmatic and argumentative text.

The Programmatic Function of Biography

These four examples from the Danish biographical tradition on Niels Stensen illustrate how the genre of scientific and scholarly biography is upheld to varying degrees by the particular interpretational patterns that a given biographer chooses to superimpose upon his subject. It is the effect of these explicitly and implicitly controlling thoughts – which may be of a contemporary nature (relating, for example, to research policy or national or religious policy), or of a more timeless nature (for instance, aesthetic) – that I have referred to here as the programmatic function of biography.

There can be no doubt that the strength and intensity of a biography are dependent on its being carried forward by a general narrative technique or a train of thought that ensure the progression of the account, while simultaneously creating cohesion and consistency in the presentation of the scientist's life. There must be an underlying intention. I hope these four examples can serve to encourage further contemplation and discussion of the frameworks that can apply to such fundamental, biography-controlling intentions.

Bibliography

Aggebo, Anker (1937), *Danmarks store Søn Niels Steensen, 1638–1938: Danmarks største Anatom, Geologiens Grundlægger, Katolsk Biskop. Et Mindeskrift af Anker Aggebo,* Aarhus: Universitetsforlaget i Aarhus.

Albeck, Ulla (1961), *Stilistik*, 4th edn, Copenhagen: Gyldendal 1939.

Anon. (1938), 'Niels Steensen: Anatom, Geolog, Biskop', *Nordisk Ugeblad for katholske Kristne*, **2**.

Aristotle (1948), *Poetics*, trans. S.H. Butcher, New York: Liberal Arts Press.

Bech, Svend Cedergreen (ed.) (1979–84), *Dansk biografisk leksikon*, 3rd edn, Copenhagen: Gyldendal.

Beck, Peter (1972), *Steno: På rejse med Niels Stensen*, Copenhagen: Gjellerup, Mono-serien.

Berg, Jaques (1993), *Himmelstigen: En roman om den lærde Niels Stensen, der blev Guds og troens lidenskabelige tjener*, Copenhagen: Forum.

Bjarnhof, Karl (1972), *Støv skal du blive: På spor af Niels Steensen*, Copenhagen: Gyldendal.

Børup, Marinus (1938), *Steno: Hans Livs Historie fremstillet i Versbilleder*, Copenhagen: Gyldendal.

Bricka, C.F. (ed.) (1887–1905), *Dansk biografisk Lexikon, tillige omfattende Norge for Tidsrummet 1537–1814*, 1st edn, Copenhagen.

Churchill, Frederick B. (1982), 'Darwin and the historian', in R.J. Berry (ed.), *Charles Darwin: A Commemoration 1882–1982: Happy is the Man that Findeth Wisdom*, London and New York: Academic Press, pp. 45–68.

Dahl, Svend and Povl Engelstoft (eds) (1920–26), *Dansk biografisk Haandleksikon*, Copenhagen and Kristiania: Gyldendalske Boghandel, Nordisk Forlag.

Dansk Litteraturhistorie (1983–85), vols 1–9, Copenhagen: Gyldendal.

Engelstoft, Povl and Svend Dahl (eds) (1933–44), *Dansk biografisk Leksikon: Grundlagt af C. F. Bricka; red. af Povl Engelstoft under Medv. af Svend Dahl*, 2nd edn, Copenhagen: J.H. Schultz Forlag.

Fabroni, Angelo (1779), *Vitæ Italorum doctrina excellentium, qui saeculis XVII et XVIII floruerunt*, vol. III, Pisis, pp. 1–63.

Guerlac, Henry (1954), 'Lavosier and His Biograhers', in *Isis*, 45, pp. 51–62.

Hansen, Jens Morten (2000), *Stregen i sandet, bølgen på vandet: Stenos teori om naturens sprog og erkendelsens grænser*, Copenhagen: Fremad.

Hansen, Signe Lindskov (forthcoming), *Den danske forskerbiografi fra 1750 til i dag*, Copenhagen: Museum Tusculanum Press.

Jørgensen, A.D. (1884), *Nils Stensen: Et Mindeskrift*, Copenhagen: Bianco Lunos Kgl. Hof-Bogtrykkeri.

Jørgensen, A.D. (1958), *Nils Stensen. Et Mindeskrift*, 2nd rev. edn, Copenhagen: Munksgaard.

Koch, Carl Henrik (2003), Niels Stensen og naturiagttagelsen, Lyngby: Polyteknisk Forlag.

Kongsted, Ole, Jørgen Nybo Rasmussen and August Ziggelaar (eds) (1981), *Niels Stensen og Danmark: kirkens fattige tjener*, Copenhagen: Katolsk Forlag.

Larsen, Knud (1933), 'Johann von Rosen: La vie de Nicolas Stenon', in Meisen, Valdemar and Knud Larsen (eds), *Stenoniana*, vol. 1, Copenhagen: Hafnia, pp. 126ff.

Lund, Jørn (ed.) (1994–2001), *Den Store Danske Encyklopædi*, Copenhagen: Danmarks Nationalleksikon.

Manni, Domenico Maria (1775), *Vita del letteratissimo monsig. Niccolò Stenone di Danimarca, vescovo di Titopoli e vicario apostolico*, Florence.

Meisen, Valdemar (1932), 'Første Niels Steensen Forelæsning holdt i Medicinsk Selskab 11. Jan. 1932', special edition of *Hospitalstidende*, [Copenhagen], pp. 242–69.

Metzler, Johannes (1928), *Niels Steensen: En stor Naturforsker og Biskop. Et kort Livsbillede*, Copenhagen; translated from German (1911).

Moe, Harald (1988), *Niels Stensen: En billedbiografi. Hans utrættelige søgen, hans geni, hans stræben mod det absolutte. Skrevet til erindring om Niels Stensen ved 350-året for hans fødsel*, Copenhagen: Rhodos.

Mortensen, Miriam (1993), *Skønnest af alt. En biografi om Niels Steensen*, Aarhus: Katolsk Forlag.

Nygaard, Jens Jørgen (1989), *Niels Stensen*, Copenhagen: Forum.

Outram, Dorinda (1976), 'Scientific biography and the case of George Cuvier. With a critical bibliography', *History of Science*, **14**, 101–37.

Petersen, Carl S. and Vilhelm Andersen (1924–34), *Illustreret dansk Litteraturhistorie*, Copenhagen: Gyldendal.

Plovgaard, Karen (1953), *Niels Steensen: Anatom, geolog og biskop*, Copenhagen: Pauluskredsens skrifter.

Pouillon, Jean (1946), *Temps et roman*, Paris: Gallimard.

Rosen, Johannes von (1906), *Nicolaus Stenos Liv og Død*, trans. Vilhelm Maar, Copenhagen.

Scherz, Gustav (1956), *Vom Wege Niels Stensens: Beiträge zu seiner Naturwissenschaftlichen Entwicklung*, Copenhagen: Munksgaard.

Scherz, Gustav (1963), *Niels Stensen*, Copenhagen: Gads biografiserie.

Scherz, Gustav (1987–88), *Niels Stensen. Eine Biographie*, vols I–II, Leipzig: St. Benno-Verlag.

Scherz, Gustav (1988), *Niels Steensen (Nicolaus Steno) (1638–1686): Guldsmedesønnen fra København, der indlagde sig verdensberømmelse som banebrydende naturforsker, men ofrede videnskaben for at blive en Guds tjener af legendarisk ry*, Copenhagen: Den Katolske Kirke i Danmark.

Snorrason, Egill (1966), *Niels Stensen: En students notater fra 1659*, Copenhagen: Mölnlycke A/S, Sygehusafdelingen.

Söderqvist, Thomas (1996), 'Existential projects and existential choice in science: Science biography as an edifying genre', in Michael Shortland and Richard Yeo (eds), *Telling Lives in Science: Essays on Scientific Biography*, Cambridge: Cambridge University Press, pp. 45–84.

Söderqvist, Thomas (2006), 'What's the use of writing lives of recent scientists?', in Ronald E. Doel and Thomas Söderqvist (eds), *The Historiography of Contemporary Science, Technology and Medicine: Writing Recent Science*, London: Routledge, pp. 99–127.

Suhr, J.T. (1942), *Niels Steensen som Menneske: Det medicinske Selskabs i Københavns 10. Niels Steensen-Forelæsning afholdt d. 17/2 1942*, Helsingfors.

Theerman, Paul (1985), 'Unaccustomed Role: The Scientist as Historical Biographer – Two Nineteenth-century Portrayals of Newton', *Biography*, **8** (2), 145–62.

Wichfeld, J. (1865), *Erindringer om den danske Videnskabsmand Niels Stensen – Nicolaus Steno med Portrait og Facsimile*, special edition of *Historisk Tidsskrift*, **3** (4).

Worm, Jens (1771–84), *Forsøg til et Lexicon over danske, norske og islandske lærde Mænd, som ved trykte Skrifter have giort sig bekiendte, saavelsom andre Ustuderede, som noget have skrevet, hvorudi deres Fødsel, betydeligste Levnets*

Omstændigheder og Død ved Aarstal kortelig erindres, og deres Skrifter, saavidt mueligt, fuldstændig anføres, Helsingor and Copenhagen

Young, Robert M. (1987), 'Darwin and the genre of biography', in George Levine (ed.), *One Culture: Essays in Science and Literature*, Madison, WI: University of Wisconsin Press, pp. 203–24.

Chapter 9

Discriminating Days? Partiality and Impartiality in Nineteenth-Century Biographies of Newton

Rebekah Higgitt

Introduction

The 1830s–1850s saw an active debate regarding the life and character of Isaac Newton, generated and sustained by a series of publications that included or discussed previously unknown source materials. While the fundamental importance of Newton as a scientific icon has been well established, the relationship between the use of his reputation and its presentation within contemporary biographical models has not been fully considered.[1] This chapter is therefore concerned with how the various nineteenth-century depictions of Newton were presented and made persuasive. I will discuss authors' choices in the selection of source material, and the authority they derived from an apparently 'impartial' presentation of evidence. Their primary choice was, however, between continuing the convention of Newtonian eulogy bequeathed by the eighteenth century and providing a critique of this tendency, and I argue that the attitude of the writer to Newton's position within the natural theological tradition was a key element in dictating this decision.

An examination of the forms of biographical writing used in the debates over Newton allows us to develop a more subtle understanding of nineteenth-century biography. Typically, this genre has been understood as invariably demonstrating both the attempt to establish the heroism of the subject and an unquestioning trust in documentary evidence (for example, Cockshut, 1974). However, this essay attempts to show that the former was not always the case, although it does argue that a moral impulse remained even when the hero did not, and that attitudes to archival material might vary considerably, for its inclusion was not inevitable, but contingent. The archives became a resource for both critics and defenders of an idealised image of Newton, as did claims to impartiality and accusations of partiality. This latter point is illustrated in the article to which the title of this essay refers: a review by Augustus De Morgan, Professor of Mathematics at University College London, of the 1855 *Memoirs of Sir Isaac Newton* by David Brewster, a researcher in physical optics and a College Principal at St Andrews University. De Morgan, also a biographer of Newton, believed that they were living 'not merely in sceptical days … but in discriminating days, which insist on the distinction between intellect and morals'

1 See Yeo (1988), Theerman (1985), Fara (2002) and Higgitt (forthcoming).

([De Morgan], [1855] 1914, p. 122), but that biography had not yet responded sufficiently to this development: 'Though biography be no longer an act of worship, it is not yet a solemn and impartial judgement: we are in the intermediate stage, in which advocacy is the aim, and in which the biographer, when a thought more candid than usual, avows that he is to *do his best* for his client' (De Morgan, 1914, pp. 124–5; original emphasis). De Morgan thought biographers should aim instead for the impartiality of the judge. This is close to what we now understand as 'objectivity', a concept that Lorraine Daston (1992) has shown was in the process of formation in the nineteenth century. Crucially, both she and Peter Galison have also pointed to the simultaneous 'moralisation of objectivity' (Daston and Galison, 1992, p. 81). I argue that the morality attached to De Morgan's 'impartiality' can be considered to be in antithesis to the overt moralising of the eulogistic presentation of Newton by Brewster.[2]

It has been suggested that the scientific biographer is 'limited by the thoughts and deeds of his chosen subject', by 'the demands and appraisals that his own age makes of sciences and of scientists' and also 'by what his contemporaries demand of biography' (Fullmer, 1967, p. 285). I wish to add that he is also 'limited' by his personal understanding of the aim of biography, for there were a variety of models from which to choose. Should it be 'philosophy teaching by examples'?[3] Was the aim to teach 'at the same time great moral lessons, both as to the value of labour and industry, and the necessities of *virtues*, as well as intellectual endowments, for the attainment of lasting excellence' ([Jeffrey], 1835, p. 209)? Was it the biographer's duty 'to select such portions as may illustrate the public doings of the man' ([Aytoun], 1849, p. 222), or to show 'those little personal traits which reflect the distinctive lights of a marked individuality' ([Anon.], 1879, p. 485)? For Thomas Carlyle, the 'History of the world is but the Biography of great men' who should be revered as heroes (Carlyle, 1841, p. 21). More typically, biography was viewed as a branch of history, but to what extent was the new critical historical approach adopted by biographers?

De Morgan chided Brewster for being 'too much of a biographer and too little of an historian', and his comments suggest that he thought biography inevitably biased ([De Morgan], 1914, p. 124). However, he did not eschew the genre, and he and others attempted to demonstrate critical awareness in their interpretation of sources. The translation of historical methods to biography was, as I will show, understood to render accounts more convincing and authoritative, despite the fact that problems might arise when the presentation of an individual as a moral exemplar conflicted with information brought to light by research. This essay outlines the approaches of Newton's biographers and considers how these related to their attitude to Newton as an exemplar.[4] In particular, I contrast Brewster's first biography, largely based on

2 See also Galison (1999), and for an exploration of this ethos transferred to literature, see Levine (2002).

3 On the origin and frequent use of this phrase, see Reed (1966), pp. 16–18.

4 I use 'biography' in the widest sense, including writings that focus on the life of an individual, whether straightforward narratives, critical reviews or collections of manuscript material.

secondary sources, with subsequent publications, including his own later *Memoirs of Newton*. I highlight three approaches: 'inductive' biography, which consisted almost entirely of primary sources, 'judicial' biography, which sought to judge the available sources and conflicting interpretations, and Brewster's later approach, which moderated the partisanship of his first biography by adopting aspects of the other two.

Biographies of Newton, 1806–55

From the eighteenth century, biographers had emphasised the reproduction of original materials, and this trend became something of a fetish in the following century, with the publication of lengthy *Lives and Letters*.[5] By putting as much of the biography as possible in the subject's own words, the author, often a close relation, could demonstrate both humility and a lack of bias. However, the propriety of printing private documents was always questionable. As Tennyson said: 'What business has the public to know about Byron's wildnesses? He has given them fine work and they ought to be satisfied.'[6] Many felt it was particularly despicable to speak ill of the dead, and Diogenes Laertius' maxim 'de mortuis nil nisi bonum' was frequently cited. That the dispute surrounding Newton's reputation was fuelled by revelations from the archives, which significantly altered the depiction of his life and character, suggests that the remains of an historical figure of this magnitude were treated differently from those of the recently deceased.[7] Although Brewster was shocked that Newton 'should be attacked more than a hundred years after his death … when the hand and the tongue of the accused and his contemporaries were safely mouldering in the grave' (Gordon, 1869, pp. 260–61), others felt that, as biography invaded the sphere of history, this passage of time allowed, or required, an alternative response.

The following paragraphs outline the new material relating to Newton's life published between 1800 and 1855.[8] In 1806, Edmund Turnor, an MP whose family had bought Newton's estate at Woolsthorpe, published some items from the Newtonian manuscripts owned by the Portsmouth family.[9] These included reminiscences collected by Newton's young supporters, John Conduitt and William Stukeley, utilised by all subsequent biographers (Turnor, 1806, pp. 160–86). Sixteen years later, Jean-Baptiste Biot, a follower of Laplace, wrote a life of Newton that

5 Altick (1966), Ch. 6. On eighteenth-century biography, see Stauffer (1941).

6 Quoted by Tuchman (1981), p. 147. I have been unable to locate the original source.

7 The interest in Newton's literary remains began on his death, but little new material had been incorporated into accounts of his life since Fontenelle's *Eloge* in 1727, and views of what materials were of interest had substantially changed; see Hall (1999).

8 Most of the following works are briefly discussed in Yeo (1988). A comparison of De Morgan and Brewster's biographies appears in Theerman (1985). De Morgan's historical work is discussed in Rice (1996) and Richards (1987), and Brewster's in Christie (1984).

9 The Portsmouth Papers included Newton's manuscripts and some material added by his niece, Catherine Conduitt, and her husband. On examinations of this collection in the eighteenth century, see Iliffe (1998), pp. 140–45.

contained evidence indicating he had suffered a mental breakdown in 1692–93. A translation of this biography was published by the Society for the Diffusion of Useful Knowledge (SDUK) in 1829, which included an additional source that appeared to confirm the story (Biot, 1833). Brewster's 1831 biography drew attention to the significance of previously published correspondence, for example in Lord King's *Life of John Locke* (Brewster, 1831; King, 1829). Brewster wrote to counter Biot's thesis that the effects of the breakdown were permanent, but added plausibility to its occurrence.

The *Account of the Revd. John Flamsteed, the First Astronomer Royal* (1835) by Francis Baily consisted entirely of previously unknown material, with only the editor's preface and notes. It contained Flamsteed's copious complaints about Newton and Edmund Halley, provoking robust criticism from certain quarters.[10] De Morgan's research on Newton, published by the SDUK and in various periodicals, largely relied on published sources, but his interest in bibliography meant he recovered many forgotten texts.[11] His 1846 biography collated information regarding Newton's religious beliefs, and concluded that he was Antitrinitarian. His investigations in the Royal Society library led to his theory that Newton had orchestrated the committee appointed to arbitrate the dispute with Leibniz over who had priority in the invention of the calculus, and had written sections of the anonymous *Commercium Epistolicum* that adjudicated in his favour. De Morgan's writings on the relationship between Newton's niece, Catherine Barton, and his friend and patron, Lord Halifax, were likewise mostly based on contemporary publications, but did include a new manuscript source that he considered a 'clincher' regarding his hypothesis that they had been privately married.[12]

S.P. Rigaud, the Savilian Professor of Astronomy at Oxford, has been called 'the principal instaurator of serious modern Newtonian scholarship' (Cohen, 1972, p. v). His *Historical Essay on the First Publication of the Principia* (Rigaud, 1838), which gave a detailed history of that book's production and highlighted the crucial role performed by Halley, was based on previously unpublished material.[13] Joseph Edleston, a Fellow of Trinity College Cambridge, published the college's *Correspondence of Newton and Cotes* (Edleston, 1850). This, which related to the publication of the second edition of the *Principia*, was printed with copious notes and a 'Synoptical view of Newton's life' that included details from the college records and collated references to nearly all the published evidence relating to Newton.[14] Other collections of documents added to the sources available for Newton's biography,

10 Baily (1835). This book and the debates following its publication have been discussed in Ashworth (1997) and Higgitt (2004).

11 See De Morgan (1847).

12 De Morgan to Herschel, 15 August 1856, Royal Society, Herschel Papers, 6.286. De Morgan's writings include De Morgan (1846), [De Morgan] 1914 [1855]), De Morgan (1852) and De Morgan (1885). For a list of De Morgan's writings, see De Morgan (1882). Rice (1996), pp. 208–9, has a list of De Morgan's principal historical writings.

13 Riguad's historical writings are briefly discussed in Fauvel, Flood and Wilson (2000), pp. 160–61. His part in the debates about the *Account of Flamsteed* is discussed in Ashworth (1997).

14 Edleston is little known, but is discussed most fully in Higgitt (forthcoming).

for example the *Correspondence of Scientific Men* (Rigaud, 1841), taken from the collection of Lord Macclesfield, and Dawson Turner's *Letters from Newton to Covel* (Turner, 1848).

Brewster's 1855 *Memoirs of Newton* were, like his earlier biography, conceived as a defence of Newton, this time against the accusations contained in Baily's *Account*. This had provoked Brewster to search the Portsmouth Papers in 1837 for evidence to refute Flamsteed's version of events. However, the papers revealed that De Morgan had been correct in believing that Newton's religious beliefs were Unitarian and that he had played an active role in the dispute with Leibniz. They also demonstrated the extent of Newton's interest in alchemy, already indicated in letters published by Lord King. Brewster published much of this problematic evidence, but continued to make advocate-like decisions regarding what to reveal and how to present it to the reader. His was the longest biography of Newton produced in the nineteenth century, and it long remained the standard work. John Christie has suggested that, within Newtonian biography, 'Brewsterian categories and dichotomies still infect the very framing of the questions we pose' (Christie, 1984, p. 56), and although his biases were easily perceived, the book's influence has been lasting. However, the extent to which the contents and approach of Brewster's *Memoirs* were reactive to the writings of others should not be underestimated.

Zealously Partisan: Brewster's *Life of Newton* (1831)

Brewster was able to assert the importance of even his shorter 1831 biography, the fullest treatment of Newton's life produced by this date. From this superior position, he complained of the scanty amount of material already collected, asserting that it was 'not creditable to his disciples that they have allowed a whole century to elapse without any suitable record of his life and labours' (Brewster, 1831, p. v). He highlighted his own use of archives, giving thanks to a number of individuals 'for much valuable information' from several Oxbridge libraries and a private collection (1831, p. vi). However, despite the prominence given to his search for sources, Brewster, with hindsight and in comparison with his later biography, admitted that this was 'a mere sketch, drawn up from materials known to every body'.[15] It differed little from eighteenth-century accounts, except with regard to Newton's breakdown and on several issues of peculiar interest to Brewster: the history of optical science, the neglect of science by the government, and the worthlessness of the Baconian methodology.[16]

Brewster described Newton's 'social character' as 'such as might have been expected from his intellectual attainments. He was modest, candid, and affable, and without any of the eccentricities of genius' (1831, p. 337). This judgement, designed to undermine the image of the inspired but tortured Romantic genius propagated in J. B. Biot's article, tended to conflict with Brewster's own assessment of Newton

15 Brewster to Brougham, 11 November 1854, University College London: Brougham Papers 11,248.

16 On Brewster's life, which was reflected in his works, see the essays in Morrison-Low and Christie (1984). On Brewster's anti-Baconianism, see Yeo (1985).

as a genius whose discoveries were the result of imaginative leaps rather than 'the plodding drudgery of inductive discipline' (1831, p. 336), with the published evidence regarding his illness and with John Locke's claim that he was 'a nice man to deal with', being 'apt to raise himself in suspicions where there is no ground' (King, 1829, pp. 224–5, 259). Choosing to ignore this judgement, Brewster echoed instead the eulogistic words of Conduitt, who wrote of Newton's 'meekness and sweetness', and asserted that his 'whole life was one continued series of labour, patience, charity, generosity, temperance, piety, goodness, & all other virtues, without a mixture of any vice whatsoever'.[17]

For Brewster, intellectual eminence and morality were firmly connected. The purpose of biographical writing was to display this connection and describe the subject's achievements to inspire the reader. Negative characteristics were both a mischievous example and, arguably, irrelevant to the celebration of public achievements. As John Davy wrote in the biography of his brother Humphry: 'To hold up the infirmities of a man of genius to observation is neither necessary nor useful; on the contrary, injurious, as tending to lower him as an example in the minds of posterity, and diminish the influence of his name' (Davy, 1836, vol. 2, p. 410). Brewster's reviewers, however, did not all endorse this view. In the *Edinburgh Review*, B.H. Malkin commented on Brewster's anxiety to support certain positions, warning that 'we cannot entertain any very high opinion of the accuracy with which he examines any facts which appear to bear on them' ([Malkin], 1832, p. 3).[18] Thomas Galloway discussed the *Life of Newton*, and a review of it by Biot, in the *Foreign Quarterly Review*. He suggested that Brewster had been 'animated by the spirit of a zealous partisan', and declared that this 'unhappy spirit of prejudice and intolerance' was 'alien to philosophy, and … incompatible with the impartial investigation of historical truth' ([Galloway], 1833, pp. 7, 24).

Some years later, De Morgan commented in the SDUK's *Penny Cyclopaedia* that the reader of Brewster's biography was misled, for he 'could only infer that [Newton's] moral character had suffered from no one instance of human infirmity, and that every action had been dictated by feelings of benevolence and the love of truth' ([De Morgan], 1840, p. 202).[19] Malkin and Galloway were also connected to the Utilitarian and non-Sectarian SDUK, and it is possible that all three men felt that they should write in defence of the account by Biot, published in translation by this society. However, their championing of 'impartial' history and criticism of Brewster's use of sources and his tendency to hero-worship went beyond this. They claimed superiority in their historical awareness by using Brewster's own

17 J. Conduitt, 'Memoir of Newton', King's College Cambridge: Keynes Ms. 129/1, p. 23 (punctuation added), reprinted in Turnor (1806), pp. 158–67. Conduitt sent this memoir to Fontenelle, through whose *Eloge* it was echoed in all subsequent accounts.

18 The *Wellesley Periodicals Index* attributes authorship to Malkin's father (1769–1842), but the true authorship is clear from Malkin's letters to Macvey Napier, editor of the *Edinburgh Review*: Malkin to Napier, 14 and 28 June 1832, and 14 August 1832, British Library: Napier Correspondence, Add. Ms. 34,615, fols 346, 354, 386.

19 An annotated copy of the *Penny Cyclopaedia* at the British Library gives the author of this article as 'Lecappelain'. I have found no other reference to such an individual, and Jourdain credits it to De Morgan in De Morgan (1914), p. 3.

sources to produce an image of Newton at variance with the paragon of all virtues created in the previous century. Writing within a tradition of rationalist criticism that queried the acceptance of religious, political and intellectual authority, they objected to Brewster's promotion of Newton within the framework of natural theology, a concept developed to support the political status quo and Anglican Church as much as to legitimise science.[20]

Biography by Induction: Baily, Rigaud and Edleston

De Morgan wrote that in the second quarter of the nineteenth century, a few men 'set the example of delving into one thing at a time, and presenting both the results and a full account of the sources from which they had been obtained' (De Morgan, 1885, p. 127). These included Rigaud and Baily, who both published large quantities of manuscript material. Baily was not a member of the SDUK, but he moved within the same reformist London circles and was close to De Morgan and Galloway. All three were Fellows of the Royal Astronomical Society (RAS), of which Baily was both 'parent and protector' ([Herschel], 1844, p. 119).[21] Most of the *Account* consisted of Flamsteed's unpublished correspondence and autobiography. Such materials were commonly used within biography, and three decades earlier Robert Southey had commented that this method was 'lively and entertaining, and carries with it a strong impression of authenticity'. However, he noted that frequently the 'thread of the narrative is broken, and all the proportion of length, to importance of matter, destroyed'. Southey insisted on the creative role of the biographer, and felt that this approach should only be used 'where the biographer is conscious of a paucity of materials for his own share of the work, or of some nice and delicate points in the story, upon which he does not choose to express himself with the responsibility of an author' ([Southey], 1804, p. 457).[22] Baily emphasised the fact that he presented Flamsteed's life '*in his own words*' (Baily, 1835, p. xviii; original emphasis), and Southey's latter point is relevant to this case; the papers were undeniably contentious, and Baily wished to minimise the author's role.

Baily intended that his history should carry the same values of 'objectivity' as his science, being based on 'facts' with a minimum of interpretation. As William Ashworth has written: 'History, like Baily's (and Flamsteed's) astronomy, was to be about documented facts as certain as stars passing across the wires of a micrometer'

20 See Gascoigne (1988) and Turner (1993), pp. 76–80, 101–10. Members of the SDUK were divided over the use of natural theology, but it is significant that in the early 1830s the society rejected the proposal by their Chairman, Lord Brougham, to publish a new edition of Paley's *Natural Theology*; see Topham (1992), pp. 404–7.

21 These friendships are described in De Morgan (1882), for example pp. 46–7.

22 In Southey's view, the subject's 'self-love' meant his true motivations were less likely to be revealed in his correspondence and journals than by the 'sagacious and impartial biographer' (Southey, 1804, p. 457). Opposing this, Francis Jeffrey thought 'private letters, journals, or written fragments of any kind, by the party himself' more satisfying than 'the biographer's own (generally most partial) description and estimation of his author's merits' ([Jeffrey], 1835, p. 207).

(Ashworth, 1997, p. 203). Baily's scientific persona, as described by John Herschel, was projected into his historical work: 'Far-sighted, clear-judging, and active; true, sterling, and equally unbiassed by partiality and fear; upright, undeviating, and candid, ardently attached to truth, and deeming no sacrifice too great for its attainment' ([Herschel], 1844, pp. 118–19). His scientific reports were known for their careful description of equipment and recording of results, and this was matched by his emphasis on the collection, arrangement, cataloguing and transcription of the 'vast mass' of Flamsteed's papers (Baily, 1835, p. xiv).[23] Like the SDUK reviewers, Baily implied that his unbiased approach to the sources allowed him to attack the methods of other writers, including Brewster, who had discussed Flamsteed and had done damage through 'partial statements and unfounded remarks' (Baily, 1835, p. xvii). He declared that 'a proper regard for truth and justice prevents any suppression, at the present day, of the many curious and important (though at the same time lamentable) facts which these manuscripts have, for the first time, now brought to light' regarding Newton and his circle (Baily, 1835, p. xx).

Like the SDUK men, therefore, Baily believed that historical understanding was undergoing a process of improvement. He noted that the Newton manuscripts had long been thought to be of no value because, on Newton's death, they had been declared not fit to be printed. He, however, found much 'that is *now* highly interesting, and not generally known', and hoped that they would soon be examined, preserved and at least partially published (Baily, 1835, p. xxi; original emphasis). However, despite his claim to be presenting the results of dispassionate historical enquiry, the *Account* was inevitably biased, for it reproduced evidence from only one side of the argument between Newton and Flamsteed, and it was clear that Baily's scientific and personal sympathies lay with the latter. Flamsteed was intended, every bit as much as Brewster's Newton, to be a hero worthy of emulation, but one whose achievements were more attainable, and therefore more democratic. However, Baily printed evidence demonstrating that both Newton and Flamsteed were difficult men, and at times found in favour of the former. Regarding the reputation of both men, he used original sources to undermine previous accounts.

Rigaud's *Historical Essay* (1838) was innovative in treating a particular episode in the history of science, the publication of the *Principia*, in detail. However, its reliance on manuscript sources was in keeping with his earlier historical works, including the 1832 *Works and Correspondence of James Bradley*. This, like Baily's *Account*, prefixed a biographical preface to a large amount of primary evidence. Rigaud claimed that he had 'in every instance in my power derived my information from original authorities' and done his 'best to verify the facts and dates which are taken from printed accounts' (Rigaud, 1832, [p. 4]). An explicit connection was made between this historical approach, observational astronomy and morality by the antiquary Philip Bliss, who wrote that Rigaud was 'constantly applying to his own moral improvement, the accuracy of observation, and correctness of judgment, which qualified him for mathematical pursuits, and enabled him to recover and

23 On the scientific style promoted by Baily and the RAS, see Ashworth (1994). Baily's recognition of a similar approach from Flamsteed was fundamental to his decision to publish the *Account*.

ascertain so many particulars' relating to the lives of his biographical subjects.[24] Despite serious differences in their attitude to the historical figures of Newton and Flamsteed, Rigaud and Baily, who were in frequent communication, shared scientific and historical interests and a painstaking style of research in both areas that was invested with moral value.[25]

Rigaud's exhaustive use of original sources gave his history obvious authority. The conclusions of the *Historical Essay*, apparently derived inductively from the 'facts', seemed judicial and impartial. However, the Tory and staunchly Anglican Rigaud feared for Newton's reputation, telling William Whewell that 'if Newton's character is lowered, the character of England is lowered and the cause of religion is injured'.[26] His negative reaction to the publication of Flamsteed's papers demonstrates that he believed evidence had to be carefully selected. A similar caution was exercised with regard to his discussion of Edmund Halley, about whom Flamsteed had been even more vicious. Although he was candid in admitting to Whewell that Halley was 'a harsh overbearing man' and that he saw 'no possibility of getting rid of the prophaness [*sic*] of his conversation', he thought that since he was 'not aware of his having been guilty of the same faults in print, more harm therefore than good is now done by dwelling on his faults in this respect'.[27] Instead, his publications sought to refute rumours about Halley's atheism and immorality.[28]

Edleston was another conservative attached to an Anglican stronghold who used a form of historical writing typically connected with non-partisan scholarship. However, his *Correspondence of Newton and Cotes* (1850), which consisted of historical texts and scholarly notes, was a vehicle for controversial opinions. Hidden within the densely written notes were a series of discussions on controversial aspects of Newton's life that countered the findings of Biot, Baily and De Morgan. After a brief look through these notes, De Morgan detected a 'strong Anti-Leibnitio-Flamsteedian bias', and in a subsequent essay on Leibniz he declared that he and Edleston differed significantly on matters of opinion.[29] He was, however, much more appreciative of Edleston's scholarship than of Brewster's, stating that the

24 Addition by Bliss to an obituary, possibly printed in the *Oxford Herald*, Bodleian Library: Rigaud Family Papers, MS.Eng.misc.c.807. Rigaud also researched the lives of Halley, Thomas Harriot, Jeremiah Horrox and John Hadley.

25 The manuscripts of both men testify to their research style. The Rigaud Papers are held at the Bodleian Library, and include a series of notebooks on scientific and historical subjects. Baily's manuscripts are held at both the Royal Astronomical Society and among the Royal Greenwich Observatory papers in the Cambridge University Library, and demonstrate sometimes identical means of recording data in history and experimental science.

26 Rigaud to Whewell, 25 January 1836, Trinity College Cambridge: Whewell Papers, Add.Ms.a.211/79.

27 Rigaud to Whewell, 31 January 1836, Trinity College Cambridge: Whewell Papers, Add.Ms.a.211/80.

28 The *Historical Essay* began a defence of Halley that Rigaud intended to continue in a biography. Some of the research he completed before his death in 1839 was published by his son as *A Defence of Halley Against the Charge of Religious Infidelity* (1844).

29 De Morgan to Whewell, 8 January 1851, Trinity College Cambridge: Whewell Papers, Add.Ms.a.202/123.

notes contained 'such research and digestion as make it difficult to give adequate praise to the whole without appearance of exaggeration' (De Morgan, 1852, p. 74). Once clear citations were provided, he implied, questions of opinion could be more comprehensively debated, and in this essay he could tackle Edleston's interpretation with reference to the same sources.

Judicial Biography: De Morgan

The article 'Biography' in the *Penny Cyclopaedia* claimed that it was 'properly … only a branch of history' (Long, 1833–46, vol. 4, p. 414). For De Morgan, however, it had to be treated with more caution because it was at greater risk of bias. In a letter to George Long, editor of the *Penny Cyclopaedia*, he explained his view that lives outlined in biography contained 'Details of the life [and] inflated reputation', while histories included 'No details of life … [but an] acc^{t} of labours'.[30] He stressed the need for new research by persons with knowledge of the field: 'I do not think it is safe to leave a biographical researcher who is not a mathematician to copy any old opinion, or state the impression in his own mind from old opinions.'[31] Equally, he insisted that men of science attempting a history should have adequate knowledge of the period in question. He admired men like Gugleilmo Libri and Michel Chasles, whose expertise in both areas made them, in his view, 'scholars' rather than 'technologists'.[32] He emphasised the careful collection and non-judgemental reading of sources, and was, like Baily, a member of the short-lived Historical Society of Science (founded 1840), which aimed to publish source materials for the history of science (Dickinson, 1932). However, he admitted that in his 1846 biography Brewster was his 'chief reference on matters of fact'. Nevertheless, he felt able to assess these facts in a more informed manner, for on matters 'of opinion … we must differ in some small degree from our guide' (De Morgan, 1846, p. 78). His aim was to judge the known sources, including the advocate-like accounts of Newton's disciples, and correct a long-standing bias. This had not only hidden the truth regarding Newton's religious beliefs and character flaws, but had, as Malkin and Baily claimed, done an injustice to Newton's rivals. This explains his emphasis on Newton's weaknesses: 'Justice to Leibniz, to Flamsteed, even to Whiston, called for this exposure' (De Morgan, 1882, p. 256).

De Morgan criticised other writers for not reading their sources carefully, for taking them on trust without further investigation, and for not referencing their own work adequately. For example, a footnote in his 1846 biography revealed that Brewster had 'copied the false biographer [in the *Biographia Britannica*] without verifying the reference – a common, but a dangerous practice' (De Morgan, 1846,

30 De Morgan to Long, 26 March 1844, University College London: SDUK Correspondence.

31 De Morgan to Long, 3 August 1843, University College London: SDUK Correspondence.

32 See De Morgan to Brougham, 9 November 1855, University College London: Brougham Papers 10,301. On Libri, see Maccioni-Ruju and Mostert (1995); on Chasles, see Koppelman (1971).

p. 111). He was even more severe with writers who either suppressed evidence or avoided investigation for fear of what they might find. The topics that De Morgan chose to investigate – Newton's role in the calculus dispute, his religious beliefs and the possibility that he allowed his niece to become the mistress of his patron – were areas in which Newton's defenders had feared to tread. He thought the 'reserve of biographers about the illustrious dead is a very unwise proceeding' because they might do an injustice to the subject's contemporaries, but also because reserve fuelled rumours that frequently proved worse than the facts (De Morgan, 1857, p. 41).[33] Worst of all, those who held up Newton as an unblemished hero, and suppressed contrary evidence, were guilty of hypocrisy. The propagators of what he called the 'mythical Newton' were the 'respectable' classes: 'Who does not know the smug individual of this species, as he sees him picking his way through the world? His highest model is aristocracy; his social life is silver-forkery; his main pursuit is money-grubbery; and his whole religion is Sunday-prayery' (De Morgan, 1885, p. 130)

De Morgan argued that none of his investigations could hurt Newton's scientific status: 'the theory of gravitation would not be altered in value, or in opinion of value, though its author were proved to have committed murder and robbery' (De Morgan, 1858, p. 641). Insisting on the difference between intellectual and moral greatness, he was 'the *avvocato del diavolo*' who opposed Newton's canonisation.[34] By countering those who presented Newton as a hero of the establishment, he made him available as a scientific hero to radicals and Nonconformists. The figure of Newton was, in a sense, to be secularised as education had been within De Morgan's institution, University College London. However, his biographical and historical works continued to carry a moral message, for the exemplar was transferred from the biographical subject to the historian's ability to judge his sources impartially. De Morgan's critique of Newton's 'mythical' status and call for historians to banish existing prejudices can thus be related to his position as a 'Christian unattached' (De Morgan, 1882, p. 86), and a radical who would, given the opportunity, abolish the monarchy and disestablish the Church.[35]

Under the Influence of His Time: Brewster's *Memoirs of Newton* (1855)

The situation by 1855, when Brewster published his *Memoirs of Newton*, had altered considerably from 1831. His sources now included De Morgan, whom he mentioned with some reserve, and Baily, to whom he was downright hostile. He had also viewed the Portsmouth Papers for a week in 1837, and selections from them, made on his behalf by a member of the Portsmouth family, in the comfort of his own home in

33 De Morgan discovered that a clergyman had destroyed some letters from Newton to his nephew because they contained 'vulgar phraseology'. He pointed out that by doing this, he nearly caused posterity to lose 'the argument that an uncle, who reproved his nephew's vices in strong and irritating terms' was unlikely to be an uncle who was 'a consenting party to the dishonour of his own niece' (De Morgan, 1885, p. 7).

34 De Morgan to Whewell, 26 October 1846, in De Morgan (1882), p. 198.

35 De Morgan to William Frend, n.d. [1835], in De Morgan (1895), p. 128. See also Richards (2002).

1854. The result was, he said, 'wholly different' from the *Life*, for the 'materials for the present work have been collecting for nearly 25 years, independent of the use of the Newtonian Papers'. It was these papers, however, that forced him to admit the truth of De Morgan's claim that Newton 'was virtually the author' of the *Commercium Epistolicum*.[36] He also had to accept that the manuscripts proved Newton's interest in alchemy and 'place it beyond a doubt that he was a Unitarian'.[37] Brewster looked to the papers as a means of producing a version of events with the authority to counter Baily's *Account*, but he did not shirk from what he saw as his duty to the sources. He printed enough to undermine the eulogistic pattern that the biography maintained, despite consistent attempts to interpret them in most favourable manner.

However, like Rigaud, Brewster believed suppression could be the most prudent course, and he was, apparently, prepared to accept 'credit for having withheld some of the letters in my possession'.[38] He also thought it would have been better had the manuscripts of Flamsteed never been published. He accused Flamsteed and Baily, 'two English astronomers, the one a contemporary and the other a disciple' of a cross-century conspiracy to 'misrepresent and calumniate their illustrious countryman' (Brewster, 1855, vol. 1, p. xi). It is difficult to ascertain the degree to which Brewster suppressed his findings, but there are several instances of 'misrepresentation', and it is clear that he searched the archive for evidence to counter 'the injurious effects which [the *Account*] may have produced upon the public mind'.[39] While the amount of problematic evidence he did publish gave some foundation to his claim that 'not one of [Newton's] biographers has found so much fault with him as I have', it seems likely that he would have remained silent were not for the fact that this material was likely to be made public by critics if not first by himself.[40] Regarding his inclusion of Newton's heterodox theological writings, he explained that he did not feel justified in concealing that which the public 'have long suspected, and must sooner or later have known' (Brewster, 1855, vol. 1, p. xv).

Brewster seems to have had exclusive use of the Portsmouth Papers until the *Memoirs* were published, and it appears that he was chosen by the family as a more suitable researcher than Baily, who also considered publishing a selection from the manuscripts.[41] The attitude of the custodians of this collection is indicated by a report from Ada Lovelace to her mother Lady Byron that Henry Fellowes, Lord

---

36 Brewster to Brougham, 11 November 1854, University College London: Brougham Papers 11,248.

37 Brewster to Brougham, n.d. [1854?], University College London: Brougham Papers 31,833; Brewster to Rigaud, 21 May 1837, Bodleian Library: Rigaud Papers 60/86.

38 Brewster to Brougham, 18 March 1857, University College London: Brougham Papers 26,697. Brewster was referring to a review in the *Revue des deux mondes*.

39 Brewster to Brougham, 8 October 1847, University College London: Brougham Papers 26,633.

40 Brewster to Brougham, 30 May 1856, University College London: Brougham Papers 26,692.

41 Baily's interest in publishing the papers, and the family's displeasure, is suggested by Newton Fellowes to Baily, 24 November 1835 and W.S. Stratford to Baily, 30 December 1836, Cambridge University Library: Baily Papers, RGO 60/3. Brewster indicated his exclusive use of the papers in a letter to *The Times*, 28 June 1837 [5b].

Portsmouth's nephew, 'surprised me much by saying that "of theological papers, only such will be published as are sufficient to *prove that Newton believed strictly* in the Trinity, so as completely to answer the *Unitarians and Deists*, who had hitherto gloried in his authority, & appropriated him to themselves" …' (quoted in Toole, 1998, p. 70; original emphasis). De Morgan noted in his review the convenience of 'the first disclosure of these documents coming from an advocate' ([De Morgan], 1855, p. 126), but suggested that scientific biography was progressing, and that the changing times had made themselves felt on Brewster:

> Our generation, with no lack of idols of its own, has rudely invaded the temples in which science worships its founders: and we have before us a biographer who feels that he must abandon the demigod, and admit the impugners of the man to argument without one cry of blasphemy. To do him justice, he is more under the influence of his time, than under its fear: but very great is the difference between the writer of the present volumes and that of the shorter *Life* in the 'Family Library' in 1831; though, if there be any truth in metaphysics, they are the same person. ([De Morgan], 1855, pp. 122–3)

Brewster was indeed a member of an older generation, and had been forced, in his defensive strategy, to respond to developments in Newtonian biography. However, circumstances dictated that he should produce the century's fullest account of Newton's life, based on a larger range of primary sources than any until the late twentieth century.

Conclusion

I have outlined the biographical strategies used by those who wrote about Newton in a period when his life and character was the subject of debate. Brewster, Rigaud and Edleston fought to retain an idealised image of Newton, seeing this as 'a sacred duty to the memory of that great man, to the feelings of his countrymen, and to the interests of Christianity itself' (Brewster, 1831, p. 227). They opposed those such as Baily and De Morgan who sought to undermine this mythical creation of the previous century. Despite their varying aims, all stressed their reliance on, or ability to interpret, the historical sources as a means of giving their accounts authority and the appearance of impartiality. It is particularly clear in the cases of Rigaud and Baily that the values connected with their scientific personae were appropriated, by themselves and their friends, into their 'inductive' historical style. De Morgan used fewer unpublished sources, but he attempted to produce apparently dispassionate, judicial analyses of the available 'facts', the 'impartiality' of which was to be demonstrated by his willingness to face the worst about Newton. He wrote, however, as a critic of the belief that intellect and morality coincided, a position that can be seen to correspond to his position as a religious free-thinker and political reformer. While wishing to retain Newton the scientific hero, he could reveal the imperfections that establishment interests wished to suppress: 'Let a flaw be a flaw: Newton is not the less Newton' ([De Morgan], [1855] 1914, p. 182). Brewster, however, was driven to the archives by Baily's *Account*, as he searched for a means of counteracting the mass of material this work contained. Faced with the contents of the Portsmouth

Papers, it appears that he decided that partial exposure at the hands of a Newtonian advocate was preferable to full exposure by a critic.

Nearly forty years ago, A.N.L. Munby noted that it was during the 1830s and 1840s that sources for the history of science were first put into print, and speculated as to why astronomers such as Baily and Rigaud were prominent in this development. 'Worship of Newton', he thought, provided part of the answer (Munby, 1968, p. 5). While Newton did generate unique interest, it cannot be claimed that 'worship' was the impetus behind the publications of Baily and De Morgan, except to the extent that they wished to put hero-worship to the test. It was the writings of men on opposing sides of the debate about Newton that ensured that by the 1860s the gap between lay and expert knowledge of Newton had widened dramatically. This was noticed by a writer in the *Edinburgh Review*, who wondered what were the 'laws which govern the award of fame', since they were clearly not influenced by abstruse historical research. Two key factors in Newton's case were thought to be national prejudice and the 'all-pervading power of theological sentiment' which 'works with facts, or in spite of facts' ([Hemming], 1861, pp. 401, 402–3). It was, however, Newton's importance to Church and state interests that drove historians and biographers to search for these 'facts' in their attempts either to undermine or to defend the status he had been accorded by the majority of the population.

Acknowledgements

I would like to thank Andrew Warwick, Rob Iliffe and John Higgitt for their comments on earlier drafts of this paper, and the Syndicates of Cambridge University Library, the Bodleian Library, the British Library, the Royal Society, the Master and Fellows of Trinity College Cambridge and University College London for permission to quote from and refer to manuscripts in their possession.

Bibliography

Altick, Richard D. (1966), *Lives and Letters: A History of Literary Biography in England and America*, New York: Alfred A. Knopf.

[Anon.] (1879), 'Contemporary Literature', *Blackwood's Edinburgh Magazine*, **125**, 482–506, reprinted in Ira Bruce Nadel (ed.) (1986), *Victorian Biography: A Collection of Essays from the Period*, New York and London: Garland.

Ashworth, William J. (1994), 'The calculating eye: Baily, Herschel, Babbage and the business of astronomy', *British Journal for the History of Science*, **27**, 409–41.

Ashworth, William J. (1997), '"Labour harder than thrashing": John Flamsteed, property and intellectual labour in nineteenth-century England', in Frances Willmoth (ed.), *Flamsteed's Stars: New Perspectives on the Life and Work of the First Astronomer Royal (1646–1719)*, Woodbridge: Boydell Press, pp. 199–216.

[Aytoun, W.E.] (1849), 'Modern biography – Beattie's Life of Campbell', *Blackwood's Edinburgh Magazine*, **65**, 219–34, reprinted in Ira Bruce Nadel (ed.) (1986), *Victorian Biography: A Collection of Essays from the Period*, New York and London: Garland.

Baily, Francis (1835), *An Account of the Revd. John Flamsteed, the First Astronomer Royal; Compiled from his Own Manuscripts, and Other Authentic Documents, Never Before Published*, London.

Biot, Jean-Baptiste (1832), 'Revue de The Life of Sir Isaac Newton', *Journal des savants*, April, 199–203, May, 263–74, June, 321–39.

Biot, Jean-Baptiste (1833), *Life of Newton*, trans. H. Elphinstone, London: Baldwin & Craddock.

Brewster, David (1831), *The Life of Sir Isaac Newton*, London: John Murray.

Brewster, David (1855), *Memoirs of the Life, Writings and Discoveries of Sir Isaac Newton*, 2 vols, Edinburgh: Constable.

Carlyle, Thomas (1841), *On Heroes, Hero-Worship, and the Heroic in History*, London: James Fraser.

Christie, J.R.R. (1984), 'Sir David Brewster as an historian of science', in A.D. Morrison-Low and J.R.R. Christie (eds), *'Martyr of Science': Sir David Brewster 1781–1868*, Edinburgh: Royal Scottish Museum, pp. 53–6.

Cockshut, A.J.O. (1974), *Truth to Life: The Art of Biography in the Nineteenth Century*, London: Collins.

Cohen, I. Bernard (1972), 'Introduction', in reprint of S.P. Rigaud, *Historical Essay on the First Publication of Sir Isaac Newton's Principia*, New York: Johnson Reprint Corp., pp. i–xi.

Daston, Lorraine (1992), 'Objectivity and the escape from perspective', *Social Studies of Science*, **22**, 597–618.

Daston, Lorraine and Peter Galison (1992), 'The image of objectivity', *Representations*, **40**, 81–128.

Davy, John (1836), *Memoirs of the Life of Sir Humphrey Davy*, 2 vols, London: Longman, Rees, Orme, Brown, Green & Longman.

[De Morgan, Augustus] (1840), 'Newton', in George Long (ed.) (1833–46), *The Penny Cyclopaedia of the Society for the Diffusion of Useful Knowledge*, vol. 16, London: Charles Knight, pp. 197–203.

De Morgan, Augustus (1846), 'Newton', in Charles Knight (ed.), *Cabinet Portrait Gallery of British Worthies*, reprinted in Augustus De Morgan (1914), *Essays on the Life and Work of Newton*, ed. P.E.B. Jourdain, Chicago, IL and London: Open Court Publishing, pp. 3–63.

De Morgan, Augustus (1847), *Arithmetical Books from the Invention of Printing to the Present Time, Being Brief Notices of a Large Number of Works Drawn from Actual Inspection*, London: Taylor & Walton.

De Morgan, Augustus (1852), 'A short account of some recent discoveries in England and Germany relative to the controversy on the invention of the fluxions' Companion to the Almanac', reprinted in Augustus De Morgan (1914), *Essays on the Life and Work of Newton*, ed. P.E.B. Jourdain, Chicago, IL and London: Open Court Publishing, pp. 67–101.

[De Morgan, Augustus] ([1855] 1914) 'Review of Brewster's Memoirs of Newton', *North British Review*, **23**, 307–38, reprinted in Augustus De Morgan (1914), *Essays on the Life and Work of Newton*, ed. P.E.B. Jourdain, Chicago, IL and London: Open Court Publishing, pp. 119–82.

De Morgan, Augustus (1857), 'Newton's nephew, the Rev. B. Smith', *Notes & Queries*, **3**, 2nd Series (55), 41–42.

De Morgan, Augustus (1858), 'Review of King's Biographical Sketch of Newton and Brougham's Addresses on Popular Literature', *Athenaeum*, **1,621**, November, 641–2.

De Morgan, Augustus (1885), *Newton: His Friend, and His Niece*, ed. S.E. De Morgan and A. Cowper Ranyard, London: Elliot Stock.

De Morgan, Augustus (1914), *Essays on the Life and Work of Newton*, ed. P.E.B. Jourdain, Chicago, IL and London: Open Court Publishing.

De Morgan, S.E. (1882), *Memoir of Augustus De Morgan*, London.

De Morgan, S.E. (1895), *Threescore Years and Ten: Reminiscences of the Late Sophia Elizabeth De Morgan, and Others*, ed. M.A. De Morgan, London: Richard Bentley & Son.

Dickinson, H.W. (1932), 'J. O. Halliwell and the Historical Society of Science', *Isis*, **18**, 127–32.

Edleston, Joseph (1850), *Correspondence of Sir Isaac Newton and Professor Cotes, including Letters of Other Eminent Men, Now first Published from the Originals in the Library of Trinity College, Cambridge: Together with an Appendix, Containing other Unpublished Letters and Papers by Newton; with Notes, Synoptical View of the Philosopher's Life, and a Variety of Details Illustrative of his History*, Cambridge and London.

Fara, Patricia (2002), *Newton: The Making of Genius*, London: Macmillan.

Fauvel, John, Raymond Flood and Robin Wilson (eds) (2000), *Oxford Figures: 800 Years of the Mathematical Sciences*, Oxford: Oxford University Press.

Fullmer, J.Z. (1967), 'Davy's biographers: Notes on scientific biography', *Science*, **155** (3,760), January, 285–91.

Galison, Peter (1999), 'Objectivity is romantic', in Jerome Freidman, Peter Galison and Susan Haack (eds), *The Humanities and the Sciences*, New York: American Council of Learned Societies, pp. 15–43.

[Galloway, Thomas] (1833), 'French and English biographies of Newton', *Foreign Quarterly Review*, **12**, July, 1–27.

Gascoigne, John (1988), 'From Bentley to the Victorians: The rise and fall of British Newtonian natural theology', *Science in Context*, **2**, 219–56 .

Gordon, M.M. (1869), *The Home Life of Sir David Brewster*, Edinburgh: Edmonston & Douglas.

Hall, A. Rupert (1999), *Isaac Newton: Eighteenth-century Perspectives*, Oxford: Oxford University Press.

[Hemming, G.W.] (1861) 'Newton as a scientific discoverer', *Quarterly Review*, **110**, 401–35.

[Herschel, J.F.W.] (1844) 'Memoir of Francis Baily', *Monthly Notices of the Royal Astronomical Society*, **6**, 89–128.

Higgitt, Rebekah (2004), 'Astronomers against Newton', *Endeavour*, **28** (1), March, 20–24.

Higgitt, Rebekah (forthcoming), *Recreating Newton? Biographies of Newton and the Making of Nineteenth-century History of Science*, London: Pickering & Chatto.

Iliffe, Robert (1998), 'A "connected system"? The snare of a beautiful hand and the unity of Newton's archive', in Michael Hunter (ed.), *Archives of the Scientific Revolution: the Formation and Exchange of Ideas in Seventeenth-century Europe*, Woodbridge: Boydell, pp. 137–57.

[Jeffrey, Francis] (1835), 'Memoirs of Sir James Mackintosh', *Edinburgh Review*, **62**, 205–55.

King, Peter (1829), *The Life of John Locke, with Extracts from his Correspondence, Journals, and Common-Place Books*, London: H. Colburn.

Koppelman, Elaine (1971), 'Chasles, Michel', in Charles Coulston Gillispie (ed.), *Dictionary of Scientific Biography*, New York: Charles Scribner's Sons, vol. 3, 212–15.

Levine, George (2002), *Dying to Know: Scientific Epistemology and Narrative in Victorian England*, Chicago, IL and London: University of Chicago Press.

Long, George (ed.) (1833–46), *The Penny Cyclopaedia of the Society for the Diffusion of Useful Knowledge*, 29 vols, London: Charles Knight.

Maccioni-Ruju, P. Alessandra and Marco Mostert (1995), *The Life and Times of Guglielmo Libri (1802–1869): Scientist, Patriot, Scholar, Journalist and Thief. A Nineteenth-Century Story*, Hilversum: Verloren.

[Malkin, B.H.] (1832), 'Brewster's Life of Newton', *Edinburgh Review*, **56**, 1–37.

Miller, D.P. (2000), '"Puffing Jamie": The commercial and ideological importance of being a "philosopher" in the case of the reputation of James Watt (1736–1819)', *History of Science*, **38**, 1–24.

Munby, A.N.L. (1968), *The History and Bibliography of Science in England: The First Phase, 1833–1845*, Berkeley, CA and Los Angeles, CA: Berkeley School of Librarianship and Los Angeles Graduate School of Library Service.

Reed, J.W. (1966), *English Biography in the Early Nineteenth Century, 1801–1838*, New Haven, CT and London: Yale University Press.

Rice, Adrian (1996), 'Augustus De Morgan: Historian of Science', *History of Science*, **34**, 201–40.

Richards, Joan L. (1987), 'Augustus De Morgan, the history of mathematics, and the foundations of algebra', *Isis*, **78**, 7–30.

Richards, Joan L. (2002), '"In a rational world all radicals would be exterminated": Mathematics, logic and secular thinking in Augustus De Morgan's England', *Science in Context*, **15**, 137–64.

Rigaud, S.P. (ed.) (1832), *Miscellaneous Works and Correspondence of the Rev. James Bradley, D.D., F.R.S., Astronomer Royal, Savilian Professor of Astronomy at the University of Oxford …*, Oxford: Oxford University Press.

Rigaud, S.P. (1838), *Historical Essay on the First Publication of Sir Isaac Newton's Principia*, Oxford.

Rigaud, S.P. (ed.) (1841), *Correspondence of Scientific Men of the Seventeenth Century; Including Letters of Barrow, Flamsteed, Wallis, and Newton, Printed from the Originals in the Collections of the Right Hon. The Earl of Macclesfield, with introduction by S. J. Rigaud*, 2 vols, Oxford.

[Southey, Robert] (1804), 'Hayley's Life and Posthumous Writings of William Cowper', *Annual Review*, **2**, 457–62.

Stauffer, Donald A. (1941), *The Art of Biography in Eighteenth-century England*, Princeton, NJ: Princeton University Press.

Theerman, Paul (1985), 'Unaccustomed role: The scientist as historical biographer – two nineteenth-century portrayals of Newton', *Biography*, **8**, 145–62.

Toole, Betty A. (1998), *Ada: The Enchantress of Numbers, Prophet of the Computer Age*, Mill Valley, CA: Strawberry Press.

Topham, Jonathan (1992), 'Science and popular education in the 1830s: The role of the Bridgewater Treatises', British *Journal for the History of Science*, **25**, 397–430.

Tuchman, Barbara W. (1981), 'Biography as a Prism of History', in Marc Pachter (ed.), *Telling Lives: The Biographer's Art*, Philadelphia, PA: University of Pennsylvania Press, pp. 132–47.

Turner, Dawson (1848), *Thirteen Letters from Sir Isaac Newton to John Covel D.D.: From Original Manuscripts in the Library of Dawson Turner Esq.*, Norwich.

Turner, Frank M. (1993), *Contesting Cultural Authority: Essays in Victorian Intellectual Life*, Cambridge: Cambridge University Press.

Turnor, Edmund (1806), *Collections for the History of the Town and Soke of Grantham, containing authentic memoirs of Sir Isaac Newton, now first published from the original MSS. in possession of the Earl of Portsmouth*, London: W. Miller.

Yeo, Richard (1985), 'An idol of the marketplace: Baconianism in nineteenth-century Britain', *History of Science*, **23**, 251–98.

Yeo, Richard (1988), 'Genius, method and morality': Images of Newton in Britain 1760–1860', *Science in Context*, **2** (2), 257–84.

Chapter 10

Biographies as Mediators between Memory and History in Science

Bernadette Bensaude-Vincent

This chapter ventures some reflections based on my own experience as a biographer of Paul Langevin, a physicist of the early twentieth century, and Antoine-Laurent Lavoisier, a chemist of the late eighteenth century. Lavoisier and Langevin are still present in the French urban landscape. There are many streets, squares, public elementary schools or high schools named after them. To these names a number of popular images are attached. How to deal with these popular clichés when writing a biography? Shall we try to debunk them one after the other, or should we include them as part of the history of our subject?

When I started writing a biography of Paul Langevin (1872–1946) based on his personal archives and interviews with his family and former students, I quickly realised that I was confronted with a robust and self-consistent portrait shaped during his lifetime by his colleagues and companions on several occasions. Since it was the first biography written by someone who did not belong to Langevin's circle, I thought I had to give a more realistic portrait. As a historian of science, I focused on his style of scientific creativity and tried to elucidate how and why Langevin became a prominent social activist. I chose the title *Paul Langevin: Science and Vigilance*, to emphasise the dual aspects of his activities as a physicist enjoying an international reputation and as a leader of peace movements and author of educational reforms during the inter-war period (Bensaude-Vincent, 1987). However, to my astonishment, most readers and reviewers perceived this biography as a tribute to the memory of a neglected figure of twentieth-century science rather than as an attempt to develop a historiographical approach to Langevin. Although I was aware that writings escape their authors, that their meaning is reconstructed by the readership, I was somewhat disappointed that my intentions were partly misunderstood.

In order to avoid similar misreadings when I started writing about Antoine-Laurent Lavoisier (1743–94), I made my intentions more explicit, from the outset. Hence the title of my book: *Lavoisier: Mémoires d'une révolution* ('Lavoisier: Memories of a Revolution') (1993). I am not sure that this volume can be called a biography, especially as two biographies were published in English and French a few months before its issue. Anyway, I had to renounce one standard requirement of the biographical genre – the minute description of the key events in the life, in favour of a *longue durée* perspective. Starting long before Lavoisier, with eighteenth-century chemists, I also included the various layers of the construction of the founder myth by nineteenth-century chemists. This approach proceeded from my gradual realisation that biographies are essentially mediators between memory and history.

By 'memory', I mean a collective maintenance of past events in the present through monuments, rituals, commemorative events. The study of memories belongs to ethnography (Abir-Am, 1992, 1998). By history, I mean a written narrative of past events clearly characterised as belonging to the past. The former is a re-enactment of the past, while the latter is the reconstruction of the past by the use of narrative devices and conventions. Writing can kill memory, as was suggested by the myth of Thoth told by Socrates in *Phaedrus* (274C–275b). However memory can also be maintained and transmitted from one generation to the next ones through writings, especially through fictions and biographies.

If lives of scientists belong both to the realm of memory and to the realm of history, does it mean that biographers have to choose their side, to fight memories for the sake of history? Or should they rather invent a kind of hybrid narrative style?

Langevin's Popular Image in French Collective Memory

Langevin's name is most usually found in cities held by a communist or leftist municipality. This early twentieth-century physicist is mainly remembered as a glamorous figure of the French Communist Party (although he became a member only two years before his death). Several commemorative events have spread a canonical portrait of him. Already in his lifetime, Langevin received many honours and inspired respect and admiration even among his political enemies. After his election to the Académie des sciences in 1934, Romain Rolland, a French novelist and the president of an international committee for peace and against fascism, honoured Langevin as 'a leading figure of science who walks at the head of the working classes'.[1] This image was magnified after Langevin's death. In a special issue of the journal *La Pensée*, published in 1947, Jacques Hadamard, a young colleague, regretted the loss of 'a great savant and a great man' (Hadamard, 1947, p. 31); Aimé Cotton emphasised an exceptional alliance of experimental skills, theoretical inclination and human generosity (Cotton, 1947); Einstein praised 'this pure and enlightened savant'(Einstein, 1947, p. 14), and John D. Bernal emphasised the unity of Langevin's views, which 'created a link between the inorganic domain and the organic, and even the social world and expressed the unity of dialectical materialism' (Bernal, 1947, p. 17).

Two major features emerge from innumerable homages to Langevin: scientific excellence combined with dedication and generosity. Although Langevin was never awarded a Nobel Prize like his close friends Jean Perrin and Albert Einstein, he enjoyed a high international reputation crowned by his election as Chairman of the Physics Committee of the Solvay Institute in 1927 (Bensaude-Vincent, 1999). He is depicted as a brilliant and disinterested scientist who energetically advocated the 'new physics' – that is, atomic physics and Einstein's theory of relativity – against

1 Romain Rolland, letter to Paul Langevin, 24 January 1937, Langevin Archives, Ecole Supérieure de Physique et de Chimie Industrielles, Box 31.

the conservatism of the French physics community. Many collaborators insisted that he so little cared for his own interest that he did not publish a large number of his own results, which were later published by others. Einstein wrote: 'It seems certain that he would have developed the theory of special relativity had it not been done elsewhere because he had caught all its essential points' (Einstein, 1947, p. 14).

After the First World War, Langevin reduced his own research activities because he was increasingly involved in several social movements that led him to travel a lot. In the inter-war period, he more often showed up on the public stage than in the two laboratories that were his responsibility at the Ecole Municipale de Physique et de Chimie Industrielles and at the Collège de France, in Paris. Although he kept aware of the most recent developments in physics until the end of his life, he was no longer personally at the forefront of research, and became more a public lecturer.

However, in the memory of his collaborators there is no hint of reproach. Instead, they assume that Langevin was driven into social activism by his generosity and goodness. The emphasis on his personal moral virtues, such as goodness, disinterestedness and generosity, necessarily occults a more complex character. Langevin, like most humans, was a complex character with sometimes contradictory aspects in his behaviour. For instance, although he advocated pure and disinterested science, he patented an ultrasound device and spent ten years in patent litigations. This champion of peace movements pursued his collaboration with military research far beyond wartime in the 1920s. In addition, the 'saint' who was married and the father of four children created a public scandal in 1911 because of his liaison with Marie Curie, and later he had an illegitimate son with a younger woman.

In pointing to such contradictions, I provided a much more human portrait than the heroic image forged by Langevin's collaborators. My main intention was to emphasise that Langevin has been dignified as a kind of 'laïc saint', a typical ideal of the anti-clerical mythology of the Third Republic. The stereotype is interesting in itself because it shows how much the image of science and scientists is shaped by the cultural context of the biographer. In fact, in Langevin's career three of the values which were the moral pillars of the Third Republic can be emphasised: education, the people and science. Throughout his life he participated in the various reforms of the French educational system. He saw education as an instrument of social equality and as the main remedy to prevent future wars; he often defended proletarians and minorities against the government along with the Communist Party, and finally, science was more than his professional activity, since he placed his faith in it and assumed that social justice could be achieved through its progress. Langevin was driven into activism not so much by moral rules as by his conviction that science would save mankind from all injustices and the danger of collapse. He presented science as a cultural activity and praised it as the noblest production of human civilisation. In the name of science he urged his fellow scientists to devote themselves to peace movements and popularising tasks because he thought scientists had a responsibility to reach out to the people and lead society on the path of justice and progress. Thus Langevin appears as a key figure in the scientistic movement that dominated French culture in the twentieth century. He epitomised the historical alliance between the educated elite and social concern that is the essence of what we call in France the 'intellectuals' (Charle, 1993; Julliard and Winock, 1996, pp. 1,239–55). The mixture of activism and idealism of the generation of intellectuals formed during the Dreyfus Affair in the 1890 led many of them to support the French Communist Party in the inter-war period without systematically following the line decided in Moscow. The close connection between socialism and scientism in twentieth-century France remains to be investigated by cultural historians; however, it is an example of the kind of contribution that a biography attentive to the significance of popular images transmitted by the collective memory could do for cultural history.

Lavoisier as a Founding Hero

The popular image of Lavoisier also combines the two themes of science and politics. But whereas Langevin did not seek glory in science and was highly esteemed for his social activism, Lavoisier, by contrast, was a loser on the political stage and a winner on the scientific stage.

A number of legends surround Lavoisier's life and work (Bensaude-Vincent, 1993). French high school students learn that Lavoisier was the sole founder of modern chemistry. He was a true Saviour who rescued chemistry from the obscurantism of alchemy; he overthrew the speculative and qualitative chemical tradition by introducing the use of the balance in the chemical laboratory; he supposedly authored the law of the conservation of matter.

Without entering into a detailed survey of Lavoisier's scholarship, a quick glance at any recent history of chemistry would suffice to falsify all these legends (Smeaton, 1989; Donovan, 1993; Poirier, 1996). First, prior to the chemical revolution, chemistry was already a well-established discipline, illustrated by a long tradition of public lectures and textbooks, cultivated by a large number of enlightened amateurs or by pharmacists and medical doctors. In contrast to the traditional view of eighteenth-century chemistry as an immature, obscurantist collection of artisans' recipes, it was indeed already an academic discipline, distinguished from the chemical arts and promoted in a special class of the Académie des sciences (in contrast, there was no class for experimental physics) (Crosland, 1980; Meinel, 1983; Holmes, 1989). The disciplinary coherence of pre-Lavoisierian chemistry is not a recent discovery by professional historians; it was assumed by Pierre Duhem as early as 1902, and clearly established by Hélène Metzger's works in the 1930s (1930 [1974], 1935; Mieli, 1932). Moreover Lavoisier was not a lonely genius, a marginal who opposed the scientific orthodoxy. He belonged to the most prestigious institution of his time, the Académie des sciences, and most of his works were performed and publicised on the stage and with the support of the Academy. The reform of language, like many of his campaigns of experiments, were collaborative enterprises (Goupil 1992; Grison, Goupil and Bret, 1994; Crosland, 1994; Bensaude-Vincent and Abbri, 1995). Scales and balances had been part of the equipment of chemical laboratories at least since the seventeenth century. Not only the instruments but the balance sheet method were in use long before Lavoisier. The so-called Lavoisier's Law ('Nothing is lost, nothing is created') is not a law and was not invented by Lavoisier; it had been a tacit assumption in physics and chemistry for centuries.

Despite historians' repeated attempts to spread a more realistic portrait of Lavoisier, the cliché of the founding father survives untouched. It has never been more vivid than in 1994, the bicentennial of Lavoisier's death (Bensaude-Vincent, 1996). The Academy energetically revitalised the cult of the founder. A ceremony took place under its cupola in the same building that was formerly the Collège Mazarin attended by the young Lavoisier, and the closing session was held at the Château de Blois, following visits to Lavoisier's former properties in this region. In connection with the 'pilgrimage', an international conference was held in May 1994. It was a strange juxtaposition of amateur papers presented by retired professional chemists who spread the canonical clichés about Lavoisier's work and career, and papers given by professional historians who tried to revisit the chemical revolution. The first three lectures delivered under the cupola successively presented Lavoisier as a revolutionary chemist, as the founder of the part of physiology called bioenergetics, and as the precursor of agricultural science and agricultural policy. In a fourth and final lecture, the Vice-president of the Academy portrayed Lavoisier as the heir to Archimedes and Newton, the founder of the modern scientific experimental method. So fervent was the cult of the founding father that the President of the Lavoisier Committee suggested that had Lavoisier survived the turmoil of the revolution and lived twenty-five more years, he would have developed the atomic theory, and even structural chemistry (Kagan, 1995).

The persistence of the national cult may be linked to Lavoisier's tragic death on the guillotine. In the midst of the Terror, on 19 Floréal of Year II (8 May 1794), the revolutionary tribunal sentenced him to death together with 27 other tax farmers who were supposedly guilty 'of a conspiracy intended to encourage by all possible means the success of the enemies of France' (Grimaux, 1896, p. 305). After a show trial, they were all guillotined on the same afternoon. To add a dramatic touch to the portrayal of the founding father of chemistry, another legend is attached to Lavoisier's death. It is said that Lavoisier asked for the execution to be temporarily suspended for him to complete his scientific work, but received the following reply from the president of the revolutionary tribunal: 'La république n'a pas besoin de savants' ('The Republic has no need of scientists'). Although nothing in the minutes of the trial confirms it, these words remain deeply rooted in the French collective memory, and they are referred to over and over again in conflicts between the state and the scientific community.

To sum up briefly, the heroic image of Lavoisier derives its strength from revolutionary mythology. Lavoisier is roughly portrayed as the hero of a scientific revolution and the victim of a political revolution. This image includes two major components: on the one hand, the notion of a radical break with the past that led to the foundations of modern chemistry; on the other hand, the conflict between two systems of values – the values of politicians, representatives of the temporal power, and those of scientists, representatives of a new, non-clerical, spiritual power.

Two Alternatives for Biographers

When confronted with such persistent myths, a biographer has two alternatives: either discard them and then go back to the sources in order to reconstruct a more faithful, more complex portrait, or take the myths into account in order to build an intertextual narrative. In outlining briefly the two methods, I will explain why I rather favour the latter.

The Private Fabric of Science

Archival sources allow a detailed reconstruction of the daily life and activity of a scientist that is a remarkable antidote to the heroic image. For instance, Larry Holmes' minute study of Lavoisier's laboratory notebooks is clearly intended as a method to get rid of the public mask by focusing on the private face of science:

> The public scientist is important, and we must elucidate the ways in which scientists have assumed their public roles; but privacy is still valued, and the private science of those whose public performances have made them memorable is still worth pursuing. I believe that we still care about the person behind the mask, that we still seek the reality of external truths, but the reality accessible to ordinary people who try to understand those around them and those who lived before them as deeply individual human beings. (Holmes, 1998, pp. 145–6)

Unveiling the private side of scientific activity is a very powerful method for undermining the clichés that surround scientific creativity without depreciating the scientist. In the case of Lavoisier's revolution, the canonical accounts emphasise the coherence of the research programme he started in 1772 and conducted methodically with the use of his quantitative balance sheet method. In his detailed study of the 1773 laboratory notebooks, Larry Holmes gave an alternative picture of the contingent nature of Lavoisier's research pathway –of a young and impatient scientist whose experiments often failed, and who learnt his famous balance sheet method at the bench, through hits and misses. In pointing out the day-to-day connections between Lavoisier's experiments on various subjects, Holmes undermines all attempts to identify a key discovery, the 'hardcore' of the revolutionary process. (Holmes, 1985).

In my view, narratives of this kind offer two major advantages. Firstly, they shift the attention from theoretical breaks towards experimental practices in scientific creativity. They reveal the variety of uses of experiments. When the results reach the public stage, experiments are essentially presented as 'instruments of persuasion' or as authorities, and their exploratory functions are ignored (Gooding, Pinch and Schaffer, 1989). The trials and errors and the frequent failures in a scientist's life provide windows into 'science in action'. Secondly, and more importantly, Holmes's kind of narrative avoids all presentist biases. In constructing his narrative of a scientist's daily life, Holmes never anticipates the future. The 'past future' is treated as a future that may or may not happen: 'A story should retain its meaning up until any point to which it has been told, even if we imagine that it had historically been interrupted at that point. I intend for the story to unfold, insofar as possible, "as it happened"' (Holmes, 1991, vol. 1, p. xviii).

This method requires both a great deal of patience and an immoderate empathy with the scientist under study. These are rare qualities which are difficult to develop, especially in the case of Lavoisier, who was hardly a sympathetic character. Moreover, the density of detail confines the analysis to the micro level of the daily experimental space. It is readable for a short period in a scientist's life. A minute analysis of all 12 volumes of Lavoisier's notebooks is not only unfeasible for a single historian, but also difficult to read, and consequently, fine-grained reconstruction of the daily practices of science seems unsuitable to undermine the robust mythologies that surround many scientists' lives.

The quotation marks surrounding Holmes's 'as it happened' methodology point to another major difficulty. Objectivity is unattainable through such micro-level studies, for at least two reasons. First, the experimental space is artificially isolated from other spaces where the scientist acted in his or her daily life. Lavoisier's chemical works were just one aspect of his life, which was divided between his main professional activity as a tax farmer, his appointment in the Gunpowder Administration, his public role on the stage of the Academy, and finally, his private life. Second, the objectivity of such detailed narratives may be illusory because the macro level constantly interferes with the micro level. In order to build a narrative of a short campaign of experiments, one has to set up the background not only of past investigations, but also of international competition on the topic and related events in public. Inevitably, one has to more or less arbitrarily select and prioritise a few events among the many intricate aspects of daily life. Finally, the emphasis on the private fabric of science in the mind of individual scientists tends to minimise one essential feature of science: its publicity. The public face of science also matters. It is an essential piece in the construction of scientific facts.

Biographies as Intertextual Narratives

The interferences between micro and macro, between private and public, were the main motives for the method that I tried to develop in my own book on Lavoisier. Because I was convinced that cultural and scientific traditions transmitted by readings and by institutions create an expectation that shapes the individual way of doing science, I looked more closely at the kind of influence that the Académie des sciences had on Lavoisier. He entered the Academy at the young age of 25, and subsequently shaped his style in order to suit this prestigious institution. Not only did he gain the confidence of being a member of the scientific elite, but in reporting his experiments in public memoirs, he also adopted the Academy's prevailing 'literary technology'. Writing an academic memoir for the Academy required that the author follow a strict and stereotyped reasoning, with hypothesis, experiments, results, conclusions. Towards the end of the eighteenth century there was a strong emphasis on the accuracy of measurements rather than on the reproducibility of experiments (Licoppe, 1996, pp. 243–310). The academic code also incorporated a number of tacit rules of civility, especially for priority claims. In addition, the impression that Lavoisier methodically conducted a research programme over fifteen years in order to overthrow the phlogiston theory may well be induced by the long tradition of research programmes carried out in the chemistry section of the Academy since the early eighteenth century.

Most scholars have been very impressed by Lavoisier's revolutionary intentions, made explicit as early as February 1773, when he wrote in his laboratory notebook that his 'experiments would occasion a revolution in physics and chemistry' (Guerlac, 1961, pp. 228–9). However, this all-too-famous remark has been misunderstood as the expression of a research programme intended to debunk the phlogiston theory. In fact, in 1773, the modern meaning of scientific revolution as the overthrow of a dominant paradigm did not exist in the French language. Both lexicographical research and I. Bernard Cohen's study of the eighteenth-century origins of the concept of scientific revolution indicate that this phrase became rather frequent in the eighteenth-century French language, especially in Diderot's *Encyclopédie* (Reichardt and Lüsebrink, 1988; Rey, 1989; Cohen, 1985). It referred to any epoch-making event, although the term still retained its astronomical meaning of redoing the same trajectory. In this context, Lavoisier's famous note cannot be understood as the overthrow of a paradigm. Historians not only tend to forget that the concept of revolution has its own history, but we also often overlook the impact of the linguistic context. The term 'revolution' proliferated in eighteenth-century French scientific literature, each science having his revolutionary hero, thus creating an expectation of such big events. More precisely, the author of the 'chemistry' entry in the *Encyclopédie* explicitly called for 'a new Paracelsus', an enthusiastic revolutionary hero who, 'finding himself in a favourable position would take the advantage of some profitable circumstances', would be able to promote chemistry to the dignity of physics and get rid of the current prejudices against it (Venel, 1753, p. 409). I do not mean to indicate that Lavoisier had a clear memory of this sentence when, twenty years later, he noted that his experiments 'could occasion a revolution in physics and chemistry'. But he and his contemporaries were aware that a revolution was fulfilling a long-standing expectation. More broadly, revolutionary heroes like Charles Augustin Coulomb in electricity, Laplace in mechanics and Viq d'Azyr in comparative anatomy were proliferating on the academic stage in late eighteenth-century France, claiming they founded their science on the solid ground of a sound method and thus portraying themselves as founding heroes. In my view, such claims in conjunction with the advent of the French revolution played a key role in the construction of the modern concept of scientific revolutions as a radical and irreversible break with the past.[2]

Up to now, I have emphasised how Lavoisier's own interpretation of his achievements helped shape the image of a founding hero. However, the attitudes and writings of the main actor are but one of many factors in the construction of his public image. Lavoisier's colleagues – his disciples and his enemies as well – contributed to the emergence of the mythology of the founding hero. Louis-Bernard Guyton de Morveau in particular played a key role. Shortly after having converted to Lavoisier's anti-phlogistonist views, he compared Lavoisier to 'le grand Descartes',

2 It is not surprising that Thomas Kuhn shaped his theory of scientific change while he taught at Harvard in James Conant's programme, where the chemical revolution with the overthrow of the phlogiston theory was the favourite case study in the history of science.

who submitted the foundations of chemistry to a radical doubt and built up a new, solid and eternal edifice (Guyton de Morveau, 1786). The temptation of spreading mythical portrays was especially enhanced among those disciples who might have felt guilty for not saving Lavoisier from the guillotine. For instance, Antoine de Fourcroy's homage forged the picture of a founding hero who had become the victim of bloodthirsty barbarians. Later, in the 1830s, the French chemist Jean-Baptiste Dumas reconfigured the view of Lavoisier as a lonely genius who relied on his balance sheet method. In the second half of the nineteenth century, when nationalistic tensions between France and Germany increased, the portrait of the positivistic hero was repainted with the traits of a nationalistic hero. This is not the place to survey the various individual contributions to the mythology surrounding Lavoisier; I have analysed them elsewhere (Bensaude-Vincent, 1983, 1993). Rather, I would like to end up with a few methodological remarks.

First, the Lavoisier case lead me to give up the naive belief that somewhere one could find the 'correct' interpretation of his life. The 'authentic' Lavoisier lies neither in his public papers nor in his private notes. Lavoisier himself did not see the whole picture of his life. Langevin presumably had no better control over his public image, but he intentionally used his scientific reputation as a political activist in the public arena.

Second, Lavoisier's narrative of his own achievements was shaped by his view of science. Although he claimed that the history of science should be discarded from science textbooks, his scientific practice and his research papers were based on a tacit and implicit view of the history of chemistry. In the preliminary discourse of *Elements of Chemistry*, Lavoisier presented history as an encumbering pile of errors and prejudices. Although he did a lot of collaborative work throughout his career, he envisaged the foundation of chemistry as an individual enterprise. By contrast, Langevin, whose major contributions to physics were rather personal achievements, was convinced that science is a collective effort. He claimed that science was part of mankind's attempt to adapt to its environment, and tended to minimise the role of individuals in the advancement of science.

Third, Lavoisier's and Langevin's own narratives have no privilege over those written by their contemporaries and followers. All of them have been shaped by the perspective, the rhetorics and the tensions of their authors. All of them are interesting in so far as they are not detached from their local and historical context. They all shed a new light on Lavoisier, even those that spread legends such as 'The republic has no need of scientists.' Legends are meaningful, and we can learn a lot from them.

Further, the true meaning of Lavoisier's revolution has never been unearthed because past and future generations never finish judging and re-judging the past. It is impossible to exhaust the meaning of any life because of so many layers of meanings. Only through cross-perspectives can we attempt to approach the density of meanings embedded in these great figures of science. This approach is not far from what semioticians, following Julia Kristeva, call 'intertextuality'.[3]

Finally, I do not use 'intertextuality' in its full technical meaning which includes the notion of hypertext. The key notion, which I find useful for biographers and historians, is that texts are framed by other texts and by the writer's and readers' cultural contexts. Intertextuality emancipates science biographers from the dull and vain task of fighting against myths and distorted memories. It is an invitation to a more positive attitude of a hermeneutical reconstruction of historical realities by displaying the wide variety of their potential meanings.

In conclusion, I have been struggling with the difficulty of taking into account the impact of the public image of the individual under study. This is by no means a problem specific to biographies of scientists. Rather, it is a general difficulty raised by the lives of poets, artists, politicians and so on. In the case of scientists, however, it seems even more important to stress the interplay between the individual achievements and public images in the collective memory of scientific communities and nations. In this respect, biographies of scientists are a very efficient instrument to emphasise the cultural meanings of scientific activities and their part in the construction of national memories.

Bibliography

Abir-Am, Pnina (1992), 'A historical ethnography of a scientific anniversary in molecular biology: The first protein x-Ray photograph', *Social Epistemology*, **6**, 323–54.

Abir-Am, Pnina (ed.) (1998), *La mise en mémoire e la science: Pour une ethnographie historique des rites commémoratifs*, Paris: Editions archives contemporaines.

Bensaude-Vincent, Bernadette (1983), 'A founder myth in the history of sciences? The Lavoisier case', in Loren Graham, Wolf Lepenies and Peter Weingart (eds), *Functions and Uses of Diciplinary Histories*, Dordrecht: Reidel, pp. 53–78.

Bensaude-Vincent, Bernadette (1987), *Paul Langevin: Science et vigilance*, Paris: Belin.

Bensaude-Vincent, Bernadette (1993), *Lavoisier: Mémoires d'une révolution*, Paris: Flammarion.

Bensaude-Vincent, Bernadette (1996), 'Between History and Memory: Centennial and Bicentennial Images of Lavoisier', *Isis*, **87**, 481–99.

Bensaude-Vincent, Bernadette (1999), 'Paul Langevin and the French scientists at the Solvay Conferences', in Pierre Marage and Grégoire Walenborn (eds), *The Solvay Councils and the Birth of Modern Physics*, Basle: Birkhäuser, pp. 35–47.

3 This term, derived from the Latin *intertexto*, was coined by Julia Kristeva (1974) to refer to a transposition of one or more system of signs into another one. Her aim was to break with traditional notions of the author's 'influence' and the text's 'sources', and to view texts as systems of transformation of earlier signifying systems. A text is a 'mosaic' of multiple texts, and the internal connections outlined by the footnotes make a visible forum which can be expanded by each subsequent reader.

Bensaude-Vincent, Bernadette and F. Abbri (eds) (1995), *Lavoisier in European Context: Negotiating a New Language for Chemistry*, Canton, MA: Science History Publications.

Bernal, John (1947), 'Paul Langevin et l'Angleterre', *La Pensée*, **12**, May–June, 17–20.

Charle, Christophe (1993), 'Academics or Intellectuals? The professors of the University of Paris and Political debate in France from the Dreyfus Affair to the Algerian war', in J. Jennings (ed.), *Intellectuals in Twentieth-century France: Mandarins or Samurai?*, London: Macmillan.

Charle, Christophe (1994), *La république des universitaires 1870–1940*, Paris: Seuil

Cohen, I. Bernard (1985), *Revolution in Science*, Cambridge, MA: Harvard University Press.

Cotton, Aimé (1947), 'L'oeuvre scientifique de Paul Langevin', *La Pensée*, **12**, May–June, 21–30.

Crosland, Maurice P. (1980), 'Chemistry and the chemical revolution', in G.S. Rousseau and Roy Porter (eds), *The Ferment of Knowledge: Studies in the Historiography of Eighteenth-century Science*, Cambridge: Cambridge University Press, pp. 389–418.

Crosland, Maurice P. (1994), *In the Shadow of Lavoisier: The Annales de Chimie and the Establishment of a New Science*, Oxford: Alden Press.

Donovan, Arthur (1993), *Antoine Lavoisier, Science, Administration and Revolution*, Cambridge, MA: Blackwell.

Duhem, Pierre (1902), *Le Mixte et la combinaison chimique*, reprint Paris: Fayard, 1985; trans. Paul Needham, (2002), *Mixture and Chemical Combination, and Related Essays*, Dordrecht: Kluwer.

Einstein, Albert (1947), 'Paul Langevin', *La Pensée*, **12**, May–June, 13–14.

Gooding D., T. Pinch and S. Schaffer (1989), *The Uses of Experiment: Studies in The Natural Sciences,* Cambridge: Cambridge University Press

Goupil, M. (ed.) (1992), *Lavoisier et la Révolution chimique*, Palaiseau : Sabix, Ecole polytechnique.

Grimaux, Edouard (1896), *Lavoisier, 1743–1794, d'après sa correspondance, ses manuscrits, ses papiers de famille et d'autres documents inédits*, Paris : F. Alcan.

Grison, E., M. Goupil and P. Bret (1994), *A Scientific Correspondence During the Chemical Revolution*, Berkeley, CA: University of California Press.

Guerlac, Henry (1961), *Lavoisier, the Crucial Year: The Background and Origin of His First Experiments on Combustion in 1772*, Ithaca, NY: Cornell University Press.

Guyton de Morveau, Louis-Bernard (1786), 'Second avertissement', in *Encyclopédie méthodique, Dictionnaire de chymie*, vol. 1, pp. 628–34.

Hadamard, Jacques (1947), 'Paul Langevin au Collège de France', *La Pensée*, **12**, May–June, 31–2.

Holmes, Frederic L. (1985), *Lavoisier and the Chemistry of Life: An Exploration of Scientific Creativity*, Madison, WI: Madison University Press.

Holmes, Frederic L. (1989), *Eighteenth-century Chemistry as an Investigative Enterprise*, Berkeley, CA: University of California Press.

Holmes, Frederic L. (1991), *Hans Krebs*, New York: Oxford University Press.

Holmes, Frederic L. (1998), *Antoine Lavoisier: The Next Crucial Year*, Princeton, NJ: Princeton University Press.

Julliard, J. and M. Winock (1996), *Dictionnaire des intellectuels français*, Paris: Seuil.

Kagan, Henri (1995), 'Lavoisier, chimiste', in *Il y a deux cents ans, Lavoisier*, Actes du colloque, Paris-Blois, 3–6 mai 1994, Paris: Académie des sciences, pp. 9–11.

Kristeva, Julia (1974), 'La revolution du langage poétique', trans. in Léon Roudiez (ed.) (1984), *Desire in Language: A Semiotic Approach to Literature and Art*, Oxford: Blackwell.

Licoppe, Christian (1996), *La formation de la pratique scientifique: le discours de l'expérience en France et en Angleterre, 1630*, Paris: La Découverte.

Meinel, Christoph (1983), 'Theory or practice? The eighteenth-century debate over the scientific status of chemistry', *Ambix*, **30**, 68–103.

Metzger, Hélène ([1930] 1974), *Newton, Stahl, Boerhaave et la doctrine chimique*, Paris: Blanchard.

Metzger, Hélène (1935), *La Philosophie de la matière chez Lavoisier*, Paris: Hermann.

Mieli, Aldo (1932), 'Le rôle de Lavoisier dans l'histoire des sciences', *Archeion*, **14**, 51–6.

Poirier, Jean-Pierre (1996), *Antoine-Laurent de Lavoisier, Chemist, Biologist, Economist*, Philadelphia, PA: Pennsylvania University Press.

Reichardt, R. and H.-J. Lüsebrink (1988), 'Révolution à la fin du XVIIIe siècle. Pour une relecture d'un concept clé du siècle des Lumières', *Mots*, **16**, pp. 35–68.

Rey, Alain (1989), *Révolution: histoire d'un mot*, Paris: Gallimard.

Smeaton, W.A. (1989), 'The legacy of Lavoisier', *Bulletin for the History of Chemistry*, **5**: 4–10.

Venel, Gabriel-François (1753), 'Chymie', in D. Diderot and J. D'Alembert (eds), *Encyclopédie ou dictionnaire raisonné des sciences, des arts et des techniques*, vol. 3, Paris, pp. 408–37.

Chapter 11

'*La Mauvaise Herbe*': Unwanted Biographies Both Great and Small

Jacalyn Duffin

> La biographie est une mauvaise herbe, dont le foisonnement dans les champs de l'histoire est funeste à l'avancement de la science véritable. (de Bertier de Sauvigny, 1948, p. vii)

I have written two biographies of dead, white, male doctors. From the academic perspective of our time, my choice is quixotic, perhaps even risky, especially for a woman who began her study of history only twenty-five years ago.

In what follows, I will briefly examine the status of biography in popular culture and scholarly literature. Then I will present an outline or 'mini-biography' of each of my biographies. One is the life of a 'great doctor', the other of his complete opposite, an 'everyman'. For each, I will describe what I think was original in my work – in other words, what quality should have justified the publication. Then I will recount the journey I took with various reviewers and editors before the books were published. In both cases, the fact that I was writing biography posed the greatest barrier to publication with an academic press. I will conclude with why I think biography still belongs, and will remain, in the panoply of resources produced and used by historians of medicine. My conclusions return to the aesthetic of pleasure.

Biography in Popular Culture and Scholarly Literature

First, I have a confession. I *like* biographies to read, as well as to write. And I know a lot of others do, too. Biography is immensely popular in North America, and I suspect in Europe, too – for pleasure, entertainment, storytelling, television, and movie-making. For many people, history *is* biography. A desultory survey of the 'History' sections of bookstores in train stations of western Europe during my sabbatical travels of 2001–2002 revealed that at least half the so-called 'History' books were biographies; in Rome, the figure was close to three-quarters.

People like to map their own personal histories onto the lives of others; and they like to gossip. We prepare our CVs with reverence and read others' with voyeuristic curiosity. Biography is painless, accessible, irrefutable. It can be sympathetic or hostile – and we do not have to agree with the interpretations. One can find examples of contrasting biographies on controversial figures, such as Théophraste Renaudot or Grigori Rasputin or Florence Nightingale. The great differences in the final summation of these lives do not refute each other; rather, they enhance the enjoyment and fascination of trying to understand experience in other times and places. Most of my fellow contributors in this volume wrap their scholarly work around one or two biographical subjects; the life-story parts of their essays are my favourites, and the easiest to digest. Virgil-like, the biographical bits subtend and exemplify the analytic points. Biography serves up history in human doses, which is the only dimension to which we can truly relate as individuals. Biography, therefore, is personal, intensely personal, for the subject, the author and the reader. And not only is biography popular, it is ancient. We have the examples of Plutarch, Boswell and Voltaire, to which we could add the biblical prophets and the evangelists without stretching a point.

So where is the problem? Why might the liking of biography be 'confessed' like a sin? Why would such a popular topic merit a symposium and a transatlantic voyage? Many of the authors in this volume, including Aubin, Bigg, Bensaude-Vincent and Söderqvist, address the 'problems' and 'limitations' inherent to this genre that everyone else seems to crave without apology. Are we not simply re-inventing an old wheel? Or are we reacting against a recent fad?

Certainly, biography is nothing new. In fact, one problem is its venerable age. Historians prefer new ways of approaching the past. Writing about lives cannot claim originality of method or of structure. Perhaps, too, biography suffers because it is so well liked; its appeal and accessibility can be construed as absence of erudition or rigour.

Notwithstanding the popularity of biography among the general public, however, its status has declined in the ivory tower of academe, where, as Thomas Söderqvist (Chapter 15 in this volume) aptly says, it is 'despised and rejected' (Isaiah 53:3). And the decline is not new. Back in 1948, Guillaume de Bertier de Sauvigny introduced his life of a colourful ancestor with the observation that many modern historians considered biography to be 'a weed whose proliferation in the fields of history is considered deadly for the advancement of true knowledge' (1948, p. vii, and epigraph above).

Half a century later, these sarcastic words still have a special resonance, as we confront a publishing conundrum. Popular biographies may become bestsellers and earn prizes. But scholarly biographies (the kind with footnotes, placed with academic presses) are far less celebrated now than in the past. For a time, they even became less numerous – not that scholars avoided life-writing, but because finding publishers was difficult. The decline can be understood as a by-product of the rise of social history and the passing of the 'great man' theory in academia.

Curious to see whether or not I could find statistics to confirm my impression that biography had dwindled in popularity, I used the articles indexed in Medline from 1966 to 2002. Table 11.1 displays the number of articles classified as 'biography' or as 'historical article', or as both combined (to eliminate obituaries), over the three-and-a-half decades of *Medline* indexing since 1966. When expressed as a proportion of all historical articles, biography declines from a high of nearly 60 per cent in the late 1960s to a low of 42 per cent in the mid-1980s, with some recovery up to 2002. As the total number of articles published and indexed increased through time, the *relative rate* of publishing is also important. Historical biography might be expected to represent a vanishingly small percentage of all publications in this

Table 11.1 'Biography' and 'historical articles' in *Medline*, 1966–2002.

	1966–74	*1974–81*	*1982–88*	*1989–92*	*1993–95*	*1996–2002*
No. of years	9	7	7	4	3	5.2
Biography	27,215	21,951	15,182	9,741	7,894	15,579
Historical articles	46,059	42,335	36,009	20,437	15,932	32,759
Biography + historical	25,893	23,271	15,182	9,741	7,894	15,547
Biography as % of historical	**59.1**	**51.9**	**42.2**	**47.7**	**49.5**	**47.5**
History average/year	5,118	3,136	5,144	5,109	5,311	6,300
Historical biography average/year	2,877	3,324	2,168	2,435	2,631	2,989
Medline average total/year*	220,720	255,820	277,806	337,781	380,102	474,276
Historical as % of *Medline* total	**2.32**	**1.23**	**1.9**	**1.5**	**1.4**	**1.3**
Historical biography as % of *Medline* total	**1.3**	**1.3**	**0.78**	**0.72**	**0.69**	**0.63**

Note: * *Average total number of Medline items indexed in each year. These rates were calculated on total number of references available for slightly different but overlapping periods: 1966–74, 1975–79, 1980–84, 1985–89, 1990–94, 1995–99.*

medico-scientific index; yet I was startled to find that it was as high as 1.3 per cent of the total until the mid-1980s; less surprising, it is now less than half of that, and continues to decline.[1] Beth Linker (Chapter 14 in this volume) confirms this impression; she observes that the decline is more prevalent in medical history than history of science.

If we think of history as the study of change, we can see how the problem came about. Rather than attributing change to the remarkable energy and ideas of intelligent, hard-working individuals, social history looks to notions of gender, race, class, authority and power. It is concerned with collectives, cultures, structures, inevitability, and even continuity. Long before I had written any book, one gloomy professor told me that unless I could demonstrate that my subject's ideas – notwithstanding his fame and originality – 'were held by large numbers of important people', then my work was uninteresting. It seemed a tall order.

To freshen the approach to biography and avoid charges of naive idealism, historians use several 'strategies'. The contributions to this volume by Hankins (Chapter 5) and Higgitt (Chapter 9) address some of the techniques head-on; together with Kragh (Chapter 7) and Linker (Chapter 14), they linked these methods to purpose. One strategy allows the single life under study to represent an entire era, or intellectual movement. For example, S.E.D. Shortt (1986) used the life of a nineteenth-century asylum-keeper in a medium-sized Ontario town as a lens for all psychiatry of the contemporary English-speaking world. His title? *Victorian Lunacy*.

Another approach to freshening biography is debunking. Instead of attributing change to individual geniuses, people are portrayed as obstacles to change, plagiarists and villains. Throwing respect to the wind, biographers looked for hidden motives, shameful secrets and big mistakes in the lives of their subjects. Examples of this approach include feminist attention to marriages, such as that of Phyllis Rose (1983) on nineteenth-century intellectuals, or of Andrea Gabor (1995), who showed that twentieth-century scientists sometimes extracted more from their spouses than tender loving care.

But debunking biographies appeal to even fewer readers than do the celebratory tales. People do not like them. Humans react defensively and offensively to attacks on a flawed but none the less brilliant precursor. Gerald Geison's book *The Private Science of Louis Pasteur* might fit that bill. For his uncompromising remarks about the ethics of Pasteur's research – and notwithstanding the archival evidence he used to bolster his claims – Geison was roundly attacked by the mighty pens of Bruno Latour (1995) and Nobel laureate Max Perutz (1995). As if in compensation, he was awarded the Welch Medal of the American Association for the History of Medicine.

1 All historical publishing indexed in Medline has declined, but to a lesser extent than biography; a slight rise occurred in the mid-1980s. Although Medline offers handy figures, it is not the ideal source for this assessment. It does not include books or book chapters. Furthermore, and contrary to the observations of Beth Linker in Chapter 14 of this volume, medical writers and readers may favour biography as a form of history when compared with other academic groups. If anything, I suspect that the decline is less obvious in Medline than it would be if a historical database had been used.

Did Geison resort to judgemental prose to ease his book into print? I have often wondered.

Social history has been in vogue for so long now that we may actually be witnessing a minor reaction as the biography empire strikes back. What is my evidence? First, we have seen a collection of 'compensatory' biographies that document the lives of, for example, women doctors, black doctors, black women doctors and others previously absent from the white male doctor-dominated medical history of the past. In this tradition, and in response to the late Roy Porter's famous appeal (1985), we can also include recent lives of patients, both famous and obscure, who were also excluded from histories of great doctors and great ideas.

Signs of reaction can be found in the periodical literature, too. The *Canadian Bulletin of Medical History* devoted a special issue to the topic of medical biography in 1996 with contributions by many historians. Perhaps in desperation, new journals have been founded as forums for life-writing. The *Journal of Medical Biography* is a good example. Founded back in the late 1960s, a precursor produced sparse and sporadic issues in 1977 (vol. 5) and 1988 (vol. 6). Resuscitated from this *mors interrupta* with a new Volume 1 in 1993, it has since published over 380 articles. I have no insider information, but I wonder if would-be authors, frustrated by constant rejection from mainstream history of medicine periodicals, banded together to create a venue where they could publish and read about the past in a manner that made sense to them.

Other evidence comes from the recent comeback of obituaries in magazines and newspapers (for example, 'Obituaries' are a recent addition to the *Economist*, as is the 'Lives Lived' column of the Toronto *Globe and Mail* newspaper). From January 1995 to August 1996, the *British Medical Journal* published a series of heated editorials and letters on the topic of obituaries. Finally, I have the irrefutable evidence of my own two biographies. They haven't broken any sales records, but at least they made it into print. Two editors took risks on me.

Mini-biography No. 1: My Life with Laennec

At his death, René Théophile Hyacinthe Laennec (1781–1826) was one of the most famous doctors in Europe. Breton-born, he lived through the French Revolution, the First Empire and the Bourbon Restoration. Intellectually gifted, financially challenged and chronically ill, his existence can be read (and has been written over and over again) as a romantic life in romantic times. Laennec's medical fame resides in his discovery of the phenomenon of auscultation and his invention of the stethoscope. Not only did this instrument come to symbolise medical practice and diagnosis, it also coincided with a change in the way doctors think about disease. We still use the medical paradigm endorsed by Laennec.

Laennec's achievement cannot be separated from the turbulent times in which he lived. His delicate mother died when he was five. Abandoned by his ne'er-do-well poet father, he was sent to his physician uncle, who set the adolescent on medical training during the Revolution. In Paris, Laennec studied with Corvisart, Napoleon's personal physician, who was a promoter of Auenbrugger's percussion.

In the eighteenth century, physical diagnosis was little relevant to the diagnostic process; diseases were made up of symptoms. They were subjective: to be sick, you had to feel sick. Patients were the ultimate authority on their own well-being. Attempts had been made to link diseases to the body, but detecting inner change was not possible. Laennec's stethoscope, when combined with percussion, enabled doctors to recognise the anatomical state of the lungs and heart before the patient became a cadaver. Disease concepts shifted from symptom patterns to organic changes. Doctors began finding anatomical 'causes' of all diseases, and disease names changed from symptom-based descriptions, like cough, haemoptysis, consumption, shortness of breath, to anatomic descriptions, like bronchiectasis, effusion, emphysema, tuberculosis and tumour.

Laennec could even hear signs of disease inside the bodies of people who felt completely well. Something had changed coincident with his work. From the nineteenth century forward, you no longer had to feel sick to be sick. Instead, the doctor had to detect an objective sign or proof of disease. The patient was no longer the ultimate authority on her own well-being. This conceptual shift to anatomical diagnosis and physician authority dominates practice today.

Following his discovery, the royalist Catholic Laennec found a niche in Restoration circles where he enjoyed many benefits of his monarchical contacts; yet he had enemies – an unpopularity that only to adds to the romance.

I studied Laennec because my adviser, Mirko Grmek, had suggested him to me as a thesis subject back in 1982. Grmek had just finished a catalogue of the extant manuscripts – some ten thousand pages of lecture notes and patient records held in several French archives. A vast set of letters remained within the family. For the thesis, I concentrated on the scientific papers alone. Later, for the book, I needed to find the letters.

But Laennec was so famous that a mountain of secondary literature also weighed on me too. Most authors had relied on the two-volume biography written in the early twentieth century by Nantes physiologist Alfred Rouxeau. Unabashedly hagiographic, Rouxeau made extensive use of the personal correspondence to reconstruct a gentle account of a much-maligned, good and brilliant man who lived in turmoil, poverty and illness. I needed to examine Rouxeau's papers, too.

Had I been contemplating a debunking strategy, I would have had a formidable uphill battle. Laennec has been called 'the French Hippocrates' and 'the French Sydenham', his influence has been likened to that of Galileo, Newton and Gutenberg. Because he was a devout Catholic during the period of atheism, Breton admirers are pushing for his canonisation. All this material was very daunting. How could I find anything new to say? And why bother? So intimidated was I after a year's research that I told Grmek that I needed to change my subject. He gave me a summer to explore something else – but the lure of Laennec proved too strong. Only after reading most of the scientific papers, especially the lectures, did I finally realise that I might have something new to say.

What was original in my biography of Laennec? First, Rouxeau had done little with the scientific papers and patient records; I could fill that gap through simple exposition. Second, most authors, including Foucault, reduced Laennec to a pathologist who saw disease as the product of anatomic change. But I realised that

Laennec saw himself as a physiologist too, at a time when both disciplines were being defined. Laennec acknowledged the advantages of anatomical diagnosis, but he thought that disease could come from disturbed emotions, from what he called a 'lesion in the life force'. If you were unhappy, he said, you could become physically ill.

I also found a dramatic inner consistency between this vitalistic science and his personal life. I could combine the letters with the scientific papers to prove my point. Most authors blamed Laennec's unpopularity on his politics and his religion. But I noticed that it was his apparent vitalism, his rejection of the positivist materialist trend to which he had contributed so much, that got him into the deepest trouble. Furthermore, I understood how Laennec handled evidence of his observations – sensitivity and specificity of his signs; the psychology of discovery. I could also re-personalise the man by adding what I could assess of his character, his marriage and his natural foibles to Rouxeau's sanitised version.

Finally, I could relate Laennec's life to the huge body of French medical history, beginning with Foucault and extending through Ackerknecht, Gelfand, Goubert, Grmek, Huard, Imbault-Huart, Leonard, Lesch, Maulitz, La Berge, Rey, Weiner, Weisz and many others. I would put Laennec back into French medicine, and I would put the new social history of French medicine into the life of Laennec.

The thesis was completed in 1985, and my own life intervened – a second baby, a return to Canada, the search for a job, and new research based on Canadian sources. Only after I had published the Canadian book, which I will describe next, could I return to France to find the original letters and complete my research on Laennec. But first I tried to find a publisher.

In June 1993, I circulated a prospectus to the leading university presses in North America, including Oxford, Harvard, Chicago, Cambridge, Rutgers, Hopkins, Toronto and Princeton. I had hoped that Toronto would be interested – after all, that press was publishing my first book. Toronto's rejection was immediate and blunt, and it typified the field. Editors have a difficult task when it comes to rejection, and they do it as delicately as possible. Nevertheless, the words hurt. The Toronto editor said: 'the subject-matter lies outside the areas in which we normally publish'. They wished me 'success in placing the work with an appropriate publisher'. Similar responses came from Harvard, Cambridge and several others. Now, a decade later, Oxford has still not replied. Cambridge had been urged upon me, and against my better judgement, by my friend and colleague Charles Rosenberg. The prospectus was forwarded twice, finally the reply came. 'I share you enthusiasm for biography,' said the editor, 'but … market realities concerning … recent biographies … have forced us to pull back from these areas.'

Biography again! And what a strange, paradoxical thing. Biography is a best-selling genre in the popular market, and yet for academics it sells so poorly that it doesn't deserve to be printed. To be fair, these editors might have chosen a polite way to tell me to go away: it is much easier to reject my genre than to reject me. I kept watching Cambridge for new biographies to assess whether or not I had been lied to out of 'kindness'.

Real or euphemistic, the reasons given for rejecting the Laennec book meant that my work had been put in a box before it had been read. Where did I say I was 'enthusiastic' for *biography*? Was it 'biography' that I had written? I thought I had expressed something new about the history of clinical medicine through the examination of one life. Only two of the many publishers sent my prospectus out for review. Both probably did so out of courtesy for favours I had recently done for them as a reader and translator.

One of these two, a Very Important Academic Press had me complete many pages of forms. I did so gladly; then began my wait. Months went by and I heard nothing. Finally, a phone call came from the VIAP editor, followed by a letter. The letter said that a 'straightforward narrative of the life of Laennec may be overly limiting … that in fact focusing on Laennec "the person" may prevent you from employing Laennec as a "window to an age", as a vehicle for examining the broader idea of medicine at the time' and for 'demolishing the positivist approach to medicine and science in the nineteenth century'. The VIAP editor suggested that I keep looking for a publisher because they were 'not in a position to offer [me] a contract'.

Once again, biography was the problem; and demolition was a goal. On the telephone, the editor had asked questions with candour and sympathy. 'Can't you make it look like it is not a biography? Make it appear to be about something else? Something more broad?' The suggestions were sincerely meant and gently put. But I was irritated, and replied: 'It's a bit disingenuous to pretend that the book will not be biographical when the main, primary sources are the previously unexamined papers of an individual.'

I began to wish that my French was better so that I could try in Europe with greater ease. Was biography welcome there? With sadness, I grew resigned to the idea that my book would never be published, when I heard from Princeton. Yes, they were interested! Who knew why? Perhaps they thought that the book would be purchased by many rich doctors. I had an advance contract in a short time. My gratitude was (and still is) immense.

Princeton made a good decision for me. However, as much as I am willing to stand by my scholarship and claim that my book is useful, I am aware that, perhaps, Princeton made the wrong decision for itself. Certainly, it was courageous. Against the advice of all the leading editors in the field, they published another intellectual biography about a dead, white, male doctor who turns out *not* to be a demon in the end. *To See with a Better Eye* appeared in March 1998. It is handsome but expensive, and exists in hardcover only. Sales are slow. In May 2002 it ranked 444,893 on Amazon.com's best-seller list; by May 2004 it has slipped to 579,996. Reviews are still coming in. I am patient, and mostly content. In November 2001, nearly four years after it appeared, it was honoured with the Hannah Medal of the Royal Society of Canada for 'the best work published within the preceding three years' – just under the wire! That the 2000 prize went to a book published in 1999 is a sobering comparator, the implications of which are lost neither on me, nor, I fear, on my unremunerated publisher.

Mini-biography No. 2: My Life with Langstaff

The second biography, in fact, became my first book. It is a product of a Hannah post-doctoral research project started in 1985, when I returned to Canada after defending my thesis on Laennec. By then I was in love with manuscripts and archives, and I wanted to study how those nineteenth-century scientific ideas, like percussion and auscultation, came to North America. It would be an extension of my thesis in both time and space. I began searching for sources. One day, I found a single manuscript daybook of a Dr James Miles Langstaff (1825–1889).

The opposite of Laennec, Langstaff was a complete unknown. Born one year before Laennec died, he was a Canadian-trained country doctor who practised for forty years from the timber-framed home that he had built in the village of his birth, near Toronto. Through a stroke of good fortune, I tracked down his grandson – also a doctor, well on in years and still living in the same house. In his attic lay the remainder of the forty-year trove of daybooks, and most of the financial records, too.

Langstaff had been involved in local politics, but he made no discoveries, travelled little, and his fame extended no more than ten miles from his home. Again the opposite of Laennec, he left almost no personal letters, articles or books. And bonus! – no historian had ever written about him; there really had been no reason. In many ways, this project was easier than the one on Laennec, in which hundreds of other authors and physicians (both dead and alive) seemed to peer over my shoulder, caring about what I would write. I was free. However, without Langstaff's own account of his life and work, I often wondered if he would have agreed or disagreed with my interpretations.

I decided to build a database profile of the practice. Microcomputers were just coming into widespread use. I had a huge, untapped collection of papers. Combined with this new technology that could make the records exploitable and, above all, *countable*, the daybooks and financial records promised an intimately detailed picture of an ordinary life in medicine. Langstaff was fascinating because his life was banal; he was a nobody, and could therefore be seen as an everyman.

And so began my almost monastic relationship with these beautiful papers. I read the entire set of daybooks, collecting information about all deaths, births, innovations and unusual cases. But I also divided the forty-year practice into four blocks of time, up to five years long, each representing a decade; I entered all information for these periods into database files.

Rather than a biography of a person, I called this book a 'biography of a practice'. I was awash in statistics. I learnt how many patients Langstaff saw in a day, a week, a year, and how his practice grew. I found out what diseases he diagnosed, and how diphtheria came relentlessly every winter to kill little children, but scarlatina came in waves at ten-year intervals. I know what drugs he used to treat those diseases. When his therapeutics changed, I discovered how quickly he adopted new treatments, or how reluctant he was to abandon older ones. I noticed when he first used a stethoscope, a thermometer, forceps and chloroform. I counted how many babies he delivered in forty years – about 4,000 – and how many infants perished before they breathed, and how many of their mothers died before they could suckle them. I know what operations he dared to perform: emergency amputations and tracheotomies, yes, but also rare elective procedures for hare lip, club foot and breast cancer – often, but not always, without anaesthesia. I read what caused the accidental injuries in his community: horses, alcohol, fighting and the advancing technologies of farm and railroad. I recorded how he dressed those wounds and when he turned to carbolic acid. I met his colleagues at consultations for problem cases; their encounters were not always amicable. I found out that he delivered his first wife of all their children – 11 babies, of whom only three lived to adulthood. And when this exhausted wife died of typhoid fever during a menstrual period, he blamed the slaughterhouse next door and went into local politics as a reformer seeking revenge.

The little town in which Langstaff lived had two newspapers: one politically conservative and one liberal. They offered social commentary about the doctor and his work from two different perspectives. Once again, I have no idea if he would have agreed with them or not – but they softened the monotony of my source. The Canada Census also provided another touchstone in each decade for births, deaths and personal wealth. He married again, a teacher who was an activist for women's votes, and who worked to educate her community in culture, music, art, literature and temperance. In this marriage, I found evidence to dispute the claim that doctors did not support women's rights; Langstaff, at least, seemed to do so.

In many other areas, Langstaff failed to behave in the way that social historians had suggested that nineteenth-century doctors behaved as a group. But what does it mean if he did not? This is only one practice, one life. I scarcely think this study, detailed though it may be, can refute those based on large numbers. In situating itself within the field of social history, however, it offers a model for other projects. It also challenges that gloomy professor who, long ago, dismissed my early work for its focus on quirky ideas held by one man.

Langstaff shows us that biography can be used as a lens to test theories about social, intellectual, and cultural change, those broad ideas that the VIAP editor urged me to address. Maybe this is a lesson from my project on Laennec: if the theories do not 'work' for this one person, perhaps they do not work at all. Could this powerful use of biography be a historical equivalent of Popperian falsification?

For this, my first printed book, I did not try to find a publisher before writing. Like my subject, I was completely unknown; perhaps I could not have done otherwise. In terms of my career, however, it was a big mistake.

By mid-1991, I sent the first draft to the editor of a series of short medical biographies. I had enjoyed a small grant of $2,500 Canadian from that series on the promise of the manuscript. In the covering letter, I acknowledged that the manuscript was longer and had more footnotes than specified, but I had written it the way it seemed to need to be written. I hoped that he would extend the length limits. If the work was unacceptable, I hoped that he would read it anyway for criticism and suggestions. Instead, the manuscript bounced by return mail, unread, with a cool letter asking for return of the grant, which, of course, had been spent long ago. I complied.

Then I submitted a prospectus to the University of Toronto Press. They expressed an interest and sent the full manuscript for peer review. Later, I revised and resubmitted as requested. Then this revision was submitted for a national grant-in-aid for scholarly publishing; more revisions followed. In the end, I re-wrote the book three times to satisfy different readers and granting agencies – always downplaying the life of the man, and trying to generalise from him to his century. The third version restored many things that had been removed from the first in preparation of the second. I admit that the repeated revisions made the book better, but my conclusions were unaltered, and the mechanical effort added greatly to the delay and my dismay. The time I have invested in these projects took on a life – a biography – of its own. These men essentially moved into my home and joined my family: Langstaff consumed seven years; Laennec – who did indeed overlap – fifteen.

Langstaff is available in an affordable, illustrated paperback. It received positive reviews, including Olivier Faure's rave in the *Annales ESC*, of which I am inordinately proud. Nevertheless, Amazon.com puts it even lower down the sales list, ranked at 1,432,914 in May 2002; by 2004, its position had slipped to 1,765,688. And the single review posted by an anonymous Californian in September 2000 warns that:

> you won't ['enjoy' this book] if you're not fond of the academic history style. There are charts, and frequently the text is a detailed discussion of the charts, but … you can work through/skip over those parts … What you won't get is intimately involved with the good doctor Langstaff. This isn't the volume for character psychology – it's about medicine, education, and science. (<www.amazon.com>, 5 May 2004)

In other words, for my Californian reader, it was not biographical enough!

Conclusion: Biography and Non-biography

Using the *mauvaise herbe* metaphor of de Bertier de Sauvigny, I once wrote that:

> scholars compartmentalize history into artificial though useful divisions defined by topic or period. For early nineteenth-century France, we have studies of the Revolution, the First Empire, and the Restoration; for its medicine, we have a rich literature about hospitals, practitioners, patients, pathology, physiology, and other sciences. Biography reminds us that individual lives transit these units and allows us to test, sometimes even disrupt the neat definitions of their boundaries. (Duffin, 1998, p. 4)

Life writing provides us with vigorous specimens to challenge the formal garden of theoretical history: there is art in cultivating weeds.

And yet without the rich collection of imaginative social histories, neither my Laennec nor my Langstaff would have been the books that they are. Those other works, the *non-biographies*, provided more questions, and indeed *better* questions, than I could have imagined without them. We need to maximise our approaches, accommodate all genres – even if it means including the venerable and popular, too – if we really wish to understand our medical past. I contend that labelling (and dismissing) a project as 'biography' ought to be a non-issue; it is the fragile shelter of the unimaginative who lack confidence in their own abilities to spot originality. Much more important is the historical 'work' that a biographical analysis can do.

Finally, let me return to the pleasurable aspects of biography with which I opened this paper. Why apologise for writing and reading enjoyable stories? Why succumb to the pernicious view that scholarship must always be sombre? Entertainment is only part of the pleasure. A greater, even deeper form of pleasure comes from the aesthetic of biography as an art form. The papers in this volume suggest two specific ways in which biography conforms to aesthetics; both are structural.

First, biography is a play of variations upon a theme. Like an opera, a sonnet or an altarpiece, each life has certain anticipated parts, components and rhythms. In general, everyone is born, has a mother, is a child, is educated, loves (or not), has sex (or not), becomes a parent (or not), works and is paid (or not), grows old (or not), falls ill (or not), and dies. In creating the art, a biographer can play with, emphasise, embellish or omit any of these various components.

Freud would tell us that the events of our early lives and our desires shape all that we become as individuals. Yet Thomas Söderqvist alone alludes to Freud in this volume. Only because of the conference that preceded this book was I led to an interesting idea: does the waning of Freud explain why my two publishers complained of my writing about the childhood and marriages of my subjects? (I did anyway.) Does the elegant connection of human 'personality to product', discussed by Geoffrey Cantor at the conference, mean that Laennec's stethoscope cannot really be understood without reference to his personal existence? Does it mean that the life of an obscure practitioner in rural Ontario could stand in for any isolated practice in mid-nineteenth-century North America?

Second, biography fulfils an aesthetic of pleasure by allowing authors to construct themselves as well as their subjects. Both David Aubin and Charlotte Bigg (Chapter 3 in this volume) describe 'self-fashioning' with respect to a subject's participation in his or her own biography; the influence of what they leave behind and what they destroy. Biographers also fashion themselves through their subjects and their writing, by their imagination, probing, taste, sympathies and revulsion, and by the inclusion and exclusion of materials and events. In this volume, Rebekah Higgitt (Chapter 9), Helge Kragh (Chapter 7) and Betty Smocovitis (Chapter 13) have all noticed this practice – and use words like 'suppress' or 'erase' to describe it. It is not a deliberate fictionalisation, but the product of an honest interaction between the personality of the writer and her subject. In the end, all books are the artistic creations of their authors – even the ones with footnotes.

Bibliography

de Bertier de Sauvigny, Guillaume (1948), *Le Comte Ferdinand de Bertier (1782–1864) et l'énigme de la Congrégation*, Paris: Les Presses Contintentales.

Duffin, Jacalyn (1993), *Langstaff: A Nineteenth-Century Medical Life*, Toronto and Buffalo, NY: University of Toronto Press.

Duffin, Jacalyn (1998), *To See with a Better Eye; A Life of R.T.H. Laennec*, Princeton, NJ: Princeton University Press.

Faure, Olivier (1998), [Review of Duffin, *To See with a Better Eye*, 1993], *Annales ESC*, March–April, 338–40.

Gabor, Andrea (1995), *Einstein's Wife: Work and Marriage in the Lives of Five Great Twentieth-Century Women*, Harmondsworth: Penguin.

Geison, Gerald L. (1995), *The Private Science of Louis Pasteur*, Princeton, NJ: Princeton University Press.

Latour, Bruno (1995), 'A propos de "La science privée" de Louis Pasteur', *Recherche*, **281**, November, 35–6.

Perutz, Max F. (1995), 'The pioneer defended [review of Geison, *The Private Life of Louis Pasteur*, 1995]', *The New York Review of Books*, 21 December, 54–8.

Porter, Roy (1985), 'The patient's view: doing medical history from below', *Theory and Society*, **14**, 175–98.

Rose, Phyllis (1983), *Parallel Lives: Five Victorian Marriages*, New York: Knopf.

Shortt, S.E.D. (1986), *Victorian Lunacy: Richard Maurice Bucke and the Practice of Late Nineteenth-century Psychiatry*, Cambridge: Cambridge University Press.

Teigen, P.M. (1996), 'Why historians of medicine are wary of biography', *Bulletin Canadien d'Histoire de la Medecine/Canadian Bulletin of Medical History*, **13** (2), 181–6.

Chapter 12

Primary Suspects: Reflections on Autobiography and Life Stories in the History of Molecular Biology

Rena Selya

Biographers researching scientists active in the early history of molecular biology face a particular set of challenges with these often colourful individuals. James D. Watson, Joshua Lederberg, Max Delbrück, Salvador Luria, Gunther Stent and others began the process of writing the history of molecular biology in the 1960s, as the field was maturing into an independent discipline. Several figures wrote autobiographies that frame their scientific careers in personal terms. Their versions of the history of their field reflect their priorities in ways that both help and hinder biographers of twentieth-century life scientists.

This chapter describes some of the challenges of writing biographies of scientists, who, as a group, have a strong received history, complete with heroes, occasional villains and victims. These reflections are based on my experience writing a biography of Salvador Luria, an overlooked figure in the early history of molecular biology. Luria collaborated with Delbrück in the 1940s, and was James Watson's Ph.D. advisor at Indiana University 1947–50. He shared the Nobel Prize for Physiology or Medicine with Delbrück and Alfred Hershey in 1969, and was a committed political activist throughout his life in the United States.

I first encountered Salvador Luria in his autobiography, *A Slot Machine, A Broken Test Tube*, which was published as part of the Sloan Foundation series of autobiographies of prominent scientists (Luria, 1984). Luria describes his career in the post-Second World War biological boom, and as an ardent political activist in the 1960s and 1970s. Although his life story provides an opportunity to examine changes in both the history of science and American history, historians have thus far overlooked Luria. His contributions to the development of the life sciences through his experiments with bacteriophage have been overshadowed by his role as Watson's adviser, and more aggressive politically active scientists such as Linus Pauling have captured historical attention.

Luria's autobiography inspired my own biographical work that examines the relationship between science and democracy throughout his life (Selya, 2002). Nevertheless, over the course of my project I struggled with it and many other scientific autobiographies and memorial testimonies as primary sources, as I worked to construct my own distinct narrative that would locate Luria in all of his historical contexts. I am his biographer, but I am also a historian of science, and my presentation of his life is with an eye towards a deeper understanding of the history of biology and American history. As a scientist and autobiographer, Luria did not share my intellectual and academic goals for his life story. Along with other molecular biologists who self-consciously constructed and shaped their own history over the past fifty years, Luria's case provides a particularly vivid example of the benefits and challenges of working with autobiographies, festschrifts and commemorative volumes when reconstructing and analysing the life of one particular actor.

Of course, questions about how to select and use primary sources and then construct a compelling narrative are not unique to scientific biographers. As Paula Backscheider has elegantly described, biographers of literary and political figures also face challenges in establishing a voice, working with available archival and autobiographical evidence, and making an original scholarly contribution that will be relevant to general history or literary criticism (Backscheider, 1999). Are biographies of scientists somehow different? Do historians of science face unique challenges of evidence and voice? Considerations of these questions are a part of a larger discussion of the place of biography in the history of science.

Autobiographies and other published recollections of scientists can be both a blessing and a curse for historians of science undertaking a biographical study. When scientists sit down and write out their life stories, or participate in an oral history project, they offer useful information to historians. The scientist can provide the basic facts of his or her life, and often is able to give a clear and logical explanation of his science that is accessible to non-scientists. Personal, political and philosophical reflections humanise the individual scientist and allow the historian to begin to formulate an image of the scientist as a whole person, rather than someone who is only interesting when they are in a laboratory or a scientific meeting. The subject's own words and thoughts can be invaluable for the scholar who is seeking to portray a complete, complex personality rather than merely to chronicle a series of experiments or discoveries. What better primary source for a biographer than the words of the subject himself?

Alas, these valuable tools are mere starting points for a trained historian, for they are by definition suspect. There are several related pitfalls associated with these primary sources. In addition to containing distracting factual errors that could be the result of a faulty memory, these sources are not neutral chronologies of events. Molecular biologists are often confident of their abilities to construct a plausible historical narrative, but their writings are often self-serving or clouded by the passage of time. Autobiographies and oral histories are catalogues of the stories scientists tell about themselves, and those stories inevitably have a scientific or historical agenda. This is not a criticism of scientific autobiographers, *per se*. All of us, young and old, famous and obscure, have stories that we tell about ourselves that consciously or unconsciously reflect the way we perceive ourselves. Leon Edel cautions biographers to be aware of those stories as private myths, 'the private self-concept that guides a given life, the private dreams of the self', and to take them as a starting point for biographical study, rather than as the study itself (Edel, 1984, p. 31).

These private stories can offer insights into the personality and psychology of an individual scientist, and even serve as a foil for the biographer's interpretation of her subject's life, but they cannot form the basis of an independent biographical study. They must be treated as critically as any other primary source. For a historian seeking to write a biography that provides insights into the individual *and* their historical context, these texts can be more of a hindrance than help.

These scientific myths are not only personal, but professional as well. Scientists, even those who claim to be uninterested in history, have a stake in how the history of their work is written. Whether their peers, the next generation of scientists, historians or journalists produce that history, there are strong disciplinary interests that can be served or challenged by the narrative. Festschrifts and memorial volumes are especially problematic, since they tend to skew the history around a particular individual or event. In addition, autobiographies and memoirs rarely cite letters, papers and other more reliable sources, which make it difficult for a scholar to check the historical facts. These problems are then compounded when trained historians uncritically use these documents as primary sources.

The Problem of the Origins of Molecular Biology

In the case of molecular biology, personal disciplinary histories have dictated the tenor and direction of historical analysis until very recently. The founding myth of molecular biology has not varied much since 1966, when James Watson, John Cairns and Gunther Stent edited a volume to commemorate Max Delbrück's sixtieth birthday and the twenty-first anniversary of the Cold Spring Harbor Phage Course, entitled *Phage and the Origins of Molecular Biology.* Known as *PATOOMB*, this collection of essays and reminiscences about the mythical Phage Group has served as the institutional memory of early molecular biology. As Pnina Abir-Am has pointed out, this volume presents a very specific vision of the Phage Group that reflects James Watson's conscious reshaping of his intellectual ancestry (Abir-Am, 1998; Abir-Am, 1999). Along with James Watson's *The Double Helix*, published in 1968, it has shaped historical and scientific accounts of molecular biology ever since.

According to the narrative in *PATOOMB*, in the 1930s the young and talented physicist Max Delbrück turned his mind to biological questions, hoping to follow in the footsteps of his legendary mentor, Niels Bohr, and discover a new law of nature in biology. He eventually moved to the United States to study genetics at the California Institute for Technology. Although some of his initial ideas were incorrect, many physicists and biologists were inspired by his theories equating the gene with the atom. In this group was the eminent physicist Erwin Schrödinger, who used Delbrück's model as the basis for his inspiring book, *What is Life?* In 1940, Delbrück and Luria happened to meet in Philadelphia, and discovered that they were both interested in bacteriophage. They collaborated on several important experiments, but their real historical significance lay in Delbrück's establishing the Phage Group and running the Phage Course at Cold Spring Harbor. The culmination of all of their hard work came in 1953, when Luria's graduate student James Watson discovered the double helical structure of the DNA molecule, thereby uncovering the secret of life. Watson and Delbrück are the major actors in a story that links bacteriophage and DNA.

Like all historical myths, this one is useful in that it does have some validity, but it is by no means the only possible historical narrative that can be constructed from the scientific papers and research reports of the 1940s and early 1950s. Nevertheless, since 1966 historians have uncritically treated *PATOOMB* as a primary source. Rather than examining this narrative as one interpretation of the scientific events of the 1940s, historians such as Robert Olby, Horace Judson and Lily Kay have accepted this private myth as historical fact, and constructed narratives that support it (Olby, 1974, pp. 225–47; Judson, 1979; Kay, 1985, p. 245; Kay, 1992; Kay, 1993, pp. 243–56). The Phage Group has become a fixture in the history of molecular biology, with a clear bias towards the version of events and influences offered by the *PATOOMB* contributors.

This volume has had literary effects on the field of the history of molecular biology as well. Biographies and memorial volumes tend to feature the 'ATOOMB' (and the 'Origins of Molecular Biology') suffix, which narrows the focus of the history of biology to the history of molecular biology (Stahl, 2000; Sapp, 1990; Fischer and Lipson, 1988; Summers, 1999; Lederman and Tolin, 1993). As a result of this historical trend, Salvador Luria's significant contributions to the development of biology in the United States have been overlooked by historians who have been narrowly focused on the question of the origins of molecular biology as defined by *PATOOMB*.

Personal Narratives and the Biographer's Voice

This historiographical situation is certainly problematic for my project and for other attempts to re-evaluate the history of molecular biology. However, a larger, related challenge for an historian using autobiographical and other memory-based materials is how to separate his or her voice from that of the subject. In addition to avoiding the subject's private myths, the biographer must not allow herself to rewrite the scientist's own story in his own words. As Thomas Söderqvist deftly points out, scientific biographies, especially of living or recently deceased subjects, can become a cacophony of voices, including that of the author (Söderqvist, 2003). It is a struggle to construct an engaging narrative that is true to the subject but at the same time is distinct in tone and argument from the subject's version of his own life. While one goal for the scientific biographer could be to write a biography that the subject would recognise and even agree with, that does not mean that the biographer should write the history that the scientist himself would have written (Comfort, 2001). The biographer must find a balance between sympathy for her subject and seeing the world through his eyes. This is a crucial part of writing a biography that makes a historical statement, rather than one that simply reports on a life.

James Watson presents a particularly clear example of how a scientist's public presentation of his own life can determine a biographer's narrative. Since the late 1960s, Watson has told his life story in popular books aimed at the scientist and the layman alike (Watson, 1968; Watson, 2002). While he is particularly candid about his personal life, he also is very clear about his position in the history of molecular biology. Victor McElheny's recent biography was written without access to Watson's papers or interviews with his subject, but Watson did leave a clear trail of primary sources for McElheny to use (McElheny, 2003). The book closely follows Watson's version of his life as described in his autobiographical work and speeches, and does not offer any historical analysis of Watson's place in the history of science.

Several recent biographies of molecular biologists further illustrate this problem. The subjects of these biographies – Max Delbrück, Linus Pauling and David Baltimore – are interesting, complex individuals whose lives intersected with some of the most pivotal historical and scientific threads of the twentieth century. Their biographers made little or no attempt to situate them within those trends, preferring simply to chronicle their lives, often using the subject's own words.

In the cases of Delbrück and Pauling, this problem is in part due to the closeness of the biographer and his subject. Thomas Hager has written a clear and compelling narrative detailing Linus Pauling's considerable scientific and political activities but does not locate those activities within the larger narratives of the history of science or American history (Hager, 1995). Hager's account is the most sophisticated in the Pauling literature (Serafini, 1989, Goertzel and Goertzel, 1995), which usually tumbles quickly into hagiography, although his work still suffers from lack of personal distance from his subject. Similarly, Delbrück's biographers acknowledge that they were completing a project that Delbrück started before his death, and rely as much as possible on his oral history and autobiographical sketches (Fischer and Lipson, 1988).

In the third case, Shane Crotty does not have a close personal relationship with David Baltimore, but he takes an uncritical approach to Baltimore's version of his life, which was recorded in the 1970s by another historian. For much of his book, Crotty paraphrases and cites the Baltimore oral history, even borrowing the structure of Baltimore's narrative. By not distancing his narrative from Baltimore's, Crotty turns Baltimore's personal memories into historical fact.

Due to the public interest in scientists, and programmes by organizations such as the Sloan Foundation to solicit autobiographies, there have been many published autobiographies of twentieth-century biologists over the past thirty-five years (Russell, 1988; Abir-Am, 1991). The authors of these works, which include Francis Crick (1988), Arthur Kornberg (1989) and Kary Mullis (1998), are prime candidates for biographical studies, since they not only participated in key scientific events but also played important social and cultural roles. Their autobiographical works are all quite different, but in each case, the strong personality of the scientist emerges from his self-portrait. The challenge to their current and future biographers will be to let that strong personality shine through without allowing it to take over. Historians must resist the temptation of these works, and instead turn to other primary sources such as letters and diaries, to provide a portrait of the scientist that still captures the strength of his personality without becoming his mouthpiece. With so many of the colourful figures in molecular biology still alive, it is a relatively simple task to solicit their version of events. As historians, we must be on guard not to privilege this version, or allow our voices to be overshadowed by that of the scientist. Our distinct biographical voices, when supported by archival sources, are just as legitimate as those of our subjects.

Biographers of these and other figures from twentieth-century biology do have some examples of recent works that achieve a balance between the subject's and the historian's voice. Daniel J. Kevles' article about Renato Dulbecco's contribution to the study of animal tumour viruses refers to Dulbecco's published autobiography, but relies more heavily on published scientific papers and archival sources to outline a chapter in the history of virus research (Kevles, 1993). In his recent study of Barbara McClintock, Nathaniel Comfort strikes a balance between McClintock's version of her life, as recorded in interviews, and documentary sources. He uses McClintock's words 'as pointers to and elaborations of issues verifiable with documents; and as guides to her character and personality such as could not be obtained any other way' (Comfort, 2001, 13).

Conclusion

Let us briefly return to the question I posed earlier. Is there anything particularly unusual about writing a biography of a scientist that could somehow intensify concerns about evidence and voice? Answers to this question can be taken as suggestions for a discussion of the poetics of biography in the history of science.

The task of humanising the subject, making his life more than just a list of discoveries and accomplishments, could be particularly difficult for biographers of scientists who are trying to write a more sophisticated kind of study. The autobiographical voice is strong, especially when the source is still alive, and it is often the easiest tool for making the subject into a more complex, if not sympathetic character. Given the cultural tendency to hold scientists up as heroes, the biographer/historian of science must avoid hagiography while maintaining a separate voice from her subject. This may be an additional burden that other biographers do not have to carry.

Another way to approach this question is to examine the place of biography in the history of science. The debate in the history of science community over the relevance of biographies persists even as we continue to try to produce historically nuanced, culturally and socially informed biographies of scientists (Shortland and Yeo, 1996). My discussion of different types of primary source material for scientific biographers has assumed that these works are still an important part of the literature of the history of science. The history of molecular biology and the rest of the life sciences are still in their early stages, and a biographical approach is often the easiest and most comprehensive way to examine the history of recent science. As biographers take the time to make their work more rigorously historical by relying on archival rather than autobiographical resources, the historiographical importance of biographies for the history of science will become clearer.

Bibliography

Abir-Am, Pnina G. (1991), 'Noblesse oblige: lives of molecular biologists', *Isis*, **82**, 326–43.

Abir-Am, Pnina G. (1998), 'Entre mémoire collective et histoire en biologie moléculaire: les premiers rites commémoratifs pour les groupes fondateurs', in Pnina G. Abir-Am (ed.), *La mise en mémoire de la science: pour une ethnographie historique des rites commémoratifs*, Amsterdam: Overseas Publishers Association, pp. 25–74.

Abir-Am, Pnina G. (1999), 'The first American and French commemorations in molecular biology: From collective memory to comparative history', in Pnina G. Abir-Am and Clark A. Elliot (eds), *Commemorative Practices in Science: Historical Perspectives on the Politics of Collective Memory*, *Osiris*, **14**, 324–70.

Backscheider, Paula R. (1999), *Reflections on Biography*, New York: Oxford University Press.

Cairns, John, Gunther S. Stent and James D. Watson (eds) (1966), *Phage and the Origins of Molecular Biology*, Cold Spring Harbor, NY: Cold Spring Harbor Press.

Comfort, Nathaniel (2001), *The Tangled Field: Barbara McClintock's Search for the Patterns of Genetic Control*, Cambridge, MA: Harvard University Press.

Crick, Francis (1988), *What Mad Pursuit: A Personal View of Scientific Discovery*, New York: Basic Books.

Crotty, Shane (2001), *Ahead of the Curve: David Baltimore's Life in Science,* Berkeley, CA: University of California Press.

Edel, Leon. (1984), *Writing Lives: Principia Biographica*, New York: W.W Norton and Sons.

Fischer, Ernst Peter and Carol Lipson (1988), *Thinking About Science: Max Delbrück and the Origins of Molecular Biology*, New York: W.W. Norton and Sons.

Goertzel, Ted and Ben Goertzel (1995), *Linus Pauling: A Life in Science and Politics* New York: Basic Books.

Hager, Thomas (1995), *Force of Nature: The Life of Linus Pauling*, New York: Simon & Schuster.

Judson, Horace Freeland (1979), *The Eighth Day of Creation: Makers of the Revolution in Biology*, New York: Touchstone/Simon & Schuster .

Kay, Lily E. (1985), 'Conceptual models and analytical tools: the biology of physicist Max Delbrück', *Journal of the History of Biology*, **18**, 207–46

Kay, Lily E. (1992), 'Quanta of Life: Atomic Physics and the Reincarnation of Phage', *History and Philosophy of the Life Sciences*, **14,** 3–21.

Kay, Lily E. (1993), *The Molecular Vision of Life: Caltech, The Rockefeller Foundation, and the Rise of the New Biology*, New York: Oxford University Press.

Kevles, Daniel J. (1993), 'Renato Dulbecco and the new animal virology: Medicine, methods and molecules', *Journal of the History of Biology*, **26** (3), 409–42.

Kornberg, Arthur (1989), *For the Love of Enzymes: The Odyssey of a Biochemist*, Cambridge, MA: Harvard University Press.

Lederman, Muriel and Sue A. Tolin (1993), 'Other viruses and the origins of molecular biology', *Journal of the History of Biology*, **26,** 239–54.

Lepore, Jill (2001), 'Historians who love too much: Reflections on microhistory and biography', *Journal of American History*, **88**, 129–44.

Luria, Salvador E. (1984), *A Slot Machine, A Broken Test Tube*, New York: Harper Colophon Books.

McElheny, Victor K. (2003), *Watson and DNA: Making a Scientific Revolution*, Cambridge, MA: Perseus Publishing.

Mullis, Kary B. (1998), *Dancing Naked in the Mind Field*, New York: Vintage Books.

Olby, Robert (1974), *The Path to the Double Helix: The Discovery of DNA*, New York: Dover Publications.

Olby, Robert (1989), 'From physics to biophysics', *History and Philosophy of the Life Sciences*, **11**, 305–9.

Russell, Nicholas (1988), 'Towards a history of biology in the twentieth century: Directed autobiographies as historical sources', *British Journal of the History of Science*, **21**, 77–89.

Sapp, Jan (1990), *Where the Truth Lies: Franz Moewus and the Origins of Molecular Biology*, New York: Cambridge University Press.

Selya, Rena (2002), 'Salvador Luria's unfinished experiment: The public life of a biologist in a Cold War democracy', unpublished PhD thesis, Harvard University.

Serafini, Anthony (1989), *Linus Pauling: A Man and His Science*, New York: Paragon Books.

Shortland, Michael and Richard Yeo (eds) (1996), *Telling Lives in Science: Essays on Scientific Biography*, New York: Cambridge University Press.

Söderqvist, Thomas (2003), *Science as Autobiography: The Troubled Life of Niels Jerne*, New Haven, CT: Yale University Press.

Stahl, Frank (ed.) (2000), *We Can Sleep Later: Alfred D. Hershey and the Origins of Molecular Biology*, Cold Spring Harbor, NY: Cold Spring Harbor Press.

Summers, William C. (1999), *Felix d'Herelle and the Origins of Molecular Biology*, New Haven, CT: Yale University Press.

Watson, James D. (1968), *The Double Helix: A Personal Account of the Discovery of the Structure of DNA,* New York: Atheneum Books.

Watson, James D. (2002), *Genes, Girls and Gamow: After the Double Helix*, New York: Alfred A. Knopf.

Chapter 13

Pas de Deux: The Biographer and the Living Biographical Subject

Vassiliki Betty Smocovitis

> Stuff comes out, and they'd rather be dead. It's far easier to do dead people. There's plenty of them. (Rowley, in Stright, 2004)

> It must be like marriage, with all the commitment and slog but none of the sex or comparing catty notes after dinner parties. (Handy, 2005)[1]

Introduction: The Death of the Subject

It isn't exactly customary to receive sincere condolences from family, friends, students and colleagues on the death of one's biographical subject, but that is precisely what one can expect when one chooses to work on a still living, but aged, biographical subject whose life comes inevitably to its end. So close does the identification between biographer and biographical subject become that the death of the subject (this double meaning is intentional here) is experienced not only as a personal loss by the biographer, but one that is recognised and shared by the wider community who have identified the biographer with their subject. The more famous the subject, furthermore, the more attention given to the biographer, enhancing even more the identification of the biographer with their subject.

On 19 January 2000 the calls, notes, letters of condolence and requests for interviews and personal recollections began to literally pour in following the death of G. Ledyard Stebbins. He was the American botanist, geneticist and evolutionist whose life between 1906 and 2000 spanned major developments in the twentieth century that included historical events like the 'evolutionary synthesis'. Stebbins was clearly a major figure in the history of twentieth-century science, and interest in him was further increased by the fact that his passing coincided with beginnings of the new century and millennium, which many had already hailed as the 'century of biology'. His passing was therefore an opportune time to reflect on the end of one historical epoch and the beginning of another, even more promising epoch for the history of biology.

Without surprise, his death made the front-page notices in places like the *New York Times*; the obituary that followed was about a third of their standard page

1 Bruce Handy was reviewing the new biography of actor David Niven by Graham Lord (2004).

(Yoon, 2000). Written by Carol Kaesuk Yoon – a well known science writer for the *New York Times* – it drew more than a bit from my own publications and insights. I had, in fact, been instrumental in helping to prepare it, since I had been consulted on it well before his death at the age of 94. Just one of the kernels of knowledge I gleaned from being the biographer of a living subject was the disturbing fact that major figures had such 'skeleton' obituaries prepared for them by media agencies well in advance of their actual death. Nothing had prepared me, however, for the small avalanche of notes and requests that followed within hours of his passing. The *Los Angeles Times* shortly followed with a phone-in interview, as did a number of other smaller newspaper or Internet sources responding to the breaking story. This was followed by requests for interviews or information by more focused groups interested in Stebbins, like the University of California (he was associated with that university), the California Native Plant Society, along with scientific venues that needed obituaries, like *Nature*, or more specialised scientific journals which published in his areas. With time, a virtual Stebbins 'death industry' arose that included not only obituaries, as well as tributes, memorials and retrospectives in honour of the man, his life and work, but also annual reviews, encyclopaedia and dictionary entries, and the official 'fellows' memoirs published by the National Academy of Science (he was an elected member) and the Royal Society. The death of this particular subject therefore drew considerable attention to his biographer, who had been closely associated with him for many years.

The fact of the matter is that well by 2000, I was considered the 'Stebbins person', or the 'expert' on Stebbins, having worked on him (and with him) as a biographical subject since I began graduate work in the history of science in the mid-1980s. Then approaching the fifteen-year mark, this had been a long association.[2] I had given numerous presentations on him, had published major articles about him, and had even made a number of public appearances accompanying him all through the late 1980s and 1990s. Though I bore little resemblance to the man, and had little of an intellectual pedigree traced back to him (other than a background in botany and evolutionary biology), bore no familial ties and had never engaged in any kind of conventional joint working collaboration with him, the association, and I think identification, was so close that to many in the community, the subject's name brought up the biographer's name. Even a casual attempt at Googling Stebbins drew up 'Smocovitis' (and vice versa).

More remarkable, however, was the fact that people showed sincere sympathy in expressing their condolences. I was so close to Stebbins – in their perceptions – that I must have had a serious personal loss. Thus, it was not just the number, but also the sincerity of condolences that were striking to me as biographer. The 'death of my subject' thus forced me to reflect once again on the special identification between biographer and living biographical subject, and on what this meant in historiographic terms. Elsewhere, I began to explore the advantages and disadvantages of work with

2 Biographies of major figures generally take a considerable length of time. Robert Caro's efforts to write a biography of Lyndon Baines Johnson now approach the third decade of research. Three volumes under the general title of *The Years of Lyndon Johnson* have already appeared: Caro (1982), Caro (1990) and Caro (2002).

a living biographical subject by concentrating on problems critical to the researcher pertaining to archival sources, and the reliability of oral history interviews as well as issues pertaining to distance and privacy (Smocovitis, 1999).

Here, I wish to explore more fully the personal dynamic of the interaction between the biographer and living subject with an eye to understanding the relationship that develops and the process of identification that takes place, and to then explore what the death of that subject entails. My argument will be that biographical projects on living subjects lead to an increasingly complex relationship that involves an intensified identification process. This in turn raises a set of special concerns that become even more apparent with the death of the subject. By identification, I mean the process by which we come to 'identify' with our subjects, that is, feel a 'union' with them, or empathise with them, or come to share the same kinds of values, concerns and attitudes, or perhaps even demonstrate similar behaviours as they do. As the case I explore suggests, furthermore, the process of identification involves the wider community engaged in an act of witnessing and validating the shared identification. The process does have some notable resemblance to what we might recognise as the traditional relationship known as the 'marital union', though as Bruce Handy points out in the opening epigraph above, there are some notable differences too. In my mind, a more appropriate relationship for this process draws on a metaphor borrowed from classical ballet involving the intricate choreography between two principal dancers known as the *pas de deux*.

'When Writers Wish their Subjects were Dead': The Celebrity Figure as Living Biographical Subject

Although biographers have explored the kinds of relationships that develop between them and their subjects, especially with an eye to problems of distancing and voice employed, they have only recently begun to explore the process of identification and the kinds of relationships that develop with living subjects.[3] Much of the literature exploring the special problems and challenges that emerge with work on living subjects is associated more generally with what are known as 'literary biographies', but especially – and sometimes amusingly – in the genre known as celebrity biography.[4] Thanks in part to the growing diversity of media that actively invent them, and thanks in part to a culture eager to consume them, celebrity figures seem to be increasing in number and diversifying in type. With this growing interest, there has grown a market keen to capitalise on celebrity figures. Included here are not only paparazzi and tabloid journalists, but also serious biographers following a

3 The best recent source on biographical methods and issues is Backsheider (1999); see also earlier classic works on biography, for example Edel ([1959] 1984). Biographers have suffered much anguish over the issue of subjectivity and their feelings towards even their long-dead their biographical subjects. See, for instance, Baron and Pletsch (1985) and Edel (1984); for one example from the history of science, see Westfall (1985); see also Söderqvist (1996). The peculiar identification of biographer with subject has been the subject of interest in popular works of fiction as well: see Byatt (1990).

4 For one analysis of the genre, see Collins (1998).

lucrative pursuit. Some of the best literature exploring the complex dynamic between biographer and living subject thus concentrates on celebrity figures.

Celebrities, can of course, come from all walks of life (art, sports, politics, religion and science, as well as the more obvious stage and film), but by definition they become celebrities through a process of projection or identification, playing the full range of human roles – lover, leader, saviour, healer, artist or scientist – and embodying and evoking the full range of human emotions: love, hate, fear, hope and the like. Serving at times as idealised or heroic living figures, their celebrity image oftentimes bears little resemblance to their flesh and blood existence. Biographers of such living celebrity figures thus may have an especially acute sense of their subject's allure, fame, status, notoriety, wealth or infamy as well as sheer power; the power dynamic of the relationship is thus inevitably skewed, favouring the subject. For some biographers, this very imbalance in the power dynamic may result in a happy union (especially if they are accustomed to playing passive observer in the hope of writing an 'authorised' biography the subject likes), but more often than not it leads to serious problems between the biographer and subject that must be handled with care.

So problematic are these relationships that they were the feature of a 1996 *New York Times* article with the amusing title, 'When writers wish their subjects were dead: moving from ancient lives to modern ones, biographers meet hostility and worse' (Scott, 1996). Charting the course of the relationship between living biographical subject and biographer, Marion Meade, the biographer of comedian Woody Allen described a series of appropriate emotional stages that biographers inevitably follow: first 'they adore the subject, then despise the subject, then wonder whatever possessed them to choose that subject at all'. Eventually, she noted, they 'work through their hostility and come out in the proper place'. The 'proper place', unfortunately, may be a kind of 'twilight zone' where the biographical subject exists in a kind of reality/unreality, but always linked inextricably to the biographer. Biographers of living subjects, the article notes, thus exist in a kind of 'living purgatory', where their subjects are neither 'consenting nor dead'.

The tumultuousness and volatility of this relationship is perhaps best known from the notorious 'collaboration' between sportswriter Al Stump and the controversial baseball player Ty Cobb (also known as the 'Georgia Peach'). In 1960, Stump was invited to serve as a ghostwriter for Cobb's autobiography. A difficult (in fact, probably pathological) individual famous for regularly bloodying opponents with the spikes on his shoes in his famous 'Cobb's Kiss', Cobb proved to be a ghostwriter's ultimate nightmare figure. He was abusive, manipulative and deceitful, and regularly toyed with Stump, who frequently found himself a captive audience to his celebrity figure. Stump's role, as it turned out, was to serve as witness to a range of bizarre performances, some of which were staged solely for his 'benefit'. Initially attracted to the famous pro, Stump quickly grew to despise Cobb, and by Stump's own admission, the completed book he co-authored with Cobb entitled *My Life in Baseball: The True Record* was anything but true (Cobb, 1961); it was a much-sanitised, self-serving account, the kind only an egocentric (and indeed narcissistic) figure like Cobb would produce. It took over thirty years to recover and gain both courage and distance from his biographical subject before Stump could continue with the project. It therefore took the death of his subject (and the relationship) to complete his own telling account of Cobb's life, based on what he had observed and what Cobb had told him. Describing the curious relationship that developed between them as a result of the proximity, he wrote the following:

> During the long stretches of time we spent together, my feelings for Ty Cobb were often in flux. My respect for greatness, my contempt for his vile temper and mistreatment of others, my pity for his deteriorating health, and my admiration for his stubbornness and persistence produced a frustrating mix of emotions. With so much material left over, there was need for another manuscript, but it wasn't until three decades later that I finally felt compelled to put the real Ty Cobb to rest. (Stump, 1994)

Biographies of Living Scientists: Are Scientists Different?

Celebrities are, of course, renowned for being 'difficult' people. Stars (or 'stahs', as they are referred to in Hollywood celebrity culture) are oftentimes associated with temperamental, volatile or unstable personalities. Would the same kinds of difficulties emerge with figures far outside traditional 'celebrity' culture, say in politics or religion, or especially science, areas which are traditionally not associated with the kind of glamour or high visibility usually associated with actors or sportsfigures? Should we expect the same kinds of difficulties to emerge with scientists, in particular, those supposedly rational, bloodless creatures of mind? My own sense is the answer is a strong yes. Though it is not always true, scientists selected for biographical study are usually major figures and celebrities in their own right, complete with the quirks and quarks and tendencies towards narcissism that one sees within any celebrity group; we may in fact be drawn to them precisely because of these tendencies.

Scientists are, moreover, human beings, and interactions between any two human beings that take place over a prolonged period result in what we recognise as 'a relationship', though we may ascribe different meanings to the kinds of interactions that we may have. In a close, and indeed intellectually intimate, interaction such as that between living biographer and subject, it is very likely, and I think inevitable, that it will bring out the best and worst in both sides. So close can the relationship become, and so heavily dependent can it become, that it can be viewed as classic 'co-dependency' that makes critical perspective or detachment difficult, if not impossible. Even between scientists, those famously detached or disembodied beings, and their biographers, who may be well trained to seek historical objectivity, a similar difficult and complex relationship develops. In fact, I would argue that as living biographical subjects, scientists are like any other human being – why should they be any different?

For historians especially, and not so much for science writers or journalists, the relationship may be made more problematic by the kind of temporal transgression encountered by work with living subjects. Although history, by definition, is concerned with the past, the biographer's subject is of interest historically for his or her contributions; yet the living subject is not yet a part of history, unless one can easily accept the notion of 'living' history, a fundamental contradiction in terms. This is the reason that biographers of living subjects experience discomfort: there *is* something very strange and fundamentally unnatural about facing and interacting with a historical subject. The idea is so disturbing to historians that some purists hold the view that all historical actors must be dead before one can even think of proper historical analysis. In their view, only after all the actors have died, and only after a reasonable amount of time has passed, is it possible to attain proper scholarly distance. Other historians, however, argue strenuously against this point of view. For them, living subjects are invaluable sources of information; instead of lending historical 'distance', time lapsed really means that historical documents and historical memory are irretrievably lost.[5] This is a well-known debate some historians relish; but others view such debate as wasteful because opportunities and documents are lost.

Despite the predictable reluctance from historical purists, many historians have already turned to detailed studies of living biographical subjects; and nowhere is this more prevalent than in the history of science. In recent years, a number of scholars have turned to work with scientists as biographical subjects, many of whom are still alive. The reasons for this are fairly simple. More scientists are alive today than have ever lived in the past (this was first recognised by sociologist of science Derek de Solla Price in his famous statement that '80 to 90 per cent of all scientists that have ever lived are alive now').[6] This is certainly the case for those of us following the history of modern biology, a science which is still so recent that many of its great figures are still alive. This is especially true for newer areas of the biological sciences like molecular biology, genetics, biochemistry, developmental biology and immunology, areas which became growth industries in the second half of the twentieth century (Provine, 1986; Söderqvist, 1998; Keller, 1983; Kevles, 1998; Judson, 1979).

Despite the growing interest in the recent history of science,[7]and despite growing interest in science biography as a unique genre,[8] biography of living science remains largely uncharted terrain. It is thus useful to rely on those accounts from either literary biography or celebrity biography as historiographic or methodological models for the history of science. Marion Meade's characterisation of the 'emotional stages' that biographers inevitably follow may be especially useful for us here. Additional stages might be helpful and give a more nuanced picture, especially if identification is the subject of interest. If it does not take place at the front end of Meade's emotional stages, then it might generally emerge in the middle or perhaps even towards the tail end of those same stages – it may indeed involve a gradual process, a kind of evolutionary choreography between biographer and the biographical subject. It is this complex process that I wish to explore further with some concrete grounding from my experiences as biographer of Ledyard Stebbins.

5 See, for instance, the comments in the Preface to Provine (1986).

6 He made this famous observation in 1963: de Solla Price (1986; includes the text of *Little Science, Big Science* originally published in 1963).

7 The historiography of recent or contemporary science was the subject of conferences held at Stanford University and in Göteborg, Sweden in 1994. For a discussion of some of the unique problems encountered in writing the history of recent science, see Doel and Söderqvist (2006), Söderqvist (1997) and Lindee, Speaker and Thackray (1994).

8 See Shortland and Yeo (1996).

Pas de Deux: Choreographing the Relationship with G. Ledyard Stebbins

My project did not originally begin as a biography, and I most certainly did not find Ledyard Stebbins an attractive, celebrity-like personality I was keen to explore. In my mind, immediate identification (if it existed at all) was as weak as it could possibly be. He was fifty years older, came from an entirely different class and ethnic background, and we had obvious gender and other notable personality differences. As the central figure bringing botany to the intellectual event known as 'the evolutionary synthesis', however, he played a critical role in shaping modern evolutionary biology. That made him important to me, as I was keen to understand its history and the role that botany had played in it. I therefore understood his importance and had pored over his own reflections on the subject of 'botany and the evolutionary synthesis', but I had no initial intention to concentrate on him for my research (Stebbins, 1980).

That all started to change after my first meeting with him and what amounted to three weeks of close interaction. Early in the winter of 1987 or so, Ledyard delivered a set of important lectures at Cornell University, where I was an advanced graduate student; the lectureship required him to stay in Ithaca for a three-week period. His wife Barbara was unable to accompany him, and as the student closest to his interests who was keen for the interaction, I was placed in charge of guiding him around the campus. Ledyard was fairly far along in age, and his eyesight was poor so that he required considerable attention. Up to meeting with him, the name Stebbins had existed only in my textbooks, and was especially meaningful since my first exposure to evolutionary biology at university level came in the way of his little textbook, *Processes of Organic Evolution* (Stebbins, 1966). I vividly recall the excitement I had initially felt as I anticipated my finally meeting such an object of scientific wonder.

That came in the way of a formal dinner organised by William B. Provine, but it was anything but exciting. As the evening wore on, I was horrified to see his manners at the dinner table: he was spooning the soup of the day into his eyeglasses that he kept suspended on his chest with cords. Worse, his voice was monotonous, and he spoke in what seemed like endless chains of long paragraphs with an annoyingly pretentious 'preppie' New England accent. He struck me as overbearing, interested in hearing only himself speak, and did not seem keen to actually engage anyone else at the dinner table. I was therefore not drawn to Ledyard as a person, and not only could I not identify with him, but I was nearly repulsed by that initial interaction.

Things changed gradually over the course of that initial three-week period, however. I not only guided him across the campus, drove him in my car and served as his official 'escort' to nearly all social occasions held in his honour, but I also spent my evenings with Ledyard, tape recorder in hand, interviewing him on the topic dear to both of us, botany and the evolutionary synthesis. After three weeks of listening to him answer my questions (again in those endless chains of spoken paragraphs), I grew acquainted with the details of his life and work. I also grew to appreciate Ledyard as a person, especially as I increasingly saw him as an aging and infirm individual in need of assistance and care; his 'preppie' humour actually began to make me laugh. We went for drives in upstate New York. I got a personalised tour of his ancestral home town of Cazenovia, and I touched the remnants of the peonies he once worked on at the old Saunders house next to Hamilton College. In that three-week interval of time, I began to 'connect' with Ledyard on both the personal and professional level and to relive his own history first-hand.

The sense of connection gradually grew during the next six months, as I transcribed all the interviews I had recorded into textual form, listening attentively to every inflection and intonation in his voice over and over again, as I completed what became compiled as an oral history interview. It was largely the result of this concentration on the details of this one person's life, his insights into his work, and his insights and recollections of the field as a whole, repeated over and over again on the tapes, that started to shape my project on botany and the evolutionary synthesis. I could engage those wider issues, it dawned on me, by concentrating directly on him. The project thus shifted towards a well-established genre in the history of science, whereby a central figure is chosen as an 'organisational pivot', for getting at a wider complex dynamic in scientific culture; and the subject of 'botany and the evolutionary synthesis' was indeed a hopelessly complex and indeed unworkable project until that point. With that in mind, the completed dissertation in 1988 was organised around Ledyard Stebbins, and so could be construed as 'biographical' in terms of genre, but nevertheless bore little in the way of details into the life or inner world of Ledyard Stebbins. For at least two chapters, the figure of Stebbins dropped out entirely. It was, in short, no 'real' biography, as I have come to understand that term.

The interactions continued after his departure, and unexpected elements were introduced, especially after I reciprocated the visit by being a guest in the Stebbins households in Davis and Berkeley, California. In no time, I realised that I was working on not just one biographical subject, but in fact two: included in my biography from that point on was Barbara, Ledyard's wife. Witnessing their relationship – at an uncomfortably close range – may have made some biographers dance with glee, but I felt only acute discomfort. I saw at first hand what amounted to a 'food fight' in the kitchen, and was then witness to a series of interactions resembling the breakfast scene out of the Hollywood film *Citizen Kane*. By the end of that visit, I knew more than I ever wanted to know about the both of them, and as a young scholar, that familiarity made me confront a number of issues pertaining to trust, confidence and the dignity that both had a right to (in my view), especially since I was a guest invited into their homes.[9] No less a concern were the interactions with Ledyard's other family members, and close friends – all of whom seemed to come out of the woodwork when they learned of the project. Living with a biographical subject, I quickly discovered, meant living with a family of biographical subjects, all of whom may feel some special relationship with the primary biographical subject; in a sense, all wanted to claim some kind of 'ownership' in him. Dealing with Stebbins also meant considering the needs and wishes of the Stebbins family, along with some of his colleagues and friends, which at times contradicted each other.

It was during one of my initial visits to California, moreover, that the gradual 'witnessing' began to take place that further legitimised the relationship. Ledyard took

9 I discuss some of this in Smocovitis (1999).

me to his place of work at the university, introduced me to his friends and colleagues, and importantly enough invited me to accompany him to the 'Biosystematists', an organisation of people in the Bay area with joint interests in evolutionary biology and systematics. They were the individuals who began to draw on me for expertise into their own areas of interests and to designate me as the 'Stebbins expert'. They invited me to dine with them, talk to them and to explore California botany with them. A number of them (and other colleagues who knew him outside the Bay area) allowed me to interview them with tape recorder in hand, serving both as witnesses but also participants to the process taking place. At one point, the 'Biosystematists', then meeting at Stanford University, invited me to give a formal lecture on my project, namely Ledyard. Nothing could hold Ledyard back, and even though he didn't feel altogether well, he came for the dinner, reception and talk. We all held our breath as we anticipated some kind of outburst, if not repeated 'correctives' to my interpretation of his 'life and work'. It came, of course, but in a manner so amusingly conversational that I was subsequently told by one of the audience members that 'I interrupted the speaker a bit too much.'

The longer I worked with Ledyard, the wider the circle of acquaintances that came to recognise the relationship between us. In the course of the following decade, a full-blown public 'relationship' ensued. We grew together and grew apart. I would fly to California often, and even spent two years of my life close by in the Bay area, close to my sources (and to him). Ledyard was hospitalised, and I went to visit with an armful of his beloved California wildflowers. Thinking his life was coming to end (it didn't for some eight years), he became 'miffed' with me for not doing more to preserve his memory. I grew frustrated at his meddling and interventions and the absence of proper historical documentation (he was sloppy, and threw away too much). Ernst Mayr grew to be a friend as I continued to delve into evolutionary synthesis publishing my first book. Ledyard did not approve much of my new relationship with his old rival, Ernst. He wanted more time devoted to him. Barbara died suddenly, leaving Ledyard in the care of a group of caregivers, all of whom eventually became my own friends. I was with them all, continuing my work with Ledyard, tape recorder in hand, just a couple of weeks before his end. Hospice came and went as we sorted through his papers and collected important documentation we feared would be lost. Just a few minutes before his death, I was on the phone checking up on him. Magnified by a breathing machine next to the phone, I heard him draw his last breaths.

Up until this very end, the relationship involved a complex choreography, much like a *pas de deux*. Though the identification did not exist at the outset, it came as the result of intensified interactions of the kind that can only take place between two living subjects. It isn't exactly easy to share laughter or even negative emotions like anger or frustration with the dead. Ledyard had actually been one of the better kinds of living biographical subjects. He was co-operative, interactive and talkative, and best of all, trusting enough to share his insights, his documents and some of his personal life. That did not always make it easy – he could throw temper tantrums at the drop of a hat and was at times overly helpful, but it was probably the best that it could get between two such different human beings.

The sheer length of time – some thirteen years – which allowed us time to get to know each other, combined with the focus on the understanding the minutiae of his life, no doubt intensified the process of identification further. Over the course of this long period, the narrative of our two lives had slowly begun to converge; I had become part of his memory, as he had become part of mine. As a stunning demonstration of this sharing or mingling of historical narratives, my own existence made its way eventually into the Stebbins archives at the Shields Library at the University of California, Davis. Our shared histories were now preserved and contained in the boxes that contained letters to and from each other, photographs, and drafts of my manuscripts – all traces of our shared existence. The most striking, and poignant, sign of this was the inclusion of my name in the Stebbins date-book, noting the appointment of our first meeting at Cornell. Reciprocating the relationship, some of Ledyard's own traces became part of my own life as treasured possessions: some books, some photographs, a trophy and other memorabilia are now part of my own life. Even his unpublished autobiography, eventually entitled *The Lady Slipper and I*, drew on my understanding of the critical turns and events in his life as I was writing them in my biography. Finally, the shared narratives became public, by the appearance of a number of publications on Ledyard that drew on our interactions and jointly linked our names. News of his death saddened and disoriented me, and that's probably also when I realised that the biographical project that had started in 1988 had at some point given way to a full-scale biography – a full-scale involvement with the writing of Ledyard's life.

'True to Life?' Closing Thoughts on Biographies of Living Subjects

Given that there is always some kind of interpersonal dynamic in work with living subjects, some kind of attachment and identification is inevitable. Conflicting feelings like those described by Stump, if explored and integrated into the writing of the biography, may even enhance the quality of the biographical product. It may even be the case that some of the very best biographers give way to exploring those feelings evoked by their encounters with their subject. In his recent biography of immunologist Niels K. Jerne, for instance, Thomas Söderqvist has used his biographical subject as a way of extending his own life-experiences.[10] According to Söderqvist, such identification may prove to be 'edifying' and 'may provide us with opportunities for reorienting our familiar ways of thinking about our lives in unfamiliar terms'(Söderqvist, 1996). That much appears certainly true with my experience of work with Ledyard Stebbins. Although identification with Ledyard Stebbins was not immediate, furthermore, it did happen with time, largely through the kinds of interactions that take place between two individuals. He was not my teacher, colleague, friend, nor was I his student or caregiver, yet at times we appeared to be playing with all these roles. He was my living biographical subject, the so-called 'object' of my study as it evolved. Though it may on occasion test

10 How biography can be edifying is discussed in Söderqvist (1996), Söderqvist (2003) and Söderqvist (2006).

the emotional resilience and social skills of the biographer, working on a living biographical subject in fact provides abundant – and special – gratification for the biographer. Not only can it be an opportunity to engage in a special relationship where trivial questions can be answered readily and difficult questions can at least have the possibility of an answer, but one is far less lonely knowing that there is at least one person who is more than likely as committed as the biographer to the project. Talking (and interacting) with the biographical subject also permits the historical biographer to explore the possibility of living history.

What, if anything, happens when the living subject dies? Does it aid the process of detachment so that subjective attitudes and feelings can be put aside or behind? Does it lead to a more 'objective' historical account? If some of my recent writing is any indication, Ledyard's death does not somehow allow me to gain objectivity, but it does seem easier to paint a portrait that rings true. A recently published reminiscence of Ledyard drew so heavily on my experience of him that Ernst Mayr's response to reading it was sheer delight: 'there you have him,' he said, 'Ledyard, warts and all'. Another student of Ledyard's enthusiastically wrote me to say how much I seemed to capture someone he knew. My sense from these reactions, and the project itself as it is still unfolding, is that the death of the subject does bring forth detachment, but does not necessarily lead to a more objective or neutral account. What it can do for the biographer is to permit them to paint a more vivid picture, one that may be much more 'true to life'. As Thomas Söderqvist has concluded, 'one can hardly set out to write a biography without being emotionally involved with its central figure, but on the other hand, one has to work hard on establishing distance in the process of writing. The final result should emerge as a happy divorce, a certification that the writer has freed himself from the central figure' (Söderqvist, 2003, p. xxi).

Bibliography

Baron, S.H and C. Pletsch (eds) (1985), *Introspection in Biography: The Biographer's Quest for Self-awareness*, Hillside, NJ: Analytic Press.

Backsheider, P.R. (1999), *Reflections on Biography*, Oxford: Oxford University Press.

Byatt, A.S. (1990), *Possession: A Romance*, New York: Random House.

Caro, Robert A. (1982), *Path to Power: The Years of Lyndon Johnson*, vol. 1, New York: Alfred A. Knopf.

Caro, Robert A. (1990), *Means of Ascent: The Years of Lyndon Johnson*, vol. 2, New York: Alfred A. Knopf.

Caro, Robert A. (2002), *Master of the Senate: The Years of Lyndon Johnson*, vol. 3, New York: Alfred A. Knopf.

Cobb, Ty (1961), *My Life in Baseball: The True Record*, Garden City, NY: Doubleday.

Collins, Gail (1998), *Scorpion Tongues: Gossip, Celebrity and American Politics*, New York: William Morrow.

de Solla Price, Derek J. (1986), *Little Science, Big Science ... and Beyond*, New York: Columbia University Press.

Doel, Ronald E. and Thomas Söderqvist (eds) (2006), *Writing Recent Science*, London: Routledge.

Edel, Leon ([1959] 1984), *Writing Lives: Principia Biographica*, New York: Norton.

Handy, Bruce (2005), 'Blithe spirit', *The New York Times Book Review*, Sunday 9 January, 24–5.

Judson, Horace Freeland (1979), *The Eighth Day of Creation: Makers of the Revolution in Biology*, New York: Simon & Schuster.

Keller, Evelyn Fox (1983), *A Feeling for the Organism: The Life and Work of Barbara McClintock*, New York: Freeman.

Kevles, Daniel J. (1998), *The Baltimore Case*, New York: W.W. Norton.

Lindee, M. Susan, Susan Speaker and Arnold Thackray (1994), 'Writing history while it happens', *Knowledge, Creation, Diffusion, Utilization*, **85**, 293–6.

Lord, Graham. (2005), *The Authorized Biography of David Niven*, London: St Martin's Press.

Provine, William B. (1986), *Sewall Wright and Evolutionary Biology*, Chicago, IL: University of Chicago Press.

Scott, Janny (1996), 'When writers wish their subjects were dead: Moving from ancient lives to modern ones, biographers meet hostility and worse', *The New York Times*, 6 October, 19.

Shortland, Michael and Richard Yeo (eds) (1996), *Telling Lives in Science: Essays on Scientific Biography*, Cambridge: Cambridge University Press.

Smocovitis, Vassiliki Betty (1999), 'Living with your biographical subject: Special problems of distance, privacy and trust in the biography of G. Ledyard Stebbins Jr.', *Journal of the History of Biology*, **31**, 1–18.

Söderqvist, Thomas (1996), 'Existential projects and existential choice', in Michael Shortland and Richard Yeo (eds), *Telling Lives in Science: Essays on Scientific Biography*, Cambridge: Cambridge University Press, pp. 45–84.

Söderqvist, Thomas (ed.) (1997), *The Historiography of Contemporary Science and Technology*, Amsterdam: Harwood Academic Publishers.

Söderqvist, Thomas (2003), *Science as Autobiography: The Troubled Life of Niels Jerne*, New Haven, CT: Yale University Press.

Söderqvist, Thomas (2006), 'What's the use of writing lives of recent scientists?', in Ronald E. Doel and Thomas Söderqvist (eds), *The Historiography of Contemporary Science, Technology and Medicine: Writing Recent Science*, London: Routledge, pp. 99–127.

Stebbins, G. Ledyard (1966), *Processes of Organic Evolution,* 3rd edn, Englewood Cliffs, NJ: Prentice-Hall.

Stebbins, G. Ledyard (1980), 'Botany and the synthetic theory of evolution', in Ernst Mayr and William B. Provine (eds), *The Evolutionary Synthesis: Perspectives on the Unification of Biology*, Cambridge, MA: Harvard University Press, pp. 139–52.

Stright, Caleb (2004), 'To write the lives of the living and the dead', *The Carnegie Pulse*, 17 November, 1–2.

Stump, Al (1994), *Cobb: A Biography*, Chapel Hill, NC: Algonquin Books.

Westfall, Richard S. (1985), 'Newton and his biographer', in S.H. Baron and C. Pletsch (eds), *Introspection in Biography: The Biographer's Quest for Self-awareness,* Hillside, NJ: Analytic Press, pp. 175–89.

Yoon, Carol Kaesuk (2000), 'Ledyard Stebbins, 94, dies: Applied evolution to plants', *New York Times*, Friday 21 January.

Chapter 14

Resuscitating the 'Great Doctor': The Career of Biography in Medical History

Beth Linker

The study of the history of medicine began as a practice of teaching and writing about individuals. In the first history of medicine courses taught in American universities during the late nineteenth century, instructors told stories of 'great doctors', mapping out a straight line of historical advancement. Early educators in the history of medicine (most of whom were physicians) validated their practice by claiming that historical study would have a humanising effect on their students. Progressives who believed that medicine held the key to human health and happiness worried that as medicine became more scientific, students would lose the moral and cultural foundations necessary to guide them. Early proponents of medical history hoped that by putting human faces on the increasingly abstract content of their fields, teachers could inculcate classical virtue. In such an educational atmosphere, the writing of biographies flourished.[1]

Things are very different today. The change began in the 1970s with the arrival of the 'new' social history. At that time, a cohort of professionally trained non-physician historians of medicine rejected the authority of 'great doctors' and their ideas, reacting to the biographically oriented beginnings of the history of medicine with disdain. To this day, social historians of medicine throughout the English-speaking world continue to employ a rhetoric of patricide, distinguishing their 'new' (albeit now almost thirty-year-old) context-driven method of writing history from individual-centred 'traditional' medical history, a method manifested in biographies written for, about, and often by physicians.

Yet there is an important exception to the widespread hostility among American social historians of medicine to great figures in history. As part of their effort to purge the field of 'elitist' approaches of historical study that focus on individuals, social historians of medicine have come perilously close to producing a hagiographical portrait of one particular individual: Henry E. Sigerist, a Swiss physician and professionally trained historian who is today widely regarded as one of the most important figures in beginning the turn toward a more 'sociological' approach to medical history. What is most striking about the appropriation of Sigerist as a founding father to the 'new' history is not the apparent contradiction involved in looking up to one great physician-historian as the inspiration for a method that refuses to admire great doctors. Most striking is the fact that social historians have made

1 For more on the history of medicine as a field of study, see Brown and Fee (1997), Brieger (1993), Burnham (1998), Miller (1973) and Huisman and Warner (2004).

Sigerist the leader of an ideological-methodological crusade in which he would have been, at best, a reluctant participant. A close examination of Sigerist's work shows us that, compared to the 'new' social history, Sigerist's 'older' approach to social history allowed for methodological pluralism that, despite what his inheritors have proclaimed, included medical biography.

Biography and the New Social History of Medicine

In 1979, two ambitious Boston-area graduate students, eager to move the study of medical history in a new direction, articulated a founding story of the field that would come to define the identity of professional historians of medicine for many years to come. Disenchanted with what they perceived to be medical history's celebration of great white men, Susan Reverby and David Rosner wanted to make a clean break from what they called 'traditional' medical history by publishing a collection of works in the 'new' social history of medicine. But as two 'young whippersnappers question[ing] the wisdom of their elders', they needed a precedent or a respected authority who could help to defend themselves against 'traditionalist' challenges (Reverby and Rosner, 2004, p. 167). The authority they turned to was Henry E. Sigerist (1891–1957).

In the introduction to their edited book *Health Care in America: Essays in Social History*, Reverby and Rosner defined their 'new' social history of medicine as heir to Sigerist's 'old' sociological approach to medical history. This argument of inheritance, a portrait of an intellectual family tree, rooted the 'new' social history in tradition, giving it a sense of permanence that most innovations lack. This sense of rootedness held great appeal. Once Reverby and Rosner's proposed genealogy was in print and out of their hands, it went on to become a legend reiterated by many medical historians for decades to come, a founding story told and retold in classrooms, spun and re-spun at yearly history of medicine conferences.

As a Leipzig-trained historian who intended to bring the scholarly rigour of German medical history to the United States, Sigerist was, in many ways, the perfect candidate for Reverby and Rosner's campaign. His self-proclaimed 'sociological approach' that situated medicine in economic and political history challenged those who thought that medical history was a kind of mentorship steeped in biographical accounts of great men. Indeed, Reverby and Rosner titled the introductory essay to their volume 'Beyond "the Great Doctors"', drawing on Sigerist's line that the 'history of medicine is infinitely more than the history of great doctors and their books' – an indication that their work, like his, would be an investigation specifically targeted at the social relations of health, rather than the activity and intellectual concerns of physicians alone (Reverby and Rosner, 1979, p. 3).

Treating Sigerist as a model, Reverby and Rosner pronounced the death of the old-school, biography-centred approach to the history of medicine. History of medicine, they argued, had too long been in a 'political alliance' with the practice of medicine itself. Writing about great doctors, they maintained, only perpetuated the physician's professional hegemony over medical institutions and the very definitions of health and disease. With the new social history, they intended to question not only the authority of doctors, but also their supposed greatness. The new social history promised to unmask the 'pervasive societal faith in the potential and efficacy of medical science' and ultimately to break the control that physicians exercised over the stories that lay people tell themselves about health and sickness (Reverby and Rosner, 1979, p. 4). Taking the side of the laity over power-wielding experts, Reverby and Rosner implied that writing biographies about great men was an 'unprofessional' practice, an expression of false consciousness, a trade only for whiggish historians or physicians who merely dabbled in the past.

Reverby and Rosner's book was only a small part of a much larger historiographical movement that transcended the small field of medical history. During the 1970s and 1980s, the new social history was an international movement, involving historians in Europe, Latin America, Japan, India and North America who applied theories of postcolonialism and cultural relativity to the study of everyday people and everyday life.[2] At its extreme, this new history rejected the 'traditional' methods, assumptions, and aims of historical research – that is, political history, objective history, history from 'above', events-based history – in the pursuit of creating a 'total history', in which every human activity could be studied historically, and where no one event or one society would be privileged over another. But while the new social history enjoyed a universal reach, it affected the history of medicine in a distinct way.

When the new generation of social historians first made its attack against 'traditionalists' in medical history, veteran scholars fought back. Worried that the intellectual foundation of their field would collapse under the assault of social history criticism, key historians of medicine wrote in support of 'traditional' history, arguing that their field could not exist without the history of medical ideas. In January 1980, Leonard G. Wilson, editor-in-chief of the *Journal of the History of Medicine and Allied Sciences*, claimed that if social history succeeded in taking over the field, it would lead to the end of medical history, properly understood, for it would be 'medical history without medicine' (Wilson, 1980, p. 7). In the same year, Lloyd G. Stevenson, editor of the *Bulletin of the History of Medicine*, wrote a disparaging addendum to Howard S. Berliner's (1980) positive review of Reverby and Rosner's *Health Care in America*. In his 'Second opinion', Stevenson concluded that Reverby and Rosner's social history had indeed fulfilled Wilson's prediction; since their book was not 'science-oriented or even practice-oriented', it could not rightly be called history of medicine (Stevenson, 1980, p. 136).

The battles between new social historians and those of the medical history establishment polarised the field, creating a complex map of fault lines between old guard and new guard, clinician-historians and PhD-trained historians, 'insiders' and 'outsiders'. While many old-guard historians of medicine complained that the new scholars in their field lacked medical knowledge, some of the new social historians made equally inflammatory claims, implying that 'traditional' medical history was medical history without history.[3]

2 For more on the 'new' history of the 1970s and 1980s, see Burke (1992a) and Burke (1992b). For the American context in particular, see Novick (1988).

3 Berliner implies this in (1980), pp. 131–4, esp. p. 131.

That the new social history had such a polarising affect on the history of medicine is, in many ways, unsurprising. The timeless power relationship between patient and doctor, sufferer and healer, provided a ready-made programme for studying 'history from below', a methodology deemed essential to the writing of social history.[4] Already by the mid-1970s, British social historians were calling for the 'resurrection' of the patient. In 1976, historian Charles Webster, Director of the Oxford Wellcome Unit and the newly elected President of the British Society for the History of Medicine, made patient-centred history the primary initiative of the society.[5] Ten years later, Roy Porter became one of the most articulate defenders of history from below, summoning his colleagues to combat 'traditional history of medicine [which] simply ignored the patient' by undertaking an exploration into the patient's view of health and disease (Porter, 1985, p. 175).[6]

Once social historians of medicine accepted history from below, the step towards rejecting biography, especially of 'great' doctors, seemed a sensible one to take. This step, however, led to a mixing of historical methodologies and a blurring of the past with the present. Social historians not only wrote about patients, but also identified with the powerlessness that contemporary patients felt while under the authority of physicians.[7] Although few of the younger historians of medicine were

---

4 To appreciate fully the degree to which history from below defined the social history of medicine, it is helpful to compare the fields of medical history and the history of science. Whereas the introduction of social history brought 'history from below' to medical history, historians of science followed the line of 'social construction of knowledge'. There are some key differences that could explain this divergence between the fields. First, and perhaps most obviously, the practice of science did not (and does not) have a ready-made above-and-below distinction for its historical actors as medicine does. Moreover, whereas most social historians of medicine did not have backgrounds in practising medicine, most historians of science had spent some time at the laboratory bench, since, according to Nathan Reingold, 'a majority' in the field have scientific backgrounds (1981, p. 276). In other words, social historians of science, having some common ground with even 'great' scientists, might not have felt the same struggle against power as social historians of medicine did. Little has been written comparing social history in both fields. One exception is Jordanova (1995).

5 Webster, as quoted in Porter (1995), pp. 350–51.

6 I would argue, however, that for historians of medicine, there was (and still is) an inherent difficulty in defining the 'below' in medicine as simply a category that constitutes 'patients', since illness is a shared phenomenon that cuts across all boundaries of class, gender and race. Labour historians of the 1970s were the first to employ history from below as a way to investigate the political activity of the working class. Since the division between working class and industrialists proved to be fairly consistent and concrete, labour historians could easily fit their historical actors into the appropriate hierarchical categories. The same is not true for the history of medicine. While in terms of power relations the division between physicians-as-'above', and patients-as-'below' works most of the time, the divide is nevertheless porous. Work has yet to be done on how health and disease as phenomenological experiences complicate the categories of 'above' and 'below', for even physicians, who medical historians tend to assume are 'above', are potential victims of disease. Almost every historical actor participates in the 'below' category at one point or another.

7 The best examples of how social historians began to identify with their patients as historical subjects can be found in the public health literature of the 1990s. Writing in the

trained as physicians, they did have experience as patients, which they drew upon as they wrote history from below as well as when they entered into disputes with old-guard historians of medicine (many of whom had MDs).[8] As a result, the rejection of medical biography became a measure of one's commitment to history from the bottom up, and ultimately to the new social history. At the same time, old-guard historians who still wanted to preserve 'traditional' medical history rushed to rally behind biography in the hopes of restraining the rise of the new social history.

The issue of biography created an easily caricatured artificial duality in the field of medical history, with the non-PhD-trained physician-historian who wrote and read 'cheery' biographies on the one side and the sceptical PhD-trained historian who wanted to debunk physician-centred history on the other. In reality, the field of medical history was diverse and complex, home to scholars of many different stripes and backgrounds, yet divisive disputes about biography made the field look black and white.

But while certain 'traditional' medical historians might have promoted biography in theory, fewer and fewer biographies of physicians were being written. Whereas almost 40 per cent of the articles published in English-speaking medical history journals during the 1960s and 1970s were biographical in nature, by the year 2000 the number of articles devoted to medical biography dropped to nearly 10 per cent. As Olga Amsterdamska and Anja Hiddinga have demonstrated (2004, pp. 245, 249), *both* PhD historians and physician-historians had become less likely to write medical biography and more likely to devote their attention to the history of medical practice and its institutions.

The decline in medical biographies throughout the 1980s led certain social historians to announce themselves as victors in the historiographic battles waged a decade earlier. In Britain, Andrew Wear boldly announced in 1992 that the social history of medicine had 'come of age', for it had replaced 'the history of great doctors, great discoveries and great ideas', and had proven to be the superior approach, showing how 'medicine ha[d] affected society and how society ha[d] shaped medicine' (Wear, 1992, p. 1). Wear had every reason to boast, since social history in Britain had earned a celebrated position in the medical history community, evinced most concretely by the establishment of its own scholarly periodical *Social History of Medicine* in 1988.[9]

In the United States, medical historians told similar triumphalist stories about the new social history, building on the founder story that Reverby and Rosner had established in 1979. In a 1990 *American Historical Review* essay surveying the

---

wake of the AIDS epidemic, many social historians wrote about the public health movements of the early twentieth century as a means of curtailing what they saw as potentially invasive health policies in their own time. The public health literature is quite expansive, but some of the best examples from the 1990s include Kraut (1994), Leavitt (1996), Rothman (1994) and Tomes (1998).

8 For the decline in MDs writing medical history over the last forty years, see statistics cited in Amsterdamska and Hiddinga (2004), p. 258. While very useful, this study is also limited since it only considers articles published in the field, not books.

9 For more on the evolution of the social history of medicine in Britain, and the early years of the *Social History of Medicine*, see Porter (1995).

history of medicine, Judith Walzer Leavitt categorised the authors under review as direct descendents of Sigerist, for they had all moved the focus of historical investigation away from the 'great physicians and their texts' towards a context-driven approach.[10] Not surprisingly, there were no biographies of physicians on Leavitt's list of reviewed books. Historiographic surveys of the field followed suit. In 1993, Gert Brieger discerned an emerging consensus in medical history. The Sigeristian program of study had become so prevalent, according to Brieger, that historians of medicine had come to take it 'almost for granted' (Brieger, 1993, p. 25). The question that needed to be asked was not whether the move to social history was worthwhile, but rather, why it took medical historians so long to 'heed the call to a social history voiced so eloquently by Sigerist' in the first place (Brieger, 1993, p. 26).

A common thread that ran throughout these triumphalist stories on both sides of the Atlantic was the denigration of the so-called 'Oslerian' approach of medical history.[11] To the new social historians, William Osler (1849–1919) – who held positions as a clinical physician in Canada, the United States, and Great Britain – typified 'traditional' history of medicine. From his time as a founder of the Johns Hopkins Hospital Historical Club in 1890 continuing to his position as Regius Professor of Medicine at Oxford University during the early twentieth century, Osler was known for his historically oriented speeches, lectures and publications, where he would praise the accomplishments of dead white male physicians, for he believed that 'medical biography could inspire young members of the profession to envision bright possibilities for their lives'.[12] This practical and highly optimistic view of medical history troubled many new social historians of medicine who wanted to bring about concrete changes to the way medicine was practised in the clinic, and to the way that doctors told stories about themselves.[13] As a result, Osler came to represent everything that the new social historians had fought against: intellectual history, physician worship, an overenthusiastic 'love of history', and not least, the writing of medical biography.

The gap between the new social history and the 'traditional' Oslerian approach to history became even wider in 1990 when Elizabeth E. Fee and Thomas Brown led the effort to create the 'Sigerist Circle', a section within the American Association for the History of Medicine that paralleled the older and more well-established 'Osler Society'. According to Reverby and Rosner, with the institution of the Sigerist Circle, social historians of medicine and activists scholars finally 'found a home' at the association's meetings (Reverby and Rosner, 2004, p. 176). Sigerist Circle historians

---

10 Leavitt (1990), esp. pp. 1,471–3.

11 For examples of this, see Porter (1995), Reverby and Rosner (1979) and Fee and Brown (1997), pp. 336ff. Recently, however, Fee and Brown have contended that social historians constructed an exaggerated portrayal of Osler as a amateur medical historian who only 'mined the history of medicine for inspirational messages'. See Fee and Brown (2004), pp. 139–40.

12 William Osler, as quoted in Fee and Brown (2004), p. 144.

13 For more on how social history should be motivated by present-day medical concerns, see, for example, Fee and Brown (1997) and (2004), Porter (1995) and Reverby and Rosner (1979, 2004), as well as Rothman (1994).

carved out an intellectual space where they and other like-minded scholars aimed to merge academic scholarship with political advocacy, an approach very different from the clinician-historian Osler Society and its unapologetic celebration of Osler's life and scholarship.[14]

And yet, as the Sigerist Circle gained numbers and recognition, two notable historians of medicine working in Canada – Michael Bliss and Jacalyn Duffin – appeared to buck the anti-biography trend. To be sure, 'corrective biographies' of women physicians, patients and unorthodox healers continued to be written (despite not being described or publicised in the field as biographies) since they were devoted to uncovering stories about groups of people marginalised by the medical profession.[15] Yet Bliss and Duffin (each approaching the writing of history in a unique way) chose the 'traditional' path of writing about white male physicians and extraordinary scientific discoveries.

Aware of the pressing need for a dialogue about the status of biography in medical history, J.T.H. Connor, editor of the *Bulletin of Canadian Medical History*, interviewed Bliss and Duffin, side by side, in 1996. At this time, Bliss had already published an award-winning biography of Frederick Banting (1984) and was well on his way to finishing *William Osler: A Life in Medicine*. During the interview, Bliss admitted that he had made a self-conscious effort to become a biographer, chiefly to challenge the new social history. 'It's clear', argued Bliss, 'that in our current academic culture, it's fashionable to be antibiography, but in the real world ... people want biography' (1996, p. 134). In order to legitimise the practice of biography, Bliss equated doctors who had a hagiographic need for historical role models to 'feminists' and 'ethnic groups' who were in 'search for a usable past'. As Bliss put it, feminist-oriented historians who comb the past looking to support their own political agendas 'are the same thing [as physician biographies] disguised all over again' (1996, p. 135).

Duffin, by contrast, arrived at medical biography by happenstance, largely unaware of the origins and agenda of the new social history. While she was still preparing a book manuscript on the scientific writings of René Laennec in 1985 (published in 1998), she began work on a little known nineteenth-century Ontario doctor, James Miles Langstaff. Sounding as much like a social historian as a biographer, Duffin claimed that she wrote her book *Langstaff: A Nineteenth-century Medical Life* in order to see 'what ordinary practice could tell us about the nineteenth century that hadn't already been said by very eminent people' (Duffin, 1996, p. 128). But despite having a research goal that was ostensibly in harmony with the new social history and demonstrating a rigorous method of analysing medical daybooks – a method reminiscent of Fredric L. Holmes' use of laboratory notebooks to recreate the process of scientific creativity, as well as Laurel Thatcher Ulrich's utilisation of Martha Ballard's midwife diary – Duffin was labelled an 'internalist' and had

14 For more on the Osler Society, see Fee and Brown (2004), esp. p. 140.

15 Ellen Singer More (1999) and Regina Morantz-Sanchez (1985, 1999) have written some of the most compelling histories of women physicians. Their accounts verge on biography, but do not entirely fit into the genre, since neither devotes a book to the entire life of an individual woman physician.

tremendous difficulty getting her books on Laennec and Langstaff published.[16] Tellingly, in the introduction to *Langstaff*, she resisted calling her book a biography. Instead, she wrote that it was '"biography" of a practice not of a person', putting biography in scare quotes in order to distance herself from a word that had the power to taint her career (Duffin, 1993, p. 6).

Despite the many differences between Duffin and Bliss, certain social historians lumped them together, tagging them both as apologists for 'traditional' history of medicine. Responding to the interview, Janice Dickin of the University of Calgary wrote that biography was fundamentally 'at odds with the disciplined development of writing within the history of medicine'. Whereas true academic scholarship demanded 'a great deal of suspicion', biographers, she argued, engaged in uncritical use of sources and wrote from the heart instead of the head (Dickin, 1996, p. 174).

There was, it turns out, much truth in Dickin's portrayal of medical biography. Medical biographers themselves have admitted the role that emotion sometimes plays in their work. For example, Duffin has professed in person and print that she embarked on writing biographies of Laennec and Langstaff because it was 'fun'.[17] She has also argued that good history results from an unfailing 'passion' and 'personal desire' to answer a question that leads a scholar to delve into the past (Duffin, 2004, pp. 442–3). In a similar vein, Sherwin B. Nuland, a surgeon-turned-medical-historian who won popular acclaim for his *Doctors: The Biography of Medicine*, has staunchly defended what critics call his 'gee whiz!' style of writing and his enthusiasm for great doctors of the past (Nuland, 1988b, pp. xix–xx).[18] He even urges his medical historian colleagues to adopt a similar writing style on the grounds that doing so might lead to better and more readable medical history, books that one 'may curl up with on those pleasure-filled evenings' (Nuland, 2004, p. 453).

Salvaging Sigerist and 'Great Doctors'

There are many ways to explore the question of medical biography's potential worth to social history. One way would be to attend more fully to the career of biography in the history of science, a field that has long been known as the 'sister discipline' to medical history. While certain historians of science lament the steep decline in scientific biography since the 1970s, social history did not bring scientific biography to an end as it did in medical history. Although 'social constructionism' – a school of thought that believes that specific local practices and contexts produce scientific

16 For Duffin's own compelling account of her experience as a 'biographer', and her repeated attempts at trying to get her books published, see her Chapter 11 in this volume and Duffin (2004). For more on her methodology as compared to other highly regarded contemporary historians, see Duffin (1993), Holmes (2004) and Ulrich (1990).

17 For more on Duffin's view of her own work, see Duffin (2004), esp. p. 442.

18 In the same year, Sherwin Nuland (1988a) wrote an editorial for *The Journal of the History of Medicine and Allied Sciences* advocating a kind of truce between non-PhD clinician-historians and social historians, arguing that each had something to learn from the other.

knowledge, not individual scientists – threatened scientific biography in theory, it failed to do so in practice.

For example, despite their radical claims to demonstrate that so-called experiential 'matters of fact' are entirely contingent and socially constructed, Steven Shapin and Simon Schaffer in *Leviathan and the Air-Pump* (1985) (arguably one of the most influential books to defend social constructionism) employed a rigorous historical methodology, firmly rooted in empirical evidence. Not only did they ground their argument in the authority of Boyle's *New Experiments Physio-Mechanical Touching the Spring of the Air*, but they also relied on Thomas Hobbes' *Dialogus Physicus* – a heretofore untranslated Latin text – to show other 'alternative' methods for knowledge-production. In the end, Shapin and Schaffer implied that without Boyle, the new empiricism of modern science would not have taken hold, and in so doing, they upheld the biography-centred notion that Boyle was a crucial figure to the history of science.

In a way, social constructivism helped biography to flourish in the history of science, for during the 1990s a kind of Hegelian synthesis occurred in the field, where the study of the individual and the social context merged to create a new form of biographical writing, known as 'social biography' (Söderqvist, 1996).[19] Historians such as Gerald Geison in *The Private Science of Louis Pasteur* (1995) and Mario Biagioli in *Galileo, Courtier: The Practice of Science in the Culture of Absolutism* (1993) began to redefine the very notion of scientific 'greatness', for they

19 This is a contentious claim, for several historians of science have argued that social constructionism undermined the writing of biography: that is, the context-driven approach of social history decentred the individual about whom biography is to be written. Perhaps the first defender of biography in the wake of social constructionism was Thomas Hankins, who in 1979 argued that the best way to get at both the intellectual *and* social contexts of science is to perform a life study of a single scientist. Two years later, Frederic L. Holmes (1981) made a similar attempt at defending biography against the new social historians of science, arguing that the study of individual scientists should remain at the 'heart of the discipline', not only because it allowed the writer to narrow the complexities of contexts to a manageable level, but also because the historian needed a central subject, a pivot point around which the 'various contexts surround', for without this, Holmes contended, contexts were meaningless. More recently, Thomas Söderqvist (1996) has voiced similar concerns about the decentring of the individual scientist, arguing that the sociology of science threatens to reduce the individual to a 'mere instance in contextual history', leading the social historian to catch only a glimpse of a scientist's full existential personhood. While I am sympathetic to the argument that the range and depth of biographical writing would suffer if it were completely subsumed by social history, I do not have the same dismal outlook on the effects of social history on scientific biographies. My different outlook on this point can be explained by the fact that I am engaged in a comparative study, looking at biography both in the context to the history of science and in the history of medicine. While the social history of medicine virtually eradicated biography, the social history in science (understood as social constructionism) did not fundamentally contest histories written about 'great scientists'. Moreover, given the particular professional dynamics and historiography of medical history, there appears to be no other way to incorporate biography back into the field of medical history except through social history. For more on the position that social history undermines the writing of biography in the history of science, see also Shortland and Yeo (1996).

saw that some of the greatest scientists in the past were motivated as much by career pressures and patronage as they were by the pursuit of knowledge. For these new social biographers, greatness was not so much a function of the capacity to reach truth as it was a tribute to a scientist's salesmanship.[20]

Thus, by looking to history of science, historians of medicine can see how it might be possible to incorporate the two approaches of biography and social history. Social historians of science and medicine have much in common; at base, both believe that scientific authority, medical or otherwise, is largely rooted in the practices of particular societies and cultures. But unlike most social historians of medicine, social historians in science have continued to acknowledge the importance of intellectual history and the centrality of scientific texts. As Ludmilla Jordanova and John Harley Warner have argued on separate occasions, there is much that social historians of medicine can learn from social constructionism in science, not least of which is how to salvage the history of ideas while still remaining true to social history.[21]

Another way to determine the potential value of medical biography to social history is to take a fresh look at the life and scholarship of Sigerist, a man who has come to represent the contemporary anti-biography stance in medical history. Fee and Brown, co-founders of the 'Sigerist Circle' and today's leading experts on Sigerist's life and scholarship, have called for a similar re-evaluation of Sigerist, realising that he has been 'converted into [a] symbolic representation' of himself, his differences with Osler 'exaggerated'. For the purposes of judging the merit of medical biography, then, we must approach Sigerist's life with a singular and penetrating focus, asking if he would have supported the anti-biography view ascribed to him.

By early twentieth-century standards, Sigerist was a renegade of sorts. As a Social Democrat who fled a disintegrating Weimar Germany in 1932 to become chair of the Johns Hopkins Institute of the History of Medicine, he pushed for socialised medicine in America at a time when the medical profession was overwhelmingly conservative, supporting the ideologies of big business and small government.[22] Shortly after he arrived in the United States in 1933 – a time when American anti-communist sentiments ran high and 'red baiting' was common in national politics – Sigerist wrote in his private papers: 'a social revolution [in American medicine] is needed. Under the capitalist system, preventative medicine is not possible. Russia therefore signifies the beginning of a new epoch in medicine.'[23] For Sigerist, political ideologies were not simply a private matter. He held the firm belief that he could change the practice of medicine through his historical writings, and in 1937 published *Socialized Medicine in the Soviet Union*, advocating compulsory health insurance. While the book created controversy in the United States, it received

---

20 Geison (1995), esp. pp. 10 and 278. Geison came under a significant amount of criticism for his book, for its 'debunking' qualities and its 'presentist' criticisms of Pasteur's research ethics. No matter the criticisms, though, Geison's book was still a form of scientific biographical writing, a genre that had all but disappeared in the history of medicine. For a sampling from Geison's critics, see Perutz (1995).

21 Jordanova (1995) and Warner (1995). For a useful set of articles that address social constructionism in medicine, see Wright and Treacher (1982).

22 For more on Sigerist's life history, see Fee and Brown (1997).

23 Beeson (1966), p. 83 (see Sigerist's entry for 17 April 1933).

a warmer reception in Canada, India and South Africa, where he was invited to lecture.[24] In America, however, criticisms of his work mounted throughout the 1940s. Under strain from having to constantly defend himself against the American Medical Association and Hopkins medical alumni, Sigerist left the United States in 1947 to return to Switzerland.[25]

This was the political side of Sigerist's life – the part that appears to harmonise with the uses to which he has been put by contemporary historians of medicine. But there was another side of Sigerist's life – one that many American social historians of medicine from the 1970s onward have chosen to overlook: namely, that Sigerist often engaged in intellectual history and biographical writing in his own practice of social history. For instance, anticipating his 1931 lecture tour in America, Sigerist wrote that he would 'speak of medicine and its growth, of the laws that control its development, of great doctors and great errors, and of cultural and intellectual patterns' (quoted in Beeson, 1966, p. 69). Sigerist made this notation shortly after he published *Grosse Ärzte: Eine Geschichte der Heilkunde in Lebensbildern* ([1931] 1954), which appeared two years later in English translation as *The Great Doctors: A Biographical History of Medicine* (1933).

It would be inaccurate to call Sigerist's biographical writing hero worship. Yet neither could it be described as sceptical towards the very notion of greatness, as is common among today's social historians. Sigerist treated his subjects with reverence, but he also viewed them with a critical eye, motivated by the conviction that medicine should improve the general public's health. Recognising the commonality shared by all physicians, ordinary and exceptional, Sigerist dedicated the *Great Doctors* to the 'unknown doctor', to all practitioners who do 'heroic work' on a day-to-day basis. For Sigerist, then, the physicians about whom he wrote served as inspiration for everyday doctors. As he put it: 'The fact that [the great doctors of history] were privileged to reach supreme heights makes them our masters and exemplars, the thought of whom can encourage and invigorate us when the trivialities of the daily round are tending to dim our faith in the splendor of our calling' (Sigerist, 1933, p. 18).

Sigerist's concern for biography and 'great doctors' did not wane as he became more politically motivated. Even when he began to think of himself as a Marxist in the late 1930s, arguing that doctors should 'lead struggles for the general improvement of working conditions', he also upheld the writing of medical biographies.[26] In a 1936 letter to George Sarton, Sigerist admitted frustration with 'amateur historians', who did not know that 'historical research had exact methods', just as scientific research did (Sigerist, 1936, pp. 3, 6). But he nevertheless concluded that such clinician-historians were 'enthusiasts' who should not be 'discouraged', even when they choose to write medical biographies. Indeed, Sigerist argued that physician-historian enthusiasts might be the best candidates for writing local histories – a kind

24 For Sigerist's direct involvement in the development of Canada's universal medical care system (1941–44), see Duffin and Falk (1996).

25 For the best and most concise account of Sigerist's life, see Fee and Brown (1997).

26 For Sigerist's relationship to Marxism, see Fee and Brown (2004), pp. 154–5.

of history he believed sorely lacking – for, as he put it, 'who could do it better than the physician who lives on the spot?' (Sigerist, 1936, p. 2).

Sigerist's support of biography persisted throughout his career, informing his method of 'sociological medicine' until the end. Upon his return to Switzerland in 1948 and in the midst of writing an eight-volume history of medicine from Greek antiquity to the modern era, Sigerist advocated the need to understand medical theory, claiming that the medical historian must be familiar with 'the chief actors [in medicine], their training, their contributions, and the ideals that guided their actions'.[27] Five years later, Sigerist returned to *The Great Doctors* in order to revise the German edition, which had been out of print since the Second World War. In the preface to the 1954 edition of the *Grosse Ärzte*, Sigerist voiced his initial concern that the book may have been 'too old to revise'. 'But with surprise and elation,' wrote Sigerist, 'I found that [the book] had never lost its original appeal, and that it was still capable of awakening respect for the great accomplishment of medical science and enthusiasm for its work' (Sigerist, [1931] 1954, p. 11).

Sigerist's interest in medical biography should not be overstated, for his approach to the social history of medicine demanded that historians understand the economics and politics of medicine as well as its theory and 'great men'. While he used biography in *The Great Doctors* as a possible way to write a synthetic account of medical history, he was not, first and foremost, a medical biographer; he did not, that is, give biography priority over other methods of understanding medical history.[28] Rather, he experimented and utilised very different methodological forms of historical writing throughout his career, ranging from high-culture explanations of medical discoveries to the employment of economic theory to explain poor public health. To Sigerist, biography had its uses (even if limited), and he said as much. 'Many doctors of old', wrote Sigerist in 1936, 'deserve to have their biographies written. Is there a more beautiful task than to resuscitate the life and achievements of a fellow-doctor even if he is not among the heroes of medicine?' (Sigerist, 1936, p. 3)

Here Sigerist points to several potential answers to the question 'Why biography?' As a man who wore many hats, Sigerist supported the writing of medical biography for varying, and equally compelling, reasons. At the time of his letter to Sarton, Sigerist admitted that it was not easy to keep the institute at Johns Hopkins afloat.

---

27 Sigerist, as quoted in Beeson (1966), p. 217. This quotation comes from Sigerist's discussion in 1948 of what his eight-volume history of medicine would aim to do.

28 As a contemporary of Sigerist, Owsei Temkin argued that Sigerist's *Great Doctors* was not 'biography'. Temkin contended that the book was 'not made to spring from a historical vacuum; nor [was] the history of development attributed to abstract forces of which the individual is a mere point of intersection' (1958, p. 492). Here we see how Temkin's rather narrow definition of biography stems from his background as a German-trained medical historian and his observations of lesser-trained physician-historians discussing medical ideas and theories without much concern for historical context. I would argue, however, that with a more robust definition of biography as a study of a physician's life, practice and ideas, Sigerist's *Great Doctors* does fall into the category of biography, and more precisely, collected biography. Although Temkin did not categorise *Great Doctors* as a biography, he did argue (1958, p. 493) that, as a book, it was one of Sigerist's three attempts to write a synthetic account of the history of medicine.

Acutely aware of his dependence on physician support, he concluded that biography might be the best method of maintaining financial backing from the medical school and its donors. In addition, like Osler, Sigerist saw biography as a means to recruit would-be medical students.[29]

But biography was more than a means of financial gain or a tool for popularising medical history. To write biography was, in Sigerist's mind, as much a choice of utility as it was a matter of aesthetics. In his early career, Sigerist likened William Harvey's discovery of circulation to the Baroque, demonstrating that 'medical history itself could be beautiful, approached with the esthete's eye'.[30] He brought much the same mindset to medical biography, well into his late career. When he spoke of biography, he talked about it in terms of 'beauty', a form of historical writing that could 'awaken respect' for physicians, a way to reveal the 'splendour' of doctoring, a 'calling' that included everything from the mundane tasks of caring for the chronically ill to extraordinary moments of medical discovery.[31]

Finally, because he assumed that all medical historians would possess medical degrees, as he himself did, Sigerist thought that the life stories of 'fellow-doctors' would be of inherent interest to medical historians.[32] Medical biography, in other words, would be history used as a process of self-discovery. As today's social historians readily point out, Sigerist insisted that the writing of history should be motivated by present-day concerns; for Sigerist, this also included the writing of medical biographies. 'Why', Sigerist asked, 'should a physician undertake the labour of consulting the past, of recreating it in history if it were not that he felt driven to such a task by medical considerations?' 'History', he concluded, 'is an instrument of life and medical history is an instrument of medical life' (Sigerist, 1936, p. 7). What mattered to Sigerist was not any particular historical methodology, but rather that a scholar's aim be true, with deeply felt 'historical dedication'.[33]

Biography as a Means to Plurality

There is a limit, of course, to what historians of medicine can learn from Sigerist's own views and scholarship. He lived in an era when the history of medicine was remarkably homogeneous, when an overwhelming majority of its practitioners had medical degrees. Today, the field is heterogeneous, with historians of medicine coming from a multitude of educational, cultural and socio-economic backgrounds. But perhaps historians of medicine need not look outside of their own time period to approach the question of medical biography. There is, for example, Harold Cook's *Trials of an Ordinary Doctor: Joannes Groenevelt in Seventeenth-century London*

29 For more on these points, see Sigerist (1936).

30 For Sigerist's use of art history to explain Harvey's discovery of circulation, see Sigerist (1929). On Sigerist's aesthetic sensibility, see Temkin (1958), esp. p. 490.

31 See Sigerist (1933) and (1954), as well as Sigerist (1936), p. 3.

32 For Sigerist's assumption that medical historians would also be medical doctors, see Temkin (1958), esp. 492.

33 Temkin (1958), p. 489, discusses Sigerist's belief in the importance of historical dedication.

(1994). Following traditional biographical writing, Cook begins his book with a full description of Groenevelt's Netherlands birthplace and education, and then proceeds to trace Joannes' medical career to England, where eventually, in 1694, the College of Physicians charged him with malpractice.

The fact that Cook chose to entitle his book *Trials of an Ordinary Doctor* tells us more about the profession of medical history than it does about the subject of his book. Cook takes great pains to avoid calling his study of Groenevelt a biography. Even though he employs much the same rhetoric that biographers use to introduce their subjects, Cook claims that the 'best justification for studying a person such as Groenevelt … lies in the power of the genre "microhistory"' (Cook, 1994, p. xvii). While it is true that the scope of biography, like microhistory, is inherently limited – and that both tend to draw conclusions about the general from the particular – the two genres of writing are far from equivalent. Biographies might be a kind of microhistory, but not all microhistories take the form of biography. Indeed, surveying some of the original works of microhistory written in the 1970s and early 1980s – such as Carlo Ginzburg's *The Cheese and the Worms* (1980) and Natalie Zemon Davis's *The Return of Martin Guerre* (1983) – one finds that microhistories rarely focus on one person, and instead concentrate on a local group of people, usually in small, rural towns.

It seems that Cook's motivations for calling his book a microhistory are similar to those that led him to identify Groenevelt as an 'ordinary doctor'. Cook writes that Groenevelt was 'notorious' for his 1694 malpractice trial, becoming a *cause célèbre* in so far as his vindication in the English courts 'established an important legal precedent' (Cook, 1994, p. xv). Moreover, Cook points out that Groenevelt had published several works in medicine and was 'associated with an influential group of medical dissenters, innovators and translators, including Edward Tyson, John Pechey and Thomas Sydenham' (1994, p. xiv). With such a description, one is left no other choice but to conclude that even Cook himself is not convinced of Groenevelt's 'ordinary' status. In a very real sense, *Trials of an Ordinary Doctor* is a biography of a great doctor in disguise.

With its biographical defensiveness, *Trials of an Ordinary Doctor* might give pause to the new social historians of medicine. Is the avoidance of biography and 'great' doctors detrimental to a field that aims toward scholarly sophistication?[34] Jordanova, for one, has contended that such avoidance will inevitably keep the social history of medicine in a state of infancy. As she argued in 1993, the field of medical history will never mature until it is more 'fully confident in tackling areas of major historical debate' and it broadens its 'range of genres', including, and most

34 Since I delivered the original version of this paper in May 2002, there have been two biographies written about 'great' doctors, namely Thomas Bonner's *Iconoclast* (2002) and Alan M. Kraut's *Goldberger's War* (2003). At this point, it is too soon to determine whether these two books represent the beginning of a pro-biography trend. If they do and we are entering a new era of medical biography, we need to be fully self-conscious of the profession's troubled relationship with biography throughout the twentieth century. In other words, we need to ask ourselves 'Why biography?', and begin to establish some arguments for its use in the history of medicine.

importantly, 'scholarly biographies' (Jordanova, 1993, 438). In the United States, at least, it seems that the Sigerist founding story of liberation from physician-centred history has played a significant role in narrowing social history's scope, precluding full methodological and topical pluralism.

Conclusion

Since the 1970s, social historians of medicine have been trying to rewrite the history of their field from the bottom up. Many of them have been trying to recover the patient's view, the ordinary healer's view, the people's view – basically, any view that might challenge the elitist view, or, as it is commonly put, the view of the 'great doctors'. In the process, historians of medicine have banned biography as a valid form of historical scholarship, for they have convinced themselves that 'great doctor' history is synonymous with biography. Even scholars sympathetic to the new social history, such as Cook and Duffin, who find biography a fruitful means of exploring the past, are so affected by today's professional mores that they conceal their biographical leanings.

If the new social historians decide that biography should be reintroduced into the history of medicine, its reincorporation will not be an easy task, for their very identity is inextricably bound up with biography, or rather, with opposition to it. Social historians of medicine will thus only be able to accept biography once they reassess their own assumptions about their profession's past, whether it is a passing remark denouncing Oslerian history, or a tribute in praise of Sigerist. This, along with a re-evaluation of medical and science biographies written by some of today's leading historians, might be enough to generate a much-needed debate about biography – a form of historical writing that has, for better or worse, been crucial to the making of medical history throughout the twentieth century.

Acknowledgements

I wish to thank Thomas Söderqvist for the invitation to present the original version of this paper at the 'Poetics of Biography in Science, Technology, and Medicine' workshop held in May 2002 in Copenhagen, Denmark, and more importantly, for inspiring me to explore the writing of medical biography, a research path that has proven to be highly rewarding. I also wish to thank the other participants of the workshop, particularly Jacalyn Duffin, for their helpful feedback. For their insightful comments on drafts of this paper, I am especially grateful to Daniel J. Kevles, Damon Linker, Sally Romano and John Harley Warner.

Bibliography

Amsterdamska, Olga and Anja Hiddinga (2004), 'Trading Zones or Citadels? Professionalization and Intellectual Change in the History of Medicine', in Frank Huisman and John Harley Warner (eds), *Locating Medical History: The Stories and Their Meanings*, Baltimore, MD: Johns Hopkins University Press, pp. 237–61.

Beeson, Nora Sigerist (ed. and trans.) (1966), *Henry E. Sigerist: Autobiographical Writings*, Montreal: McGill University Press.

Berliner, Howard S. (1980), 'Book Reviews', *Bulletin of the History of Medicine*, **54**, 131–4.

Biagioli, Mario (1993), *Galileo, Courtier: The Practice of Science in the Culture of Absolutism*, Chicago, IL: University of Chicago Press.

Bliss, Michael (1984), *Banting: A Biography*, Toronto: McClelland and Stewart.

Bliss, Michael (1996), 'Life Writing in Medical History', interview by J.T.H. Connor, *Canadian Bulletin of Medical History*, **13**, 123–37.

Bliss, Michael (1999), *William Osler: A Life in Medicine*, Toronto: University of Toronto Press.

Bonner, Thomas Neville (2002), *Iconoclast: Abraham Flexner and the Life of Learning*, Baltimore, MD: Johns Hopkins University Press.

Brieger, Gert (1993), 'The Historiography of Medicine', in W.F. Bynum and Roy Porter (eds), *Companion Encyclopedia of the History of Medicine*, vol. 1, London: Routledge, pp. 24–44.

Brown, Theodore M. and Elizabeth Fee (1997), 'Anything but "Amabilis": Henry Sigerist's Impact on the History of Medicine in America', in Theodore M. Brown and Elizabeth Fee (eds), *Making Medical History: The Life and Times of Henry E. Sigerist*, Baltimore, MD: Johns Hopkins University Press, pp. 333–70.

Burke, Peter (1992a), *History and Social Theory*, Cambridge: Polity Press.

Burke, Peter (1992b), *New Perspectives on Historical Writing*, University Park, PA: Pennsylvania State University Press.

Burnham, John C. (1998), *How the Idea of Profession Changed the Writing of Medical History*, London: Wellcome Institute for the History of Medicine.

Cook, Harold L. (1994), *Trials of an Ordinary Doctor: Joannes Groenevelt in Seventeenth-century London*, Baltimore, MD: Johns Hopkins University Press.

Davis, Natalie Zemon (1983), *The Return of Martin Guerre*, Cambridge, MA: Harvard University Press.

Dickin, Janice (1996), 'Life Writing versus Biography', *Canadian Bulletin of Medical History*, **13** (2), 174–6.

Duffin, Jacalyn (1993), *Langstaff: A Nineteenth-century Medical Life*, Toronto: University of Toronto Press.

Duffin, Jacalyn (1996), 'Life Writing in Medical History', interview by J.T.H. Connor, *Canadian Bulletin of Medical History*, **13** (1), 123–37.

Duffin, Jacalyn (1998), *To See with a Better Eye: A Life of R. T. H. Laennec*, Princeton, NJ: Princeton University Press.

Duffin, Jacalyn (2004), 'A Hippocratic Triangle: History, Clinician-historians, and Future Doctors', in Frank Huisman and John Harley Warner (eds), *Locating Medical History: The Stories and their Meanings*, Baltimore, MD: Johns Hopkins University Press, pp. 432–49.

Duffin, Jacalyn and Leslie A. Falk (1996), 'Sigerist in Saskatchewan: The quest for balance in social and technical medicine', *Bulletin of the History of Medicine*, **70**, 658–83.

Fee, Elizabeth and Theodore M. Brown (1997), 'The Renaissance of a Reputation', in Theodore M. Brown and Elizabeth Fee (eds), *Making Medical History: The Life and Times of Henry E. Sigerist*, Baltimore, MD: Johns Hopkins University Press, pp. 1–11.

Fee, Elizabeth and Theodore M. Brown (2004), 'Using Medical History to Shape a Profession: The Ideals of William Osler and Henry E. Sigerist', in Frank Huisman and John Harley Warner (eds), *Locating Medical History: The Stories and Their Meanings*, Baltimore, MD: Johns Hopkins University Press, pp. 139–64.

Geison, Gerald L. (1995), *The Private Science of Louis Pasteur*, Princeton, NJ: Princeton University Press.

Ginzburg, Carlo (1980), *The Cheese and the Worms: The Cosmos of a Sixteenth-century Miller*, trans. John and Anne Tedeschi, Baltimore, MD: Johns Hopkins University Press.

Hankins, Thomas (1979), 'In defense of biography: The use of biography in the history of science', *History of Science*, **17**, 1–16.

Holmes, Frederic L. (1981), 'The fine structure of scientific creativity', *History of Science*, **19**, 60–71.

Holmes, Frederic L. (2004), *Investigative Pathways: Patterns and Stages in the Careers of Experimental Scientists*, New Haven, CT: Yale University Press.

Huisman, Frank and John Harley Warner (eds) (2004), *Locating Medical History: The Stories and their Meanings*, Baltimore, MD: Johns Hopkins University Press.

Jordanova, Ludmilla (1993), 'Has the social history of medicine come of age?', *The Historical Journal*, **36**, 437–49.

Jordanova, Ludmilla (1995), 'The social construction of medical knowledge', *Social History of Medicine*, **8**, 361–81.

Kraut, Alan M. (1994), *Silent Travelers: Germs, Genes, and the 'Immigrant Menace'*, New York: Basic Books.

Kraut, Alan M. (2003), *Goldberger's War: The Life and Work of a Public Health Crusader*, New York: Hill and Wang.

Leavitt, Judith Walzer (1990), 'Medicine in context: A review essay of the history of medicine', *The American Historical Review*, **95**, 1,417–84.

Leavitt, Judith Walzer (1996), *Typhoid Mary: Captive to the Public's Health*, Boston, MA: Beacon Press.

Miller, Genevieve (1973), 'In praise of amateurs: Medical history in America before Garrison', *Bulletin of the History of Medicine*, **47**, 586–615.

Morantz-Sanchez, Regina (1985), *Sympathy and Science: Women Physicians in American Medicine*, New York: Oxford University Press.

Morantz-Sanchez, Regina (1999), *Conduct Unbecoming of a Woman: Medicine on Trial in Turn-of-the-century Brooklyn*, New York: Oxford University Press.

More, Ellen Singer (1999), *Restoring the Balance: Women Physicians and the Profession of Medicine, 1850–1995*, Cambridge, MA: Harvard University Press.

Novick, Peter (1988), *That Noble Dream: The 'Objectivity Question' and the American Historical Profession*, Cambridge: Cambridge University Press.

Nuland, Sherwin B. (1988a), 'Doctors and historians', *The Journal of the History of Medicine and Allied Sciences*, 43, April, 137–40.

Nuland, Sherwin B. (1988b), *Doctors: The Biography of Medicine*, New York: Alfred A. Knopf.

Nuland, Sherwin B. (2004), 'Medical history for the general reader', in Frank Huisman and John Harley Warner (eds), *Locating Medical History: The Stories and Their Meanings*, Baltimore, MD: Johns Hopkins University Press, pp. 450–59.

Perutz, Max F. (1995), 'The pioneer defended', *The New York Review of Books*, 21 December 1995, pp. 54–8.

Porter, Dorothy (1995), 'The mission of social history of medicine: A historical view', *Social History of Medicine*, **8**, 345–59.

Porter, Roy (1985), 'The patient's view: Doing medical history from below', *Theory and Society*, **14**, 175–98.

Reingold, Nathan (1981), 'Science, scientists, and historians of science', *History of Science*, **19**, 274–83.

Reverby, Susan and David Rosner (1979), 'Beyond "the Great Doctors"', in Susan Reverby and David Rosner (eds), *Health Care in America: Essays in Social History*, Philadelphia, PA: Temple University Press, pp. 3–16.

Reverby, Susan and David Rosner (2004), '"Beyond the Great Doctors" revisited: A generation of the "new" social history of medicine', in Frank Huisman and John Harley Warner (eds), *Locating Medical History: The Stories and Their Meanings*, Baltimore, MD: Johns Hopkins University Press, pp. 167–93.

Rothman, Sheila (1994), *Living in the Shadow of Death: Tuberculosis and the Social Experience of Illness in American History*, New York: Basic Books.

Shapin, Steven and Simon Schaffer (1985), *Leviathan and the Air-Pump: Hobbes, Boyle, and the Experimental Life*, Princeton, NJ: Princeton University Press.

Shortland, Michael and Richard Yeo (1996), 'Introduction', in Michael Shortland and Richard Yeo (eds), *Telling Lives in Science: Essays on Scientific Biography*, Cambridge: Cambridge University Press, pp. 1–44.

Sigerist, Henry E. (1929), 'William Harvey's Stellung in der europäischen Geistesgeschichte', *Archiv für Kulturgeschichte*, **19**, pp. 158–68.

Sigerist, Henry E. ([1931] 1954), *Grosse Ärzte: Eine Geschichte der Heilkunde in Lebensbildern*, 3rd edn, Munich: J.F. Lehmanns Verlag.

Sigerist, Henry E. (1933), *The Great Doctors: A Biographical History of Medicine*, 1st edn, trans. Eden and Cedar Paul, New York: W.W. Norton.

Sigerist, Henry E. (1936), 'The history of medicine and the history of science', *Bulletin of the Institute of the History of Medicine*, **4**, 1–13.

Söderqvist, Thomas (1996), 'Existential projects and existential choice in science: Science biography as an edifying genre', in Michael Shortland and Richard Yeo (eds), *Telling Lives in Science: Essays on Scientific Biography*, Cambridge: Cambridge University Press, pp. 45–84.

Stevenson, Lloyd G. (1980), 'A Second opinion', *Bulletin of the History of Medicine*, **54**, 134–40.

Temkin, Owsei (1958), 'Henry E. Sigerist and aspects of medical historiography', *Bulletin of the History of Medicine*, **22**, 485–99.

Tomes, Nancy (1998), *The Gospel of Germs: Men, Women, and the Microbe in American Life*, Cambridge, MA: Harvard University Press.

Ulrich, Laurel Thatcher (1990), *A Midwife's Tale: The Life of Martha Ballard, Based on Her Diary, 1785–1812*, New York: Random House.

Warner, John Harley (1995), 'The history of science and the sciences of medicine', *Osiris*, **10**, 164–93.

Wear, Andrew (ed.) (1992), *Medicine in Society: Historical Essays*, Cambridge: Cambridge University Press.

Wilson, Leonard G. (1980), 'Medical history without medicine', *The Journal of the History of Medicine and Allied Sciences*, **35**, 5–7.

Wright, Peter and Andrew Treacher (eds) (1982), *The Problem of Medical Knowledge: Examining the Social Construction of Medicine*, Edinburgh: Edinburgh University Press.

Chapter 15

'No Genre of History Fell Under More Odium than that of Biography': The Delicate Relations between Scientific Biography and the Historiography of Science

Thomas Söderqvist

A Cultural Poetics of Scientific Biography

The genre of scientific biography stands in want of two kinds of inquiries which, in principle, go all the way back to the Aristotelian poetical tradition. One concerns the formal properties of the genre – that is, features such as composition and style, the use of literal versus figurative language and so on – the other its historical embedment: its historicity, its cultural poetics. The two kinds of inquiries are, of course, closely related since the formal characteristics of scientific life writing have varied over time and between different social and cultural settings.

Some of the formal properties have been the object of discussion throughout most of the genre's history. For example, the use of thematic versus chronological composition – a standard issue in biography writing that goes back to the classical distinction between the 'Suetonian' and 'Plutarchian' biographical traditions – has been a recurrent dilemma for authors of scientific lives, too.[1] Patrice Debré explains why he chose to compose *Louis Pasteur* thematically instead of chronologically:

> I have opted for a methodical presentation of the various research topics (from their premises to their conclusions) rather than for a strict adherence to chronology, for it is often impossible to show the simultaneity of experiments of very different nature or importance. … A thematic narrative gains in clarity what it may lose in immediacy; above all, it allows every reader to linger over the subject that is of particular interest to him or her, whether it be crystallography or rabies. (Debré, 1998, p. xxiii)

---

1 Leo (1901) pointed out that Sueton tended to write thematically, whereas Plutarch was more strictly chronological in his *bioi paralleloi*. Arnaldo Momigliano suggests that the Plutarchian form was more useful for biographies of statesmen and military leaders, whose achievements can be described as one political action after the other, while the Suetonian form was better for the lives of authors, philosophers and artists, because the results of creative work was supposed to have an internal consistency over time, which was easier to represent systematically than chronologically (Momigliano, 1993, Ch. 1)

Debré evidently wanted his biography to present Pasteur's science as clearly as possible – and then it did not matter if Pasteur's life became more diffuse ('what it may lose in immediacy'). The opposite choice was made by Patricia Spain Ward in her study of the nineteenth-century American hydrotherapist Simon Baruch. Although it was the potential value of using Baruch's life as 'a mirror' for American medicine that first drew her to him, it was ultimately 'his human development' that captivated her. As she worked her way through his writings, she decided 'to follow a chronological structure for this book so as to portray the actual course of his life as closely as possible' (Ward, 1994, p. xiv).[2]

In these two examples, the choice of a chronological versus a thematic composition reflects the individual author's intentions to emphasise either the science or the person behind it. This choice in turn probably reflects a broader cultural division. Biographies written by scientists and historians of science have tended to focus on the scientific work and achievements, whereas family members, journalists, novelists and general writers have tended to be more interested in the life. Generally, one can expect authorial intentions (and readers' expectations) to be embedded in specific professional and cultural contexts. To the formal properties of biographical texts thus can be added a number of interesting questions concerning the cultural poetics of the genre.[3]

For example, one can think of a whole series of questions concerning the *begriffsgeschichtliche* context of scientific biography. Is the genre an expression of the rise of the idea of the modern scientific subject? Was the renaissance of biographical writing and the increasingly positive reaction to the genre among book reviewers in the 1980s and 1990s an expression of the surge of political individualism in these decades? Another set of questions concerns the dire topic of national contexts. Since the mid-nineteenth century, scientific biographies have been more frequently published in Britain than in any other country; Maurice Crosland once referred to '[t]he dearth of biographies of important French scientists compared to the relative abundance of biographies of British scientists' (Crosland 1978, p. x). There may also be certain national styles involved; Theodore Zeldin thought, for example, that French scholars did not 'excel in' biography, due partly to 'a preference for more grandiose philosophical themes than mere individuals can provide' (1974, p. 701). Although the descriptive life-and-letters were not entirely a British phenomenon – for example, Leo Königsberger's monumental three-volume *Hermann von Helmholtz* (1902–1903), one of the mastodons among German scientific biographies, was in all formal respects cast in the life-and-letters mould – the format was better at home with late nineteenth-century Victorian England than with contemporary French or German culture; Zeldin notes that 'the double-decker biography is a rarity in France' (Zeldin, 1974, p. 701). The narrative integration between life and work also occurs

2 Ward says that she took the decision against the advice of several of her peers – the historians of medicine at Johns Hopkins University, a hotbed for the social history of medicine in the 1970s and 1980s.

3 I'm using the term 'cultural poetics' in the sense of Gallagher and Greenblatt (2001) and others, as largely synonymous to a contextual (new historicist, cultural materialist) approach to literary works (including biography and historiography itself).

more frequently in the British biographical tradition, whereas French and German scientific biographies have habitually divided the text corpus into sections on *la vie* versus *son œuvre* (*Leben* versus 'work'), as in Michelle Sadoun-Goupil's *Le chimiste Claude-Louis Berthollet 1748–1822: sa vie, son œuvre* (1977) and Karl Vogeler's *August Bier: Leben und Werk* (1941), to take but two random examples. The not very voluminous but solid Dutch tradition for scientific life writing is characterised by a combination of meticulous scholarship and pronounced realism, and a corresponding lack of popular, essayistic biographies in the series format which has characterised the British, French and German biographical traditions.

The question of scientific biography and gender has only been touched upon recently. For example, Sylvana Tomaselli has pointed to a number of 'paradoxes' which arise from biographical projects 'that may initially be motivated by an attempt to show women as the equals of men'. She is especially thinking of how biographies of women scientists tend to 'reinforce images which one may well wish should cease to predominate in our culture', for example how a successful life is one which 'measures up well by the standards of individual masculine success' (Tomaselli 1988, pp. 102–4); the cultural images of Marie Curie and Dorothy Hodgkin come easily to mind. Paola Govoni has even suggested that the genre has been more suitable for women readers because they have traditionally been less educated in scientific subjects, 'and consequently more at ease when faced with histories "enriched" by the family and everyday affairs of the personage being scrutinised' (Govoni, 2000, p. 403). Should we then infer that more personal kinds of biographies, like Ward's *Simon Baruch* (1994), are implicitly aimed at a female readership, while biographies that focus on the equations – like Helge Kragh's *Dirac: A Scientific Biography* (1990), with the conclusion that theoretical physics was for Dirac 'a substitute for human emotions' (Kragh, 1990, p. 255) – are aimed at male readers? And to stretch the argument a bit further: should we infer that it is mainly the spouses of scientists and historians of science who indulge in biographies? In fact, women's biographies of other women seem to be exempted from today's universal taboo on eulogy and 'hagiography'. We used to think of eulogy as a phenomenon of the past, or at least something in the margins in the form of hero-portraits published on science presses. But eulogy is not at all dead, not even in respectable biographies; for example, Linda Lear's *Rachel Carson: Witness for Nature* (1997) is a prose encomium, a modern scientific saint's life, praised by all reviewers, indicating that women's biographies of women scientists are presently allowed to be unashamedly 'hagiographic'.

Another potentially interesting topic for a cultural poetics of scientific biography is its embedment in the genre of biography as a whole. Are biographers of scientists influenced by reading other kinds of biographies? André Maurois, French man of letters, author of novels, short stories and histories, and of biographies of Shelley, Disraeli, Byron, Lyauley, Voltaire, Chateaubriand, Proust, George Sand, Victor Hugo, Les Trois Dumas, and his own *Mémoires*, was probably aware of the biographical giants on whose shoulders he was standing when he decided to cross the border-line separating literary and scientific to craft *La vie de Sir Alexander Fleming* (Maurois, 1959). But what about the great majority of biography writer? Did the Boswellian 'revolution' have any impact on the writing of the lives of scientists in nineteenth-century Britain? Did scientific biographers in the early twentieth century pay any attention at all to Lytton Strachey's uproar against the Victorian life-and-letters? Historian of science George Sarton did not like Strachey at all (see below) – but what about less lettered authors, like professors of chemistry or mechanical engineering? Did they know about the discussion? Did they care? The prefaces and introductions to biographies rarely make any explicit bows to biographical influences, and possible intellectual debts have therefore to be inferred by more indirect means.

Scientific Biography and the Historiography of Science in the Service of Science

Yet another unexplored aspect of the cultural poetics of scientific biography is the attitudes of historians of science towards the genre. True, some biographers have claimed that their aims were not historical, but moral, didactic or literary – aims which indeed play an important role in the history of the genre, but are not considered further here (cf. Söderqvist 2006) – but otherwise a majority of biography authors have seen their work as more or less closely related to the writing of history. Book reviewers have almost invariably treated biography as just another way of writing about science in the past, and most commentators of the genre have focused on its relations to the historiography of science. Similarly the conferences and anthologies that were inspired by the genre's renaissance in the 1990s implicitly took for granted that it should be discussed in the context of its generic cousin, the historiography of science (see the Introduction to this volume).

The tradition for writing scientific biography dates much further back than writing history of science, however. Not all the way back to classical antiquity; even though some *bioi* of learned men, including mathematicians, natural philosophers and medical doctors, seem to have come forth already in the peripatetic tradition in the fourth and third centuries BCE, it is hardly meaningful to speak about a *genre* of scientific biography in the classical era (see Liba Taub's Chapter 1 in this volume). Nor can one talk about a genre in the medieval Latin West: hagiography (that is, the lives of saints) was certainly more extensive and more varied than the recent pejorative use of the term 'hagiographical' indicates, but there were almost no *vitae* of learned men or women.

The contours of a modern tradition for writing scientific biography can be traced back to the Renaissance. When the classical Greek literature was revived in the fifteenth and sixteenth centuries, the Hellenistic tradition for writing secular lives gradually became popular, probably because it combined the existing and widespread medieval tradition for hagiography with the new humanistic sensibilities. The renaissance for Plutarch and Sueton did not have any immediate consequences for the writing of learned lives, however. For example, Bernardino Baldi's collection of biographies of mathematicians from ancient time to his own time, written in the 1580s, was not published until the end of the nineteenth century (Baldi, 1998). The first lives of learned men in the Western tradition were the stereotypical and eulogistic *orationes funebres* of university professors in the sixteenth century (for example, Hizlerus, 1566), and the short eulogies and memorials published in biographical dictionaries in the seventeenth century, usually modelled on Diogenes Laertius' collection of lives of illustrious philosophers.[4] The Plutarchian and Suetonian traditions do not seem to have begun to make their mark on biographies of learned men until Gassendi's lives of Peiresc (Gassendi, 1641; see Miller, 2000) and of Tycho, Copernicus, Puerbach and Regiomontanus (Gassendi, 1654), and Viviani's now lost *Racconto istorico della vita del Sig. Galileo Galilei*.[5]

The eighteenth century saw a steadily growing number of biographies of mathematicians, medical doctors, astronomers, natural philosophers, natural historians and so forth. Some were published in the form of short articles in biographical and specialised historical dictionaries (German lexicographers were particularly industrious compilers of such dictionaries),[6] others took the form of eulogies in academy series, the best-known being the seventeenth-century *éloges* of the French Academy of Science, but the eighteenth century also saw a growing number of *vitae* in book format, like M. Menckenius' 230-page life of Girolamo Fracastoro (1731), William Burton's account of the life and writings of Hermann Boerhaave (1743) and Dietrich Heinrich Stoever's two volumes on Linnaeus (1792). Thus, by the early nineteenth century, biography was already an established kind of writing about science in the past – and this even before the history of science began to appear as a distinct genre.

As Rachel Laudan (1993) and others have pointed out, the historiography of science did not really take off until the turn of the nineteenth century. Prior to the nineteenth century, the few historical accounts of science were generally written to emphasise the connection between contemporary science and the 'venerable antiquity' (Laudan, 1993, p. 5); others, like Joseph Priestley's well-known histories of electricity and optics, which are often mentioned as examples of early historiography of science, were rather what today would be called 'review articles' than histories of science.[7] A number of German disciplinary histories of mathematics, physics, chemistry and technology appeared between the late 1790s and the first decade of the nineteenth century, reflecting the rapid institutionalisation and specialisation of science. Such disciplinary histories (to which were added histories of astronomy and zoology and, to some extent, geology and botany) thrived throughout the nineteenth century, and

4 Such as Adam (1615), a collection of short biographies of 126 German learned men who had died since the mid-fifteenth century, or Witte (1677–79), which contained 90 biographical articles of learned men, including Tycho, Scaligeri, Grotius and Descartes.

5 It has been suggested that Viviani's *vita* of Galileo (written in the mid-seventeenth century and published posthumously in a more or less edited version by Salvini in 1717) was cast in the Plutarchian mould via the example set by Giorgio Vasari's sixteenth-century *The Lives of the Artists* (Segre, 1989; Christianson, 1995). For Gassendi's Tycho *vita*, see Kragh's Chapter 7 in this volume.

6 The difference between biographical dictionaries and histories were not always clear. Some of the early 'histories' of the particular sciences were rather mixtures of biographical dictionaries and historical chronicles: for example, Johannes Friedrich Weidler's *Historia astronomiæ* (1741) was essentially a chronologically organised biographical dictionary of astronomers.

7 As David Knight has pointed out, most of the history of science from the seventeenth until well into the nineteenth century 'was written as part of a work of science, or as biography' (Knight, 1975, p. 48).

especially after 1850, as the sciences were becoming organisationally established in their modern sense. Most of these nineteenth-century histories of science was written by scientists, who felt, as Laudan points out, that they were involved in a cultural struggle; their aim was to convince their readers that their specialty was progressive and brought new cultural benefits to civilisation, particularly in contrast to the humanities. Hence much of nineteenth-century historiography of science can be understood as 'scientists pressing their claims against humanist dominance of the institutions of intellectual authority' (Laudan, 1993, p.2). History of science was supposed to spread the gospel of scientific progress.

Scientific biography thrived, too.[8] Besides biographical dictionaries of scientists, which continued to be published throughout the eighteenth century, monographical scientific lives became increasingly visible on the book market. Whereas only one or two titles per year were published in the major European languages in the first half of the nineteenth century, around five annual titles came out in the decades after 1850; the figure rose to about ten per year around the turn of the century. The book format changed, too. With few exceptions, like Stoever's life of Linnaeus, most biographies around the turn of the eighteenth century were tiny booklets of 50–100 pages, not very different from the preceding *vitae*. But towards the mid-nineteenth century, biographies began to take the form of these impressive volumes we usually think of as real books – like David Brewster's famous *Memoirs of the life, writings, and discoveries of Sir Isaac Newton* in two volumes (1855).[9] By the end of the nineteenth century, biographies of scientists were regularly published in two, or even three, volumes, often approaching, and even exceeding, one thousand printed pages. The life-and-letters of scientists – for example, Bence Jones' *Life and Letters of Faraday* (1870) in two volumes, Archibald Geikie's *Life of Roderick Murchison* in two volumes (1875), René Vallery-Radot's *La vie de Pasteur* (1900) and (as already mentioned) Leo Königsberger's monumental *Hermann von Helmholtz* (1902–1903) – significantly contributed to making scientific biography a most impressive genre of by the end of the nineteenth century.

Furthermore, in contrast to the seventeenth and eighteenth centuries, nineteenth-century life writing and historiography of science were clearly distinct genres. Both biographies and histories were almost invariably written by scientists themselves (not by historians), but there was a (probably non-conscious) division of work between authors who focused their efforts on disciplinary histories and those who preferred life writing. The German chemist Hermann Kopp, for example, who spent more than thirty years of extensive work on the history of his discipline (Kopp, 1843–47, 1869–75), never published a biography of any of his predecessors. Moritz Cantor, whose famous four-volume *Vorlesungen über Geschichte der Mathematik* (1894–1908), like many historical works of the time, was partly organised by individuals – that is, the chapters proceeded chronologically, each chapter dealing with a mathematical theme within which the works of individual mathematicians was discussed – never

8 Laudan's (1993) otherwise excellent account of eighteenth- and nineteenth-century historiography of science does not include scientific biography.

9 For Newton biographies, see Higgitt's Chapter 9 and Fara's Chapter 4 in this volume.

wrote a biography in book format. Similarly, Pierre Duhem did not supplement his history of mechanics with any biographical works, and Emanuel Rádl, the author of the first great synthesis of the history of biology (*Geschichte der biologischen Theorien*, 1905–1909), did not publish anything substantially biographical.

The relationship between the two genres was largely unproblematic throughout the nineteenth century and well into the twentieth. Even though the two genre projects had somewhat different aims, both were predominantly pursued by scientists (or family members of scientists), and both were basically about promoting science. Arnold Thackray has suggested that scientific biography, at least in England, was an effect of the fact that 'in the wake of the Industrial Revolution, scientific practitioners in the British Isles became more self-conscious about their status and functions', as reflected in the foundation of the British Association for the Advancement of Science in 1831. One of the moving spirits of the BAAS was in fact David Brewster; other BAAS-members, too, were involved in scholarship 'in which particular individuals were rescued from oblivion'. These and similar activities resulted in 'the profusion of heroic biographies religiously compiled to celebrate the lives both of earlier pioneers and of leading Victorian men of science, materials were steadily accumulated to explain and justify the role of science in the national life' (Thackray, 1980, pp. 9–10). Yet Thackray forgets that it was not biographies, but histories which were written to extol perceived national virtues, and to demonstrate the progress of science and its importance for civilisation over time. Life-and-letters could, for obvious reasons, hardly be used to demonstrate collective progress. Biography had different aims; they were primarily written by former colleagues or students as eulogies, to express gratitude to and honour the memories of those who had led the way, and not least to use life stories as paradigms (in the original sense of the word, before Kuhn co-opted it) for young men to emulate. 'How is an irresistible affinity for science created?', asked Santiago Ramón y Cajal in 1916:

> It is well known that youth show their respect for famous men by imitating them. Therefore, it would be a worthwhile contribution to the education of the will if each and every teacher would recount with genuine affection, and with the deliberate intent of suggestion, the anecdotal and more formal biographies of the scientists who have distinguished themselves most in the development of the student's chosen field. This would foster something of the spirit intended by the following authors: Comte, with his veneration of great men, Carlyle in modern times, with his book of heroes; Emerson, with his enthusiastic apologias on representative men or supermen – to whom we owe all of the progress and advantages of civilization; and finally, Ostwald with his inspiring book, *Great Men* [*Grosse Männer*]. (Cajal, 1999, pp. 140–41)

Biographies were also supposed to demonstrate that scientists were as virtuous as other learned gentlemen; scientists had perhaps castigated God's order in nature, but not God's moral order in mankind, and therefore they deserved the readers' admiration.

So, even though scientists rarely wrote *both* history and biography (but either one or the other), there was no real conflict between the two genres, at least as far as one can judge from the absence of critical remarks of biography from the side of the scientist-historians. The historical agenda and the eulogistic and moral agendas seems to have co-existed peacefully, side by side. The existence of many biographies written by family members or close relatives does not change this pattern. Their aim was, perhaps even more than the great man's students and younger colleagues, to honour their husbands, fathers and uncles, not to contribute to the promotion of science as an institution.

'To Cultivate and Accentuate the Human and Social Aspects of Science': Scientific Biography as a Humanising Genre in the First Half of the Twentieth Century

By the early twentieth century, scientific biography gradually became a more varied genre. The life-and-letters tradition gradually disappeared, giving place to an increasing number of secondary works, including shortened versions and rewritings of some of the earlier life-and-letters, often published in a growing number of series like 'Leaders in science', 'The century science series', 'English men of science', '*Geisteshelden*', '*Grosse Männer*', 'Masters of medicine' and so forth. The genre gradually began to lose its character of high literature; biographies of scientists entered popular literature. The proportion of scientists (or their family members) among authors went down and the aims of biographical writings changed, too. More and more biographies were written for pedagogical reasons, to show how exciting science was, and to show the human face of science in a world that believed that scientists were the potential producers of chemical weapons and (later) atomic bombs. In the inter-war years, the biographical novel and its accompanying stylistic innovation, the fictive dialogue, was introduced – a feature which serious life-and-letters biographers, whose duty was first and foremost to attend to the facts, would hardly have even dared dream of.

Another type of biographer – the journalist – appeared on the scene in the first half of the twentieth century. One of the most famous early journalist-biographers was the well-known Berlin journalist and music critic Alexander Moszkowski, who became Albert Einstein's Boswell in *Einstein: Einblicke in seine Gedankenwelt* (1921). Most importantly, fewer and fewer biographies were written to inculcate moral virtues. The erosion of the moral discourse in general – which Alasdair MacIntyre has pointed to in *After Virtue* (1981) – had its parallel in the biographical genre. The clear moral project that had characterised the genre until the late nineteenth century, and which had allowed it to co-exist with historiography of science, waned. Biography became amoral. It lost its long-term purpose, and hence some of its generic coherence.

The attitude of historians of science towards this ever-increasing and diversifying, sometimes novelistic, sometimes popular, biographical literature was ambivalent as the historiography of science emerged as an independent disciplinary field, with its own chairs, journals, societies, meetings and so on. The first international meeting was held in Paris in 1900, the journal *Isis* was founded in Belgium in 1913, the History of Science Society in the US a decade later. Scientist-historians slowly but steadily lost their intellectual influence. Instead, historians of science were increasingly defining themselves as a real scholarly profession with their own professional interests. They did not necessarily want to promote science; they primarily wanted to promote the history of science.

To begin with, scientific biography still enjoyed the former state of peaceful co-existence with historiography of science. The first 'modern' generation of historians of science – men like Arthur Lovejoy (1873–1962), Charles Singer (1876–1960) and George Sarton (1884–1956), who took it upon themselves to promote historiography of science and scientific ideas as an independent discipline – were either indifferent or mildly positive to biography, although none of them wrote one.[10] Sarton, who wrote frequently and with great passion about science history, also had a great and explicit interest in life writing and regularly reviewed biographies in *Isis* (Strelsky, 1957), and even made a distinction between two kinds of scholars of science's past. Whereas some were logically minded, others were psychologically minded, that is, interested in 'the whole intellectual and emotional make-up of the scientist'; biographies were written to satisfy 'that sound instinct' (Sarton, [1936] 1957, p. 42). Therefore, 'full and honest biographies should be encouraged by all means' because they 'help us to know our fellow men and ourselves better' (1957, p. 23). Being men, 'we are interested in other men' and as soon as we 'realize the great part played by individual men' in discoveries, we are anxious to know 'all their circumstances'. How did it happen that a person was devoted to science? How did his genius assert itself – was it hard or easy? Did he succeed in convincing his contemporaries? 'All these questions and many others are deeply interesting,' thought Sarton, especially for other scientists: young scientists are interested because of 'their dreams of the future and their hopes and doubts', and elderly scientists 'because of their memories of the past, and also, though in a different way, because of their hopes and doubts' (1957, p. 22). From a humanistic point of view, 'every detail' in a scientist's life might be interesting, because it was part in 'a great tragedy', in fact 'the basic tragedy of mankind –, the struggle for knowledge' (1957, p. 43). Therefore it would never suffice to describe a scientist's discovery:

> One must explain how and why he made it, and why it was he who made it, what idiosyncracies guided or handicapped him, and so forth. Every human aspect must be considered, because this is not simply a scientific matter but a human one. One must examine his whole behavior, his ways of searching, of finding, of checking and rechecking, and finally – most illuminating of all information – his ways of expressing himself, in short, his style. 'Le style, c'est l'homme.' (1957, p. 43)

But there were limits. With implicit reference to Lytton Strachey's impact on life writing in the early part of the twentieth century, Sarton warned against 'that loathsome fashion of our time, which is called "debunking"', that is, 'the dragging down of great men to the level of their meretricious biographers' (1957, pp. 23-24). He did not put a veto on personal details as such; it was rather a matter of measure: 'It is very well to show all the weaknesses of a hero, but this should be subordinated to the main purpose, the description of his genius, the explanation of the discoveries

10 Lovejoy, one of the most prolific writers on the history of (scientific) ideas in the first half of the twentieth century, not only eschewed writing biography, he did not even write a review of a biography or a comment on the genre. See the bibliography in Lovejoy (1948).

which it made possible, the contemplation of truth and beauty which it revealed'. If the great scientist's drinking had provided a welcome relief from the hard work, this should not be hidden by the biographer, but neither should it be 'unduly magnified'. In Sarton's view, the hero did not become more human and more lovable because he was shown to be a drunkard, 'though we may feel much sympathy for him in his troubles; and, of course, his genius is not at all explained by his drunkenness'. Those kinds of personal details might thus be 'interesting', but only in their 'proper place'. To insist on them would be 'a cruel distortion of the truth' (1957, p. 24).

Sarton's view of scientific biography as a humanising genre was taken up by Douglas McKie (1896-1964), who became one of the most astute spokesman for biography among the professional historians of science in the interwar years.[11] In 1936 he took the initiative to found the journal *Annals of Science* as a British counterpart to Sarton's *Isis* which had moved to the US after the war. As indicated by its subtitle 'A quarterly review of the history of science since the Renaissance', McKie wished to focus on modern science, but he also wanted to promote biography. The first editorial emphasized the need 'to cultivate and accentuate the human and social aspects of science, its pursuit, and the scientific life'. What was felt as a growing criticism in the British public debate of science as an inhuman institution had to be met: 'All Science, all Natural Philosopy is as purely human a production as Art or Literature, and is equally precious with them' (McKie et al., 1936, p. 2). In order to present science's 'human aspects' it was therefore necessary 'to study the life and work of the great masters and makers of Science to realize their own personal atmosphere and cultural environment'. Science was indeed partly 'high and dry light', but it also had 'emotional associations', like a man striving for justice had sentiments and amiable impulses:

> To recall the story – nay, the romance, of love for natural knowledge and devotion to its pursuit cannot fail to appeal to the higher emotions and help to evoke a widened sympathy and interest for the subject. … Indeed, the biographies of students of Nature furnish just as many examples of kindly feelings and loyal attachments as do those of any other type of man – churchman or man of letters, statesman or artist. (1936, p. 3)

McKie and his co-editors asked their readers to provide the new journal with articles on the personal and biographical aspects of science: 'The personal note in the history of Science is, indeed, just as interesting and just as valuable as the personal note in the history of Literature or the Arts.' After all, there were 'many inviting unworked corners in the field of scientific biography' and a 'mass of material' just waiting to be made known. The first British journal of history of science was thus founded with the intention to foster the study of biography for its potentially humanising effects.

---

11 McKie began a military career but switched to chemistry after having been badly wounded in the First World War. After completing his PhD he turned to the history of chemistry and was attached to the newly established Department for the History and Philosophy of Science at University College London in 1925, where he later became Lecturer and finally Professor.

'Biography, Romance and Antiquarian Curiosity': The Cold War Generation's Dismissal of the Individual and Personal Aspects of Science

Such benevolent, or outright promotional, attitudes to scientific biography were not uncommon among historians of science in the first half of the twentieth century. But in the period after the Second World War, sentiments shifted rather drastically. Among the early contesters were Herbert Butterfield (1900–1979), who thought that the essence of history of science was the synthetic study of historical changes in scientific thought, and that such studies should be written by historians, not left to scientists. Butterfield rarely commented on biography as a genre, but when he did (in passing), he did not think highly of it. 'Down to the twentieth century it has been necessary to fight against the assumption that the history of science consists of biography and anecdote,' he wrote in an undated manuscript (in Schweizer, 1998, p. 40). He associated biography with scientists' history at its worst, an 'anecdotal kind of history which is really a rope of sand – collecting narratives concerning great men and their incidental adventures'. This was 'rather a defective kind of antiquarianism', he stated in the first issue of the new *Bulletin of the British Society for the History of Science* (Butterfield, 1950, p. 51), responding, as it were, to McKie's programmatic call for biography more than a decade earlier. The history of science would not turn into 'genuine history' as long as its practitioners 'remain content with too biographical a mode of treatment', However, there were signs, thought Butterfield, that history of science was about to change away from this bad influence, not least thanks to late nineteenth-century historians of science like Paul Tannery. In Butterfield's view, one of the positive results of the publication of edited collections of sources and the more technical study of them was that 'biographical methods have been transcended' (Butterfield 1998, p. 50).

Butterfield's outright dismissal of biographies of scientists in 1950 signalled an upcoming general change in the attitude to the genre, at least among British and North American historians of science. The new sentiment was carried by a new generation of historians of science, born during, or right after, the First World War – I.B. Cohen (b. 1914), A.C. Crombie (b. 1915), Marshall Clagett (b. 1916), Charles C. Gillispie (b. 1918), Marie Boas (b. 1919), Erwin Hiebert (b. 1919), A. Rupert Hall (b. 1920), Duane H.D. Roller (b. 1920) and Thomas Kuhn (b. 1922) – a generation which was intellectually moulded just before and during the Second World War, and invigorated by the new philosophical sensibilities expressed by Alexandre Koyré. Gillispie's article on the émigré Frenchman in the *Dictionary of Scientific Biography* captures some of the enthusiasm with which the Cold War generation took him to their hearts. To Gillespie and his peers, Koyré's *Études galilèennes* (1939) was 'the right book becoming known at the right time':

> It was in the United States in the immediate postwar years that *Études galilèennes*, not much noticed amid the distraction of scholarship by war, found its widest and most enthusiastic public, a case of the right book becoming known at the right time. A new generation of historians of science, the first to conceive of the subject in a fully professional way, was just then finding an opportunity in the expanding American university system … Casting about through the literature in search of materials, they came upon *Études galilèennes* as upon a revelation of what exciting intellectual interest their newly found subject might

> hold, a book which was no arid tally of discoveries and obsolete technicalities, nor a sentimental glorification of the wonders of the scientific spirit, nor yet (despite the author's Platonism) a stalking horse for some philosophical system … Instead they found a patient, analytical, and still a tremendously exciting history of the battle of ideas waged by the great protagonists, Galileo and Descartes. (Gillispie, [1970] 1981a, p. 486)

Koyré was not against studying the thoughts of individuals. After all, part of *Études galilèennes* is about the intellectual development of the young Galileo. But it was the *ideas* of individuals and their disembodied *intellectual* battles that counted, not their personalities or virtues, or anecdotes or any other interesting personal details that would normally flesh out a biographical story. Similarly, the Cold War generation of American historians of science did not mind studying individual scientists, their science and their careers, as long as it shed light upon the history of scientific thought. Accordingly, the Editorial Board of the *Dictionary of Scientific Biography*, headed by Gillispie, asked their contributing authors 'to place emphasis upon the scientific accomplishments and careers of the subjects. … Personal biography has intentionally been kept to the minimum consistent with explaining the subject's place in the development of science' (Gillispie, [1970] 1981b, p. ix). The fact that several of the Cold War generation historians published studies of individual scientists' contributions therefore does not contradict the conclusion that as a generation they were basically anti-biographical.[12]

The attitude of the British/North American Cold War generation of historians of science towards scientific life writing was neatly summarised in two seminal conferences: first the two-week-long 'Critical Problems in the History of Science' meeting organised by Marshall Clagett in Madison, Wisconsin, in the summer of 1957, and the week-long symposium on 'The Structure of Scientific Change' organised by A.C. Crombie in Oxford in 1961. Biography was definitely not among the topics discussed at these two meetings. Crombie did not rule out an interest in individuals and their motivations – 'Who were the people … What were their … personal motives and opportunities' (Crombie, 1963, p. 10) – but he did not refer to biography, and when founding the journal *History of Science* a year later, he did not make any mention whatsoever to scientific life writing or to the personal and humanising aspects of science, as Sarton and McKie had done. In fact, only a single participant in the Oxford meeting mentioned biography at all. It was Henry Guerlac, then Professor in History of Science at Cornell University, who touched upon the status of the genre, although in a negative way, when expressing the opinion that the writing of history ought to focus exclusively on 'public history', and that individual actions were of interest 'only if they demonstrably affected the lives and thoughts of the social collectivities' (Guerlac, 1963, p. 798). In Guerlac's view, history was something clearly set apart from 'biography, romance and antiquarian curiosity' (1963, p. 799). Consequently, in one of his later resumés of the history of

12 For example, Gillispie later co-authored biographies of Laplace (Gillispie and Youschkevitch, 1979) and Carnot (Gillispie, Fox and Grattan-Guinness, 1997).

the historiography of science, he did not entertain the possibility that biographies constitute a significant portion of the total output on science in history.[13]

Guerlac's position at the Oxford conference in 1961 articulated a general attitude to biography among his generation of British and American historians of science. In the 1950s, 1960s and 1970s, biography seems to have had a status among professional historians of science similar to that of Christ in the libretto of Handel's *Messiah* – it was despised and rejected. Biography was simply not the thing to do for those who had professional ambitions as historians of science. For example, Duane Roller opened a collection of articles on new perspectives on the history of science and technology in 1971, by pointing out that the focus of late nineteenth- and early twentieth-century history science 'upon the achievements of individual scientists' was a result of several circumstances. First, it was due to 'a nineteenth-century heroic view of history'; second, it was supported by those 'who regarded the history as a pedagogic aid in teaching of that science wished to display its notables as exemplars to students'; furthermore, the individualist focus was reinforced by the fact that the source materials for the history of science 'have seemed to consist almost entirely of manuscripts and publications of individual scientists; and finally, 'the having of ideas is widely regarded as the prerogative of the individual'. For these reasons, he believed, early history of science had tended to be 'accounts of the work of individual scientists if the past' (Roller, 1971, p. v). But things were about to change. There was 'a growing tendency to spread beyond the confines of the writings of the scientist and to examine the culture which produced him'. And although studies of the ideas of individual scientists were still an important part of professional history writing, 'many historians now group the result of such studies in terms of the history of an idea or a point of view or an era' (1971, p. vi).

Thomas L. Hankins, an eyewitness of the 1960s American history of science scene, neatly summarises what emerging professional historians of science thought about biography:

> In the 1960s we graduate students thought that the history of science was a form of intellectual history and that a scientific biography should be about a scientist's scientific ideas. Most existing biographies tended to be anecdotal and did not try to ferret out the source of those ideas and that is why we held them in such low esteem.[14]

One can only speculate about the interests behind this widespread suspicion of biography. One possible explanation is that this was a generation of scholars who had been trained as scientists, but then turned into historians of science and therefore

---

13 Guerlac (1977), esp. Ch. 5. Guerlac's position *vis-à-vis* biography is ambiguous. In a paper on the Lavoisier biographical tradition (Guerlac, 1954), he acknowledged biography as a legitimate genre, although he was devastatingly critical of all the biographies published so far. Fifteen years later, he would try out biography in the short *Antoine-Laurent Lavoisier: Chemist and Revolutionary* (1975), aimed at a general public. Maybe biography was all right for laymen, but in his conception of what a serious, professional historiography of science ought to do, however, biography had no place.

14 Thomas Hankins to Thomas Söderqvist, 18 March 2002 (personal e-mail communication).

were eager to carve out an independent professional role *vis-à-vis* scientists; by focusing on the philosophical and non-personal aspects of science, they may have thought that they were behaving like true professionals. The few historians of science who were trained as general historians were apparently not haunted by the same anxieties. For example, A. Hunter Dupree revised his dissertation on Asa Gray written in Arthur M. Schlesinger's graduate history programme at Harvard University into a book about the 'personal story' of the great American botanist: 'The life of a man who spent most of his waking hours working at science and thinking about science includes in the personal story much that belongs also to the history of science', he stated (Dupree, [1959] 1988, p. xvii). It received laudatory reviews, even by Thomas Kuhn, who considered it to be among 'the very finest scientific biographies' he had read: 'The balance between the person and the career in science is unprecedented.'[15] Another historian, Edward Lurie, wanted his *Louis Agassiz: A Life in Science* (1960) 'to report the conditions and passions' of Agassiz's life as a naturalist; in Lurie's intepretation, Agassiz's understanding of nature was grounded in 'emotional involvement in intellectual effort' (Lurie, 1960, p. vii).

As general historians, Dupree and Lurie had probably a more relaxed attitude to the personal, even passionate, dimension of science, and could therefore freely embrace biography – in fact, the personal dimension was precisely what they, who otherwise lacked scientific training, could bring to the history of science. Another notable exception to the general dismissal of biography was L. Pearce Williams, whose *Michael Faraday: A Biography* (1965), the first serious Faraday biography for almost a century, was aimed to give 'a full picture of Faraday the man and Faraday the scientist' (Williams, 1965, p. xi). Looking back, Williams explains why he, in contrast to his thesis supervisor at Cornell University, Henry Guerlac, and a whole generation of professional, academic historians of science, grew an interest in the personal dimension of Faraday:

> Although I was Henry Guerlac's third Ph.D., I disagreed with him thoroughly on the writing of scientific biography. The reason was that I was brought up in the American theater – my father was the stage manager for such greats as Alfred Lunt and Lynn Fontaine and other such stars of Broadway, and my mother was the Subscription Manager of the Theatre Guild that led the way in making Broadway shows available to a very wide public. So, I grew up in a theatrical tradition where it was never enough to know one's lines, but fundamentally important to know why you were speaking them. That is to say, you had to try and get inside the character you were portraying or, to put it another way, to learn the motivations, prejudices and peculiarities of the person you were supposed to be. Only then could the lines you were speaking make any real sense within the framework of the play. So, when I came to study the history of science, I was not only in the what but the why of the production of new theories and ideas. This meant that I had to look at more than the published works which, quite rightly, are cleansed of all personal dimensions. What I found, to my delight, was that this dimension casts essential light on the process of science.[16]

15 Cited from the blurb of the 1988 reprint edition.

16 L. Pearce Williams to Thomas Söderqvist, 26 February 2002 (personal e-mail communication).

True, there were a few other professional historians of science who wrote scientific biographies, for example Michael Mahoney at Princeton, a student of Gillespie and Kuhn, who wrote *The Mathematical Career of Pierre de Fermat (1601–1665)* (Mahoney, 1973). As indicated by the title, however, Mahoney focused on the intellectual aspects of Fermat's 'odyssey ... in quest of self-determined goals'; it was the 'knowledge of and sensitivity to mathematics' that gave him the tool for understanding Fermat and his thoughts (1973, p. x). This and similar biographies by a few other professional historians of science emphasised the scientific ideas and their historical importance, and did not pay much attention to personal matters. They were rather what the Germans call *Ergographien* ('work-descriptions') than biographies ('life-descriptions').

Otherwise, the overwhelming majority of biographies of scientists in the 1960s and 1970s were written by people outside the professional history of science circle. Some 25–30 biographies a year were published in the English, German and French languages in the 1960s and 1970s. A new actor on the biographical marketplace in these years was the science writer – and none of them were more productive than Ronald W. Clark (1916–1987), a former war correspondent who began writing about the lives of scientists in the 1960s and then, in the next fifteen years, produced well-researched, thick and beautifully crafted biographies – of Julian Huxley, J.B.S. Haldane, Albert Einstein, Bertrand Russell, Thomas Alva Edison, Sigmund Freud, Benjamin Franklin, Charles Darwin and Ernst Chain – for the general educated reader. Clark's almost unbelievable productivity adds to the impression that scientific biography has been a most impressive (and most forgotten) genre for the popular understanding of science in the late twentieth century. So, at least in the UK and in North America, biography lived its own life on the margins of professional history of science.[17]

Scientific Biography as an *Ancilla Historiae*: The 1980s and 1990s

The Cold War generation dominated the historical outlook until the 1970s. In its wake came a more politically and socially conscious generation, intellectually formed during the 1960s. Whereas the Cold War generation had been inspired by *Études galilèennes*, the new generation's major source for enthusiasm was *The Structure of Scientific Revolutions*. This post-Kuhnian generation, to use Steve Fuller's term, began to set a new agenda for the history of science (Fuller, 2000). They did not want to promote science any more; they wanted to promote themselves. Something 'social' and something 'cultural' happened on the way to the 1980s.

My aim here is not to so much discuss how this new generation practiced history, but how it evaluated biography. With a few exceptions the post-Kuhnian generation took over the Cold War generation's fundamental scepticism to biography. 'Until recently,' Elizabeth Garber wrote in 1990, 'many of the more traditional approaches [to history of science] were pursued apologetically because of serious questions about

17 The picture is somewhat different in Germany, where the gap between biographies for the academic marketplace and a general readership has not been as great as in the UK and the USA.

their validity in light of historical writing derived from sociology and anthropology.' And she continued:

> No genre of history fell under more odium than that of biography. Belief in the priority of socially defined criteria necessarily destroys the importance of individual lives, especially of their peculiarities and eccentricities. Seeing the cognitive realm as derivative of social place, cultural values or political ideology undermines any concern with what makes an individual's contribution to the sciences or technology critical or crucial. Character and originality were lost to the crowd, suppressed as subjects for legitimate historical investigation. (Garber, 1990, p.9)

Garber's succinct characterisation of the post-Kuhnian generation's attitude towards biography was supported by many observers of the historiographical scene in the 1970s and 1980s.

But as usual, there were seemingly dissenters. 'Not quite all historians accepted this assessment,' Garber herself hastened to add a footnote (1990, p. 18) referring to a paper by Thomas Hankins in *History of Science* (1979) which many have seen as a signal of the return of biography to respectability again. 'In defense of biography', he titled it. The reason why life writing was worth defending, argued Hankins, was that it 'gives us a way to tie together the parallel currents of history at the level where the events and ideas occur', because 'we have, in the case of an individual, his scientific, philosophical, social and political ideas wrapped up in a single package' (Hankins, 1979, p. 2) In other words, Hankins seemed to see biography as a sub-genre of history, and therefore his 'defence' shall not be understood as a defence of biography as such – as we know the genre from its long tradition – but rather as a narrowing of the scope of life writing: scientific biography should be pursued simply because it is useful for a social and cultural history of science - an *ancilla historiae*, as it were.

Garber further echoed Hankins's position when she wrote that 'studies of individuals are proving invaluable in probing the values, behavior, and social life in complex societies' and that 'the idiosyncracies of the subject even help to shed light on the characteristics of the collective'. In other words, Garber could not imagine that biography could be used for any purposes other than a social, political and cultural history of science. She could not imagine that biography might also be used to shed light on the phenomenon of the individual, the person and the subject. This view of biography as an *ancilla historiae* was taken to its extreme by the American historian of medicine Charles Rosenberg. He expressed his intent of demonstrating 'the relevance of biography to social and cultural history':

> I had heard too many comments from academic historians disparaging the conceptual deficiencies of biography and the narrow vision of its practitioners. I have always thought very positively about biography and never felt that there was a necessary contradiction between the writing of an individual life and history generally.

But it was a fairly idiosyncratic concept of biography Rosenberg had in mind. He thought 'a life can be construed as a sampling device – as a controlled and internally coherent batch of data, a chronologically ordered set of realities and relationships as perceived and understood by a particular actor'. For that reason, he thought, biography, like 'the aggregated data of the historical demographer', was 'an indispensable tool for the collective historical enterprise' (Rosenberg, 1992, p. 215).

Yet another commentator on biography, who wanted to save the genre by assimilating it into a cultural history of science, was Robert M. Young, who presented himself as 'a humanistic Marxist' with an interest in psychohistory. Young did not buy into 'vulgar' Marxism *à la* Boris Hessen, nor into speculative psychohistory *à la* Frank Manuel. But instead of putting an embargo on such approaches, as many historians of science, in his opinion, had done, Young wanted to find a way of making both Marxism and psychobiography 'more sophisticated' and to integrate them into 'a single account' (1987, p. 205). He wanted to know 'how does the individual bear, mediate, and integrate the individual, intellectual, cultural, ideological, and socioeconomic forces which constitute the labor process of her or his production of what interests us about that person, be it knowledge, art, policy, heroism, great good or evil?' (1987, pp. 205–6). Young rejected a 'decontextualised' biographical approach to science which incorporated only 'personal history plus the influence of scientific discoveries and ideas'; instead, he wanted to bring biography into 'the growing body of studies which place science inside culture, inside society, inside the ideological and socioeconomic forces which shape the rest of the social world' (1987, p. 213).

Throughout the 1980s and 1990s, the history of science community at large began to adopt the view that biography is a useful tool for a contextual history of science, technology, and medicine, 'by opening out [the] individual to social and cultural contexts', as Michael Shortland put it (1996, p. 17). Biography was no longer despised or rejected. The odium was gone. Under the pretext of its usefulness for the cultural understanding of science, or for understanding how historical actors fashion their cultural identities, biography gradually became integrated into the history of science toolbox. Around the turn of the millennium, no serious historian of science rejected the genre – as long as it contributed to a socially and culturally informed history of science.

Adrian Desmond saw his two volumes on Huxley (Desmond, 1994, 1997a) as a contribution to the new contextual history of science. 'This is a story of Class, Power and Propaganda,' he wrote in the Preface to the American edition (1997b, p. xiii). The biographical 'approach' allowed him 'to paint an ashamedly *social* portrait' (1997b, p. xiv), and he attacked Disraeli's advice that one should read biography instead of history because biography is '"life without theory"':

> Good enough, perhaps, for the posing *littérateur*; but it would be a rash biographer today who ignored the new contextual and sociological approaches to science in 'Darwin's Century'. Isn't it the modern function of biography to carve a path through brambly contexts? To become a part of history? … And isn't that our ultimate aim, to understand the making of our world? (1997a, p. 235)

And he continued his *plaidoyer* for biography as history of science. *Huxley* was part and parcel of 'the new contextual history of science', and this itself was 'a reaction to the old history of ideas, which *dis*placed the person, made him or her a disembodied ghost, a flash of transcendent genius'. So only by 'embedding Huxley' was it possible to 'appreciate his role in the vast transformation that staggered our great-grandfathers' (1997a, p. 235). Desmond also gave an quasi-cultural and autobiographical explanation for how his post-Kuhnian generation (both he and Moore are the intellectual heirs of Robert Young) had contributed to the renaissance of biography in the 1990s:

> It is an irony that many historians who gravitated to the sociology of knowledge in the 1980s, trying to understand why different sorts of knowledge were produced by different social grounds, have come out seeing their subjects as individuals. For me this is particularly true. Perhaps it is because, as a beleaguered minority [*sic!*] we have been forced to follow our subjects so closely – the onus lay on us to prove that science really was socially contingent … This put the burden back on biography to explain scientific originality in terms of individual social trajectories. The upshot is a revitalized interest in scientific biography in the 1990s. (1997a, pp. 235–6)

Desmond's life of Huxley epitomises the attitudes among historians of science to biography today. The laudatory reviews of the two Huxley volumes demonstrate that this is how a scientific biography for our time should be written. But we should not forget that Desmond's volumes on Huxley – plus a few others which historians of science normally mention when the topic of biography is raised in a discussion – are just a tiny proportion of the total number of biographies written in the 1990s. There is a much larger and more diverse flora of biographical writing out there. Biographies written to entertain, biographies for young people, biographies written to explicate a scientist's technical work, biographies written by former students and family members to honour the memory of the great man (increasingly woman), biographies which want to glorify the nation, an institution or a scientific discipline, biographies which want to establish modern, secular, scientific heroes and heroines – even biographies written to inculcate moral virtues in the reader (and, of course, to make money).

In other words, today we still find the whole array of varieties of biographical writing that we can find back in the late nineteenth and early twentieth centuries. This ecologically diverse broad biographical flora in the shadow of the 'new' cultural history of science is not recognised by today's reviewers of biographies, however. In the 1950s, 1960s and 1970s, biography as such was rejected and despised by the profession; today, when a certain kind of (contextual) biographical writing has been accepted into the centre of the history of science, the large rest is just invisible.

I do not mind books being invisible as such. But I think both the historiography of science and the art of writing biographies of scientists would profit from bringing other aims and varieties into the discussion than the usual suspects.[18] It is time to look at the phenomenon of scientific biography from a truly historicist point of view and to see the genre in all its forms and varieties, without judging them with today's standards for what constitutes good history of science. To do so, would, I think, be the real defence of biography.

18 This argument is developed at length in Söderqvist (2006).

Acknowledgements

The Wellcome Trust gave a one-year travel fellowship for my studies of the holdings of scientific biographies in their own library and the Science Museum library; and the Wellcome Centre for the History of Medicine at UCL provided a most inspiring intellectual environment during a sabbatical in 2000–2001. I am grateful to Janet Browne for comments on an earlier version.

Bibliography

Adam, Melchior (1615), *Vitæ germanorum philosophorum: qui seculo superiori, et quod excurrit, philosophicis ac humanioribus literis clari floruerunt*, Heidelberg.

Baldi, Bernardino (1998), *Le vite de' matematici: Edizione annotata e commentata della parte medievale e rinascimentale*, ed. Elio Nenci, Milan: Francoangeli.

Brewster, David (1855), *Memoirs of the life, writings, and discoveries of Sir Isaac Newton*, 2 vols, Edinburgh: Thomas Constable.

Burton, William ([1743] 1746), *An Account of the Life and Writings of Hermann Boerhaave Doctor of Philosophy and Medicine Professor of the Theory and Practice of Physics and Also of Botany and Chemistry in the University of Leyden, President of the Chirurgical College in that City, Fellow of the Royal Society in London and of the Royal Academy at Paris*, 2 vols, London: Henry Lintot.

Butterfield, Herbert (1950), 'The historian and the history of science', *Bulletin of the British Society for the History of Science*, **1**, 49–58.

Butterfield, Herbert (1998), 'The importance of the history of science', in Karl W. Schweizer (ed.), *Herbert Butterfield: Essays on the History of Science*, Lewiston, NY: Edwin Mellen, pp. 31–52.

Cajal, Ramón y (1999), *Advice for a Young Investigator*, Cambridge, MA: MIT Press (trans. of: *Reglas y consejos sobre investigación biológica*, 4th edn, Madrid: Fortanet, 1916).

Cantor, Moritz (1890–1908), *Vorlesungen über Geschichte der Mathematik*, 2nd edn, Leipzig: Teubner.

Christianson, Gale E. (1995), 'On the Renaissance model of early scientific biography', *International Social Science Review*, **70**, 87–92.

Crombie, A.C. (1963), *Scientific Change: Historical Studies in the Intellectual, Social, and Technical Conditions for Scientific Discovery and Technical Invention, from Antiquity to the Present*, New York: Basic Books.

Crosland, Maurice (1978), *Gay-Lussac: Scientist and Bourgeois*, Cambridge: Cambridge University Press.

Debré, Patrice (1994), *Louis Pasteur*, Paris: Flammarion (Engl. trans.: *Louis Pasteur*, Johns Hopkins University Press, 1998).

Desmond, Adrian (1994), *Huxley: The Devil's Disciple*, London: Michael Joseph.

Desmond, Adrian (1997a), *Evolution's High Priest*, London: Michael Joseph.

Desmond, Adrian (1997b), *Huxley: From Devil's Disciple to Evolution's High Priest*, Reading, MA: Addison-Wesley.

Dupree, A. Hunter ([1959] 1988), *Asa Grey: 1810–1888*, Baltimore, MD: Johns Hopkins University Press.

Fuller, Steve (2000), *Thomas Kuhn: A Philosophical History for Our Times*, Chicago, IL: University of Chicago Press.

Gallagher, Catherine and Stephen Greenblatt (2001), *Practicing New Historicism*, Chicago, IL: University of Chicago Press.

Garber, Elizabeth (ed.) (1990), *Beyond History of Science: Essays in Honor of Robert E. Schofield*, Bethlehem, PA: Lehigh University Press.

Gassendi, Pierre (Petrus Gassendus) (1641), *Viri illustris Nicolai Claudii Fabricii de Peiresc, senatoris aqvisextiensis vita*, Paris: Sebastiani Cramoisy.

Gassendi, Pierre [Petrus Gassendus] (1654), *Tychonis Brahei, equitis Dani, astronomorum coryphæi vita*, Hagæcomitum [The Hague]: Adriani Vlaco.

Geikie, Archibald (1875), *Life of Sir Roderick I. Murchison, Bart.; K.C.B.; F.R.S.; Sometime Director-General of the Geological Survey of the United Kingdom: Based on His Journals and Letters, With Notices of His Scientific Contemporaries and a Sketch of the Rise and Growth of Palæozoic Geology in Britain*, 2 vols, London: John Murray.

Gillispie, Charles Coulston ([1970] 1981a), 'Koyré', in C.C. Gillispie (ed.), *Dictionary of Scientific Biography*, vols 7–8. New York: Charles Scribner's Sons, pp. 482–90.

[Gillispie, Charles Coulston et al.] ([1970] 1981b), 'Preface', in C.C. Gillispie (ed.), *Dictionary of Scientific Biography*, vols 1–2, New York: Charles Scribner's Sons, pp. ix–x.

Gillispie, Charles Coulston and A.P. Youschkevitch (1979), *Lazare Carnot, savant, et sa contribution a la theorie de l'infini mathematique: avec trois memoires inedits de Carnot*, Paris: Vrin.

Gillispie, Charles Coulston, R. Fox and I. Grattan-Guinness (1997), *Pierre-Simon Laplace, 1749–1827: A Life in Exact Science*, Princeton, NJ: Princeton University Press.

Govoni, Paula (2000), 'Biography: A critical tool to bridge the history of science and the history of women in science: Report on a conference at Newnham College, Cambridge, 10–12 September 1999', *Nuncius*, **15**, 399–409.

Guerlac, Henry (1954), 'Lavoisier and his biographers', *Isis*, **45**, 51–62.

Guerlac, Henry (1963), 'Some historical assumptions of the history of science', in A.C. Crombie (ed.), *Scientific Change: Historical Studies in the Intellectual, Social, and Technical Conditions for Scientific Discovery and Technical Invention, from Antiquity to the Present*, New York: Basic Books, pp. 797–812.

Guerlac, Henry (1975), *Antoine-Laurent Lavoisier: Chemist and Revolutionary*, New York: Charles Scribner's Sons.

Guerlac, Henry (1977), *Essays and Papers in the History of Modern Science*, Baltimore, MD: Johns Hopkins University Press.

Hankins, Thomas L. (1979), 'In defence of biography: The use of biography in the history of science', *History of Science*, **17**, 1–16.

Hizlerus, Georgius M. (1566), *Oratio de vita et morte clarissimi viri, medici et philosophi praestantissimi, D. Leonharti Fuchsii, artis medendi in Academia Tubingensi Professoris eximii*, Tübingen.

Jones, Bence (1870), *The Life and Letters of Faraday*, 2 vols, London: Longmans, Green.

Knight, David (1975), *Sources for the History of Science 1660–1914*. London: Hodder and Stoughton.

Königsberger, Leo (1902–1903), *Hermann von Helmholtz*, 3 vols, Braunschweig: Vieweg (Engl. trans. *Hermann von Helmholtz*, Oxford: Clarendon Press, 1906).

Kopp, Hermann (1843–47), *Geschichte der Chemie*, vols 1–4, Braunschweig: Friedrich Vieweg,

Kopp, Hermann (1869–75), *Beiträge zur Geschichte der Chemie*, vols 1–3, Braunschweig: Friedrich Vieweg.

Koyré, Alexandre (1939), *Études galilèennes*, Paris: Hermann.

Kragh, Helge (1990), *Dirac: A Scientific Biography*, Cambridge: Cambridge University Press.

Laudan, Rachel (1993), 'Histories of the sciences and their uses: A review to 1913', *History of Science*, **31**, 1–34.

Lear, Linda (1997), *Rachel Carson: Witness for Nature*, New York: Henry Holt.

Leo, Friederic (1901), *Die griechisch-römische Biographie nach ihrer literarischen Form*, Leipzig: Teubner.

Lovejoy, Arthur O. (1948), *Essays in the History of Ideas*, Baltimore, MD: Johns Hopkins University Press.

Lurie, Edward (1960), *Louis Agassiz: A Life in Science*, Chicago, IL: University of Chicago Press.

MacIntyre, Alasdair (1981), *After Virtue: A Study in Moral Theory*, Notre Dame, IN: University of Notre Dame Press.

Mahoney, Michael S. (1973), *The Mathematical Career of Pierre de Fermat (1601–1665)*, Princeton, NJ: Princeton University Press.

Maurois, André (1959), *La vie de Sir Alexander Fleming*, Paris: Hachette 1959 (Engl. trans. *The Life of Sir Alexander Fleming, Discoverer of Penicillin*, London: Cape, 1959).

[McKie, D., Harcourt Brown, and H.W. Robinson] (1936), 'Editorial', *Annals of Science*, **1**, 1–3.

Menckenius, M. (1731), *De vita, moribus, scriptis, meritisque in omne literarum genus prorsus singularibus Hieronymi Fracastorii*, Leipzig.

Miller, Peter N. (2000), *Peiresc's Europe: Learning and Virtue in the Seventeenth Century*, New Haven, CT: Yale University Press.

Momigliano, Arnaldo (1993), *The Development of Greek Biography*, 2nd edn, Cambridge, MA: Harvard University Press.

Moszkowski, Alexander (1921), *Einstein: Einblicke in seine Gedankenwelt. Gemeinverständliche Betrachtungen über die Relativitätstheorie und ein neues Weltsystem entwickelt aus Gesprächen mit Einstein*, Hamburg (Engl. trans. *Einstein the Searcher: His Work Explained From Dialogues with Einstein*, London: Methuen, 1921).

Rádl, Emanuel (1905–1909), *Geschichte der biologischen Theorien seit dem Ende des siebzehnten Jahrhunderts*, 2 vols, Leipzig: Engelmann.

Roller, Duane H.D. (1971), 'Foreword', in D. Roller (ed.), *Perspectives in the History of Science and Technology*, Norman, OK: University of Oklahoma Press.

Rosenberg, Charles E. (1992), *Explaining Epidemics and Other Studies in the History of Medicine*, New York: Cambridge University Press.

Sadoun-Goupil, Michelle (1977), *Le chimiste Claude-Louis Berthollet 1748–1822: sa vie, son œuvre*, Paris: J. Vrin.

Sarton, George, ([1936] 1957), *The Study of the History of Mathematics and the Study of the History of Science*, New York: Dover.

Schweizer, Karl W. (ed.) (1998), *Herbert Butterfield: Essays on the History of Science*, Lewiston. NY: Edwin Mellen.

Segre, Michael (1989), 'Viviani's life of Galileo', *Isis*, **80**, 207–31.

Shortland, Michael (1996), 'Bonneted mechanic and narrative hero: The self-modelling of Hugh Miller', in M. Shortland (ed.), *Hugh Miller and the Controversies of Victorian Science*, Oxford: Clarendon Press, 1996, pp. 14–75.

Söderqvist, T. (2006), 'What's the use of writing lives of recent scientists?', in R.E. Doel and T. Söderqvist (eds), *The Historiography of Recent Science, Medicine and Technology: Writing Recent Science*, London: Routledge.

Stoever, Dietrich Heinrich (1792), *Leben des Ritters Carl von Linné. Nebst den biographischen Merkwürdigkeiten seines Sohnes, des Professors Carl von Linné und einem vollständigen Verzeichnisse seiner Schriften, deren Ausgaben, Uebersetzungen, Auszüge und Commentare*, 2 vols, Hamburg.

Strelsky, Katharine (1957), 'Bibliography of the publications of George Sarton', *Isis*, **48**, 336–46.

Thackray, Arnold (1980), 'History of science', in Paul T. Durbin, *A Guide to the Culture of Science, Technology, and Medicine*, New York: Free Press, pp. 3–69.

Tomaselli, Sylvana (1988), 'Collecting women: The female in scientific biography, *Science as Culture*, **4**, 95–106.

Vallery-Radot, René (1900), *La vie de Pasteur*, Paris: Libraire Hachette (Engl. trans. *The Life of Pasteur*, 2 vols, London: Archibald Constable, 1902).

Vogeler, Karl (1941), *August Bier: Leben und Werk*, Munich and Berlin: J.F. Lehmanns Verlag.

Ward, Patricia Spain (1994), *Simon Baruch: Rebel in the Ranks of Medicine, 1840–1921*, Tuscaloosa, AL: University of Alabama Press.

Weidler, Johann Friedrich (1741), *Historia astronomiae sive de ortu et progresso astronomiae liber singularis*, Wittenberg: Schwartz.

Wiegleb, Johann Christian (1790–91), *Geschichte des Wachstums und der Erfindungen in der Chemie*, 2 vols, Berlin and Stettin: Friedrich Nicolai.

Williams, L. Pearce (1965), *Michael Faraday: A Biography*, London: Chapman and Hall.

Witte, Henning (1677–79), *Memoriæ philosophorum, oratorum, poetarum, historicorum, et philologorum*, 2 vols, Frankfurt.

Young, Robert M. (1987), 'Darwin and the genre of biography', in G. Levine (ed.), *One Culture: Essays in Science and Literature*, Madison, WI: University of Wisconsin Press, pp. 203–24.

Zeldin, Theodore (1974), [Review of Robert Reid, *Marie Curie*, 1974], *The Times Literary Supplement*, 5 July: 701–2.

Index